FACTORIAL ECOLOGY

for Rhiannon

Factorial Ecology

WAYNE K. D. DAVIES, B.SC., PH.D.
Professor of Geography,
Faculty of Graduate Studies,
The University of Calgary

Gower

© W.K.D. Davies 1984

Published by
Gower Publishing Company Limited
Gower House, Croft Road, Aldershot, Hants GU11 3HR, England

ISBN   0 566 00599 9

British Library Cataloguing in Publication Data

Davies, Wayne K. D.
   Factorial ecology.
   1. Social sciences — Statistical methods
   2. Factor analysis
   I. Title
   519.5'354'0234   HA29

Library of Congress Cataloguing in Publication Data

Davies, Wayne Kenneth David.
   Factorial ecology.
   Includes bibliographical references and index.
   1. Ecology — Statistical methods. 2. Factor analysis.
I. Title.
QH541.15.S72D38   1984   574.5'072   83-20617

Printed in Great Britain by
Biddles Ltd, Guildford, Surrey

CONTENTS

TABLES AND FIGURES

PREFACE

Factor analysis is a family of methods and techniques which have been among the most popular of the multivariate procedures applied to human geography and human ecology in the last twenty years. Nevertheless, geographers and ecologists have far less experience with the methods than many of their fellow social scientists, such as psychologists, who, in conjunction with applied mathematicians and statisticians, have been responsible for inventing and developing the basic procedures in the field. Several texts on the statistical basis and variety of the methods, usually in the language of matrix algebra, have been devised, whilst introductions to the methods can also be found in various social science fields or in standard quantitative texts in human geography. Given these works it is legitimate to ask why there is a need for another book, this time in the field of human geography and ecology. The answer is the same as for any quantitative text in geography or any disciplinary field. It is necessary to explain the fundamentals of the procedures to students and potential investigators in the field within the context of the particular problems faced by adherents to the discipline, in this case areally based data. But the introduction to the principles and problems of factorial methods in human geography only represents the first two of the three objectives of this study. It is also designed to summarize the practical results and potential use of the methods in the areas of regional and urban studies, with particular reference to the methodological value, in particular the synthesizing characteristics of the procedures.

As with all studies attempting to combine technical procedures and substantive results, research workers can argue the study provides too few mathematical details of both the basic techniques and the substance of the field. Yet it must be remembered that the book is designed as an introduction to the field, not to the latest research papers. In this latter context, of course, it is now traditional to express the differences between the various factorial methods in the 'language' of matrix algebra. To do so here would mean one would have to assume that students of human geography will have already acquired these fundamentals. If the assumption can be made there would be no need for the first third of the book, for students could go straight to the standard statistical texts. Instead, this study is designed to describe the basis of the techniques in as simple a way as possible, gradually making the arguments more and more complex. The hope is that any reader planning advanced work in the field will, through this introductory survey, acquire sufficient understanding of the field to go on to tackle the advanced texts such as Horst (1965), Rummel (1970),

Harman (1976) and especially Cattell (1978). Many deliberate references to these texts as well as basic research papers are made in this study simply to help guide future investigations. This means that there seemed little point in focusing upon the computational details of the various factorial procedures since college and university students now have access to sophisticated computer package programmes. This fact, plus the speed of modern computers, has meant that an important shift in the allocation of time spent on a factor analysis example has occurred. The speed of calculations means that much more time can be spent on the more important - at least to those interested in substantive results, not purely techniques - issues of problem formulation, data grooming and assessing the utility of various sets of results. This explains why this book has such an emphasis upon the scope of the field and the problems to be faced in carrying out a study (Chapters III to VI) as well as upon the particular difficulties demonstrated by published factorial studies in human geography and ecology.

As with all studies of this nature I must thank the various authors and journals who have provided permission to reproduce various illustrative tables and figures used in the text, the sources of which are identified at the appropriate place in the tables or text. On occasion revisions of the original tables seemed to be useful in clarifying or extending the arguments. Such revisions are, of course, the interpretation and responsibility of this author not the original factorial ecologist. I also wish to acknowledge the debt I owe to a number of my former graduate students who worked with me in this area over the years, especially Bev Borden, Darina Healey, Graham Barrow, Dave Clark, Tim Gyuse, John Lawson, Caroline Mills, Rod Thompson, Steve Tapper and Steve Welling. I must also thank the patience of my family and publisher in waiting for this study particularly through the last set of delays caused by my added administrative responsibilities in the Graduate Faculty during 1982/83. In addition I must thank Marta Styck and Roger Wheate for drawing the maps, Lee Fischer for typing some of the first drafts of the manuscript, and praise the expertise of Bonnie Harvey for producing the final version of the study.

W.K.D. Davies
University of Calgary

1 FACTOR ANALYSIS IN HUMAN GEOGRAPHY AND HUMAN ECOLOGY

1. INTRODUCTION

Human geographers and human ecologists study the spatial variations in the human patterns, processes and organizations found on the earth. Given the many sources of variation that intertwine and overlap in areas to produce the complex spatial mosaic that is the subject matter of these academic areas it is obvious that the field deals with what is best described as a multivariate reality. An important part of the geographical and ecological approach is to describe and interpret the multivariate patterns. Since this often involves the synthesis of relationships it is a difficult task. Moreover in trying to interpret and understand this spatial mosaic, human geographers and human ecologists are at several disadvantages compared with physical scientists.

(i) One major difficulty is that they do not have a small number of key indicators which can be used to summarize the basic variations in the human occupance of space – such as the humidity, temperature, and pressure variables basic in climatological study. By and large human geographers are still struggling to identify – let alone precisely measure – the fundamental sources of variation in a veritable jungle of overlapping relationships and to pick out indicators for these patterns. This means they are in constant danger of being overwhelmed by the variety and complexity of real world facts and observations.

(ii) A second problem is that they are rarely able to control and manipulate their variables and observations by the standard scientific analytical approach; namely, of breaking problems into smaller component parts to ease the study of interrelationships. Part of the difficulty is that they are dealing with phenomena that are larger than life – with things that cannot be reduced to laboratory size and studied under controlled scientific conditions. Hence it is more difficult to standardize descriptions or to derive unequivocal conclusions. In addition, this interest in human phenomena means that they can rarely experiment on regions or towns to test out alternative ideas under controlled conditions since this implies totalitarianism and manipulation. Moreover, their data frequently comes from secondary sources such as the census. Like sociologists they often do not have the resources to mount the full-scale, problem-specific data gathering investigations carried out by many of their colleagues in the physical sciences.

1

(iii) A third handicap is that they find it difficult to generalize beyond the single case, to relate the results obtained in one study to all areas. By contrast, many of the relationships found in physical science hold true in any area, since they are based on determinate physical properties. In part, of course, this is a function of the very fact of spatial variation in human phenomena which produces different patterns in the complex world. But it is also a scale problem, related to the fact that geographers are dealing with aggregations based on areas. Generalizations from aggregations at one scale do not necessarily hold true at another scale. When combined with the difficulty of relating the generalizations derived from one area to those of another serious problems are posed for the investigator.

(iv) Finally, the mere fact of dealing with human phenomena means that human geographers are not studying concentrations of inert objects, but areas containing beings with feelings and consciousness who often have positive desires to change their environment. Psychologists have long recognized that a single mode of explanation cannot account for all aspects of behaviour; as Hitt (1968) has reminded us in his review of the relevance of behaviourism and existentialism in psychology. By the same token it is most unlikely that all aspects of the spatial patterning of the world can be dealt with by the normally accepted <u>analytical</u> method of science. The scientific approach searches for generalizations, rather than unique events, and arranges its conclusions in systematically related statements. It also attempts to provide a dispassionate view of the relationships between variables or phenomena. In the human context the concern for unique, individual characteristics frequently takes precedence over generalities, and it is increasingly recognized that all views of the real world are biased to some extent by the viewpoint and prejudices of the observer. Complete objectivity or detachment is difficult to achieve and, for those of an existential turn of mind, it is a pointless goal as each person is considered to view the world in very different ways. Not surprisingly, therefore, many human geographers have rejected the modern scientific, positivist method shared with physical geographers. This group have found satisfaction in a variety of alternative philosophies, ranging from a generally humanistic approach, to the specifically phenomenological methods in a descriptive context, or to the prescriptive concerns of the radical or the marxist. But we must not assume these are new trends. Few of these alternative approaches (Johnston 1979) that have caused such recent impact on the philosophy of human geography are really new. What is different is that it is only in the last decade that there has been such a <u>conscious</u> orientation of geographical interpretations to these different philosophical values. The result is that the pluralism in the content of human geography is complemented by a pluralism in geographical philosophy.

These alternative philosophical approaches must be welcomed since they throw additional light upon the spatial complexity of the world. But for students interested in building up a corpus of <u>precise, systematically related knowledge</u> in human geography they are of limited value. They deal with quite different problems to the systematization of knowledge. In any case the very tide of quantification that carried the methods of modern science to geography has its own limitations for those interested in studying the multivariate complexity of the world. The most obvious is the ever-present problem of reductionism in

science; the need to isolate and abstract a limited set of relationships for scientific analysis out of the variety of alternative
linkages. Any scrutiny of the scientific literature will reveal that
the relationships studied are basically <u>univariate,</u> or at most <u>bi-
variate,</u> in which variables are studied in <u>isolation,</u> or are explained
in terms of a limited set of indicators. When this reductionist and
univariate approach is added to the need to understand the new language
of statistics, it is not surprising that many of those interested in
the general human mosaic rejected the modern scientific approach for
its apparently narrow focus. These problems were compounded by the
availability and utility of the alternative philosophies noted above.

This partial rejection of the scientific approach is unfortunate, for
it obscures the fact that scientific methods exist for dealing with
what Cattell (1952) referred to as the other half of the methodological
realm: the one that deals with the <u>simultaneous variation in many</u>
<u>variables</u> rather than the controlled, univariate, experimental
approach. These so-called 'multivariate' methods – of which factor
analysis is perhaps the most versatile set of procedures – focus
directly on the multivariate situation that causes so many problems for
the human geographer and ecologist. Fortunately, factor analysis
procedures provide investigators with a set of methods that help reduce
the impact of many of the difficulties noted above. They provide ways
of: identifying patterns of relationships in data sets; of rewriting
data sets to produce parsimonious descriptions; and separating general
patterns of variation from specific local patterns. In many ways,
therefore, the factorial methods are <u>synthesizing</u> procedures. As such
they differ from the analytical approach of most quantitative methods.

This book attempts to demonstrate the utility of factor analysis
methods for human geographers and ecologists. It stems from a decade
of fascination with the utility of these methods in developing both
generalizations and a substantive literature in urban and regional
geography. Given the synthesizing nature of factor techniques, the
procedures provide a quantitative approach to the type of generalizing
functions provided by other, more intuitive methods. Yet it must be
emphasized that this book is <u>not</u> a technical, statistical study in the
sense of focussing upon the statistical derivation of factors.
Algebraic, computational methods demand a knowlege of matrix algebra so
the derivation of new factor procedures is a branch of applied
mathematics not geography, although there is nothing to prevent
geographers contributing to this body of knowledge. If investigators
have such a knowledge it makes more sense to proceed directly to the
basic statistical texts on factor analysis such as Cattell (1978) or
Harman (1976). Similarly, it is not designed as a manual enabling
investigators to produce their own factor programmes. This would
demand a knowledge of an appropriate computer language. In any case
useful factor analysis package programmes already exist, such as those
in the various editions of the manual Statistical Package for the
Social Sciences (Nie, Brent et al 1975). Other excellent sources have
been available for some time, such as Mather's (1976) survey of
computational methods in multivariate analysis for physical geographers
or Cooley and Lohnes' (1968) description of multivariate procedures.

This book has quite different objectives to the rigidly statistical,
or computational factor derivation approach. It is designed as an

introduction to the use of factor analysis methods in human geography and human ecology, in which emphasis is placed upon the problems as well as the utility of the procedures. As such it does not assume any knowledge of matrix algebra or computer languages. Instead, its focus is upon the applied or 'user' tradition established in social science by Child (1970) and Rummel (1970), even though the latter provides an extensive discussion of matrix algebra concepts. Lying between these two studies in scope, the book is organized in such a way that the basic concepts are introduced before more advanced work on problems or more sophisticated methods are discussed. In addition, the study is designed to go beyond these applied texts since it attempts to demonstrate how factor methods can contribute to the development of geographical generalizations and to the emergence of a systematic literature in regional and urban studies or human geography and ecology. This objective cannot be attained only by illustrating factor methods with simple examples – the standard approach used by geographical texts which describe multivariate quantitative methods (King 1969; Johnston 1978).

Obviously the introductory surveys of the basics of factor analysis methods, and the discussions of the variety of procedures and problems involved, have to follow this route. This study, however, goes further. It attempts to show how the methods can be used to derive a systematic literature in part of the human ecology field, namely urban and regional studies. It must be added that within the past decade some of the earlier enthusiasm for statistical methods among human ecologists has declined. So interest in factor analysis procedures has waned. Yet during this same period interest in the deep seated patterns or mechanisms of society, expressed in the 'structuralism' school (Levi-Strauss 1952; Piaget 1970; Harvey 1973; Gregory 1978), has blossomed. To date most of the literature in this field is cast in the subjective descriptive mould. Factor analysis, with its emphasis upon quantitatively derived generalizations of the underlying dimensions or latent structure of society, provides one of the ways in which the patterns suggested by structuralists can be identified and confirmed, although it seems less useful in dealing with the mechanisms of change. Thus Blau (1976) observed that these structures represent the building blocks of theory, the collectivities not directly observable. So it can be argued (Davies 1983) that the latent structures identified by Shevky and Bell (1955) for urban social areas, provide one of the first examples of the identification of the deep seated patterns lying behind the empirically observable variables. This study provided the stimulus for many of the earliest factorial ecologies. Once these structures are identified, the mechanisms for change need to be derived.

In many ways, therefore, this study can be thought of as a partial bridge between the early interest in statistical applications and the new frontier of 'structuralism'. However, since the results of any study depend upon the statistical procedures adopted, the emphasis of this study is placed upon the former. The book falls into three parts.

(i) The first three chapters introduces the field that has been called factorial ecology. A summary of the utility of the approach in Chapter I is followed by a discussion of the basis of factor analysis

(Chapter II) and a survey of the content and relationships dealt with by the approach (Chapter III).

(ii) The next three chapters successively deal with data problems (Chapter IV), the variety of alternative factorial procedures (Chapter V), and the problems of interpretation faced by investigators (Chapter VI).

(iii) The final four chapters survey the contribution of factor analysis to a limited number of fields: first, in the development of generalizations in regional geography and the derivation of functional regions (Chapter VII); second, in the study of inter-urban (Chapter VIII) and intra-urban dimensionality (Chapter IX) both of which have links to the 'structuralist' literature. A final chapter (Chapter X) summarizes the study, evaluates the utility of the procedures for geographical and ecological work and anticipates the development of a more comprehensive approach to urban and regional ecology.

A constantly worrying issue in all applied work using quantitative methods is whether methods derived by statisticians can legitimately be applied in other areas as Gould (1970) has succinctly observed over a decade ago in geography. Much of the proof required by statisticians before a particular set of results are accepted relates to the fact that the observations are drawn from a sample of homogeneous cases. The assumption is made that the observations used to measure the variables form an unbiased sample drawn from a population. Given measurement errors, there is always the possibility of the result produced from the study of the samples being obtained by chance. Any result, therefore, has to be linked to the size of the sample from which it is drawn, and is related to the degree of probability of the result, for example, whether the result for the sample could have occurred one in one hundred times (the 1% level) or once in a thousand times (the 0.1% level). All too frequently human geographers are dealing with complete, not sample populations in their areal studies, so the significance tests are not really appropriate. In any case statisticians have to make assumptions about the distribution of variables before their significance tables can be derived. Frequently these are breached by variables used in human geography. Moreover, in the specific multivariate area of factor analysis, even the authors of basic texts, such as Harman (1976), Cattell (1978), point to the ever-present problem of the difference between 'statistical' and 'practical or content area' significance, and provide illustrations of the way a statistical test may produce factors that are additional to those the investigator is able to interpret in any meaningful way (Harman 1976 p.5,213). These two problems demonstrate that there is a very real difference between the worlds of the theoretical statistician and the applied worker. Yet users of factor analysis are fortunate in that the procedure does not have to be used in an inferential context. Cattell's (1978) recent and exhaustive survey on the 'Scientific Use of Factor Analysis' vividly deals with this general point. He points to the danger of blindly following the dictates of mathematicians with their preference for abstract, logical and deductive relationships. Cattell (1978) contrasts this with the applied worker who recognizes the need for what he calls 'more complex and flexible scientific models' in which solutions are found that are only 'approximate and often look relatively clumsy to the mathematician' but help produce

5

answers to the problems dealt with by psychologists, etc. These opinions are even more relevant in fields such as human geography. But given the fact that psychologists have had over a half a century of interest in factor analysis, it is not surprising that their level of sophistication exceeds that of other social scientists and geographers – at least at this time. After all the latter have had barely 15 or 20 years of experience in the field. Inevitably the systematic geographical literature derived from research factor analysis methods is much more limited in scope and in sophistication of result or method.

One of the really vital elements in the creation of a systematic literature is, of course, the derivation of conclusions that are acceptable to other workers, in the sense that investigators having the same problem and using the same data and techniques would produce the same generalizations. If factor analysis helps investigators achieve this basic scientific end then a great deal has been achieved. All too frequently, however, this type of conclusion cannot be reached because of technical problems or ambiguities in the study. Hence a great deal of the utility of the factor methods is lost. If, however, researchers can reach this position of agreement between alternative investigators, then standardization of results has been achieved. This is one of the essential requirements of scientific method. Of course, exponents of a strictly statistical approach will not be impressed by such arguments. But given the way many of the assumptions of the inferential statistician are breached in so much work in human geography – because of the nature of the problem being dealt with – it is difficult to see how a strictly statistical or rather inferential approach can be used universally. It applies in rather limited circumstances. This, of course, may change as the particular problems facing human geographers and ecologists are resolved. This does not, however, restrict the use of factor analysis since the method can be used in an exploratory, data reducing context, as well as in its strictly inferential mode. At this stage in our use of factorial procedures – and given the type of problems dealt with and the data sets being analyzed, or rather synthesized – it is likely that most studies will continue to adopt this more flexible approach. Inevitably this study reflects such a current characteristic of the literature. Nevertheless, given the more powerful properties of the common factor approach in the justification phase of an enquiry, the hope is that factorial investigators will work towards the day in which the inferential approach can be adopted more widely. Therefore, this work is designed to focus upon the range of variation in the data inputs and factorial methods, so as to expose the basic problems faced by an investigator in human geography or human ecology. By exposing these problems future factorial studies will be able to resolve them more easily. Then interested students can be made aware of three requirements for future work in the field.

(i) The first is to go beyond the mainly verbal descriptions of this study into the language of matrix algebra from which factorial studies have been developed by statisticians.

(ii) The second is to use the methods as additional statistical tools for deriving generalizations about the geographical and ecological variety in the complex reality of the world. In this context the case for factorial methods must not be overstated. After all Davies (1983) has shown that the factorial approach may be part of a wider methodological approach called the multivariate-structural approach, in which additional multivariate statistical methods can be applied. Factorial ecology may prove, therefore, to be a stage towards this goal: one that now needs a comprehensive review.

(iii) Finally, the hope is that factorial studies may contribute to a substantive body of literature in human geography and ecology, one linked, in part at least, to the emerging 'structuralist' approach. Obviously the structuralist approach in urban and regional studies go beyond the areas in which factor analysis can be applied. Nevertheless, the methods described in this book provide one of the ways in which the deep seated patterns sought after by structuralists can be identified.

2. FACTORIAL ECOLOGY

Most applications of factorial methods in human geography in the past 15 years have been described as factorial ecology. In essence this is a shorthand term for describing the results obtained from the application of factor analysis methods to human ecology, or more generally human geography. Originally coined by the social scientist Sweetser (1965) to describe a new approach to the study of urban dimensions and the classification of social areas, it was generalized by a geographer (Rees 1971) to describe the application of factor analysis to any set of data based on spatial or ecological units. As the product of interaction between two formerly separate streams of enquiry, factor analysis and human ecology, it must be emphasized that factorial ecology is only one of the ways in which human ecological relationships can be identified and understood, and it is primarily concerned with the study of the dimensionality, scaling, and classification of these areally based phenomena. As such the emphasis is upon the synthesis of information to provide generalizations. Factor analysis, therefore, provides geographers and social scientists with techniques for solving one of their oldest tasks - the synthesis of information - in an age in which analysis has been the standard scientific procedure.

In many areas of human ecology or human geography, such as the study of urban classifications (Berry 1972), regional perception (Gould and White 1968), international patterns (Russett 1967), and inter-regional flows (Black 1973; Davies and Thompson 1980), the factorial approach provided a major advance over older methods based on single variables. It led initially to quite euphoric feelings about the merits of the procedure (Berry 1969). Inevitably a reaction set in during the early 1970s, with critics pointing to the 'shallow reward' of the approach in urban ecology (Palm and Caruso 1972 p.132), 'a dangerous swamp where political geographers would soon exhaust their energy' (Prescott 1972 p.40), or to its 'lack of relevance' to the problems of the real world and its 'arid statistical summaries' (Harvey 1973 p.161). These and other criticisms simply remind us that one cannot justify the factorial approach - or any other descriptive statistical method for that matter - in geography or human ecology simply be reference to its technical sophistication. If the method is to be used in a geographical, as opposed to a statistical, context then it must be shown to produce

substantive results that are a distinct advantage over older or more
simple ways of achieving the same end. Taken to its logical conclusion
it could be argued that if the substantive ecological or geographical
results of the factorial approach are more important than the
techniques by which the spatially based variables are analysed, then
there is little point in dignifying the approach with such a specific
title. This argument has merit, but ignores the fact that the
continued use of the term 'factorial ecology' has two major advantages.
In the first place it provides a common focus for the problem faced by
investigators of such diverse geographical or areally based phenomena
as urban characteristics, regional interaction, or agricultural
patterns. This focus enables investigators to build up a common body
of knowledge on the pitfalls and relative advantages of particular
methods. In the second place it provides an efficient way of dealing
with similar ecological problems, namely the study of the dimensions of
urban systems or intra-urban structures, and draws attention to the
essential interdependence of ecological features, for example, the
relationships between various types of urban structures and urban
flows. Although these advantages may be considered to be sufficiently
important to continue using the term 'factorial ecology' in the way
Rees (1970) suggested, like all shorthand terms there are dangers.
Thus, once investigators begin to use cluster analysis or other
multivariate methods to group factor scores the investigator goes
beyond the purely 'factorial' context; whilst a lot of the behavioural
aspects of human ecology are not appropriately dealt with by factorial
or multivariate methods. Davies (1983) has pointed out that the
approach may be more appropriately described as multivariate-structural: 'multivariate' allows the use of other multivariate methods
instead of only factor analysis, and 'structural' pins down the
descriptions to 'structural' considerations (Blau 1976).

3. FACTOR ANALYSIS IN HUMAN ECOLOGY AND GEOGRAPHY

Factor analysis is not a single technique. It is a generic term for a
family of statistical techniques which are flexible enough to be
applied in an inferential as well as a non-inferential context. Indeed
Rummel (1970) goes as far as suggesting that factor analysis is really
a 'calculus' of the social scientist. It is necessary to emphasize
that the term 'factor' is used in a general sense in this study to
describe the axes derived from both the common factor and component
models (Chapter V); in the former model the objective is to test the
adequacy of some previously made hypothesis; in the latter, the data is
simply rewritten in a more concise form. The factoring procedure,
therefore, produces either a parsimonious description of the data - by
components in the case of the component method - or is a summary model
of the common dimensions of data in the common factor procedure. The
effect in both cases is one of data reduction. The difference is that
the components represent a new summary description of the data, whilst
the common factors are used to test the adequacy of some hypothesis
about the amount of common variance contained in the data.

Strictly speaking, factor analytical techniques are a branch of
applied mathematics, but the history of their development shows that
they have always been very closely linked with the social sciences in

8

general. In his latest authoritative survey of modern factor techniques Harman (1976) showed how the methods were originally extensively developed by mathematically minded psychologists, such as Spearman (1904, 1927), Pearson (1901), Burt (1917), and Thurnstone (1931). The major objective of these pioneers was a method for orderly simplification, to isolate the basic or underlying components or distinctive features of human ability from a series of specific intellectual traits they could individually measure from psychological tests. The factor techniques, therefore, were used to synthesize the data of the psychological tests to obtain the basic sources of variation in ability, not only to build a theory of human ability, but also to express this theory in a testable mathematical framework in the form of common factors or basic sources of variation.

By contrast, human ecology and human geography have been traditionally developed in a non-mathematical context. The term 'ecology' stemming from the Greek word 'oikos', meaning 'home' or 'habitat' was first used by 19th century biologists to describe the scientific study of the relationships between organisms and their local environment, and primarily involved plants and animals. Then the social darwinists applied Darwin's ideas of selection to the study of man-environment relationships, and geographers in their study of areal distribution speculated on the respective causal merits of environmentalist 'free will' or possibilist mechanisms. But it was not until the 1920s that the term 'human ecology' was used by a group of University of Chicago researchers to define a distinctive social science. According to Park, who, along with Burgess and Mackenzie became the major figures in the field:

> 'there are forces at work within the limits of the human community ... which tend to bring about an orderly and typical grouping of its population and institutions. The science which seeks to isolate these factors ... is what we call human, as distinct from plant or animal ecology.' (Park 1925 pp.1-2)

Three features of this statement should be noted: the first, is the emphasis upon 'orderly and typical grouping' which implies the search for order in spatial distributions; the second is the search for the reasons for the groupings; the third is the development of a science, a discipline of scientific endeavour. Despite the importance attached to the Chicago school of human ecology by the standard texts in the field, for example Timms (1971), Robson (1969, 1975), Johnston (1971) and Berry and Karsada (1977), there can be little doubt that only the third of these features was really new. For example, the study of these 'orderly and typical groupings' was hardly as important a breakthrough as has been implied. Booth's (1903) survey of the habitat and characteristics of London's poor in the 1880s had a similar objective, even if his inductive and factual methodology is not fashionable today (Davies 1978). Moreover, Henry Mayhew (1851, 1862) in the 1850s and 1860s pioneered the type of investigative reporting used by the Chicago ecologists. Also, the basic structure of Burgess's concentric zones, if not his precise zones, can be seen in the descriptions of Manchester in 1844 produced by Engels (1845), although these patterns were not integrated into a formal model. Last, but certainly not least, human geographers (albeit at a regional level, as opposed to the city scale

favoured by the Chicago ecologists) have been describing and interpreting human patterns on the face of the earth for many centuries.
Many people accept Berry's definition of geography as dealing with:

> '... the spatial arrangements and distributions, to spatial
> interactions, organizations and processes ... (in) ... the world
> wide ecosystem of which man is the dominant part.' (Berry 1964
> p.2, 3)

Hence it is clear that there is a considerable overlap between the
fields of human geography and human ecology. This introduction is
hardly the place to provide a protracted survey of the similarities and
differences between human geography and human ecology. Instead,
for the purposes of this book it is proposed that human ecology, by
reason of its concern with spatial patterns, is a subset of human
geography, perhaps with a more behavioural orientation that reflects
the sociological origins of its principal exponents.

The first applications of factor analysis techniques to problems
dealt with by human ecology and human geography were made over 40 years
ago. The statistician M.G. Kendall (1939) studied the distribution of
crop yields in England using factorial techniques, and the sociologist
Price (1942) employed similar methods to interpret the inter-urban
character of American cities. But despite these initial breakthroughs
it is only in the last 15 years that factorial methods have been
extensively used by geographers and human ecologists. This recent
flood of interest is obviously connected with the general tide of
quanitification that has swept over the social sciences in recent
years, with the fact that large scale computers and package computer
programmes were easily available in most colleges. But it is important
to note that the application of factorial techniques has been primarily
in the data reducing field, in other words isolating the basic patterns
of variation in a data set. This contrasts with the more usual
application of factorial techniques in psychology, where they are more
usually used in an inferential context to test the adequacy of
particular hypotheses. This contrast, linked to the difference between
the component and the common factor approach, is a fundamental one in
factorial procedures and can best be explained within the context of
the alternative procedures of the factorial approach in Chapter II.
The rest of this introductory chapter provides a summary review of the
utility of factorial methods with examples drawn from urban ecology.

4. FACTOR ANALYSIS AND THE SYNTHESIS OF DATA SETS

Human geography and ecology deals with the multivariate human reality
found on the earth's surface: with a complex areal variety of
overlapping sources of variation. In attempting to describe and
interpret the basic human patterns of areas, 19th century geographers
usually produced a long series of inventories describing the content of
regions or areas. Although these inventories could be organized or
classified in different ways, the development of a systematic
discipline, or of systematic ways of thought, demanded the integration
of these spatially based facts. In other words, it became necessary to
develop a series of principles, or generalizations to encompass sets of
these geographical facts. This need to interrelate objects and

phenomena, to produce generalizations which reduce the complexity of the real world to manageable proportions capable of description is, of course, one of the fundamental requirements of any academic enquiry. The problem, therefore, is one of finding ways to synthesize the spatially distributed information into summary descriptions or generalizations. To revert to the objective for human ecology (Park 1925 p.2), the problem is to identify the 'orderly, typical groupings.'

In the early stages of the development of both human geography and human ecology, this synthesis was carried out subjectively. Few attempts were made to analyze the way in which the summary statements were created; they were left as inspired guesses of inspiration. In most cases this approach usually emphasized the spatial variations and areal integrations, rather than dealing with the interrelationships of the variables. However the results certainly provided important insights and even breakthroughs in understanding. For example, Fisher (1970), in a defence of traditional regional geography, pointed to the 'breadth of vision' and perception of many of the early twentieth century geographers; the vision that presumably led the geographer Mackinder (1902) to develop the 'heartland' and 'city region' ideas, and the human ecologist, E.W. Burgess (1925), to derive his concentric zones. But any recognition of the immense value of these generalizations in the early development of knowledge should not blind us to the fact that they are methodologically unsystematic. Four separate problems in this subjective approach can be recognized.

(i) There is little order in the way the generalizations are produced and there is no procedure for relating them to either adequate causal explanations, or to apply them to other areas. Without these procedures even perceptive ideas such as the 'heartland', or 'city region', cannot be considered as scientific concepts. This is a problem for the next section of the chapter.

(ii) The effective consequence of the subjectivity of the approach is that the resultant generalizations are only likely to be produced by prepared and incisive minds; minds that also have to be skilled in dealing with these ecological patterns. As there are comparatively few minds in this category, progress is likely to be minimal. Hence it is difficult to produce the kind of systematic, carefully interrelated building blocks or generalizations that compose a scientific body of knowledge.

(iii) There is no way of knowing how many of the features present in any area have been incorporated into the generalizations, so it is difficult to establish either the relevance or the limits of the description.

(iv) The generalizations are usually put forward on a 'take it or leave it' basis and are rarely tested for their adequacy in summarizing patterns of variation. Since they are difficult to modify in any systematic fashion, the generalizations either break down upon challenge or quickly outlive their usefulness.

Obviously one cannot deny the importance of the flash of intuition in any scientific endeavour, but this breakthrough needs to be supported by rigorous testing before it is accepted by the scientific community.

At this stage it is worth describing an example which will clarify the difficulty of the subjective approach to synthesis. Suppose that our objective is to uncover the basic social dimensions and patterns of western Canadian cities and there is evidence on the spatial distribution of six variables for 28 census tracts (Figure 1.1), describing the social variations in one of these areas, the metropolitan area of Calgary in 1961 (Table 1.1). It has already been shown that one approach to the problem would not even need this information. The insppration that lies at the heart of the subjective approach would allow some people to produce the basic generalizations directly from their knowledge of the city. Others would use the evidence of the six maps as a catalyst to integrate the variations into yet another subjective generalization. Both methods do <u>not</u> apply systematic ordering to the data, and all the problems described above limit the effectiveness of the approach. The reader is left to make his own subjective judgment of the most appropriate summary of these maps. It will soon be apparent that the task is not easy.

Table 1.1
Definitions of variables in the Calgary 1961 study:

1. Occupation: The percentage of the working population engaged in manual and semi-skilled occupations.

2. Education: The percentage of the adult population with at least one year or more of high school education or above.

3. Fertility: The ratio of the numbers of children under four to the total fertile female population.

4. Women at work: The percentage of working women (in paid employment) in the total adult female population.

5. Single detached dwellings: The percentage of separate single detached dwellings (single family dwelling units).

6. Ethnic: The percentage of the population <u>not</u> of British or French stock.

Fortunately there are ways around the problem. Ordering methods have been invented which enable investigators to reorganize the data into a form from which the essential patterns are more clearly recognizeable. The factorial method is one of the most successful of these approaches, but its utility can best be appreciated by initially looking at two alternative and frequently used procedures.

(i) Human geographers trained in the pre-quantification era often wanted to summarize the basic variations in several different types of distribution. Usually this was achieved by some type of map synthesis using the <u>overlay</u> method. The essential features of each map were copied on to individual sheets. By overlaying these sheets, and viewing the patterns in combination, it was frequently possible to

12

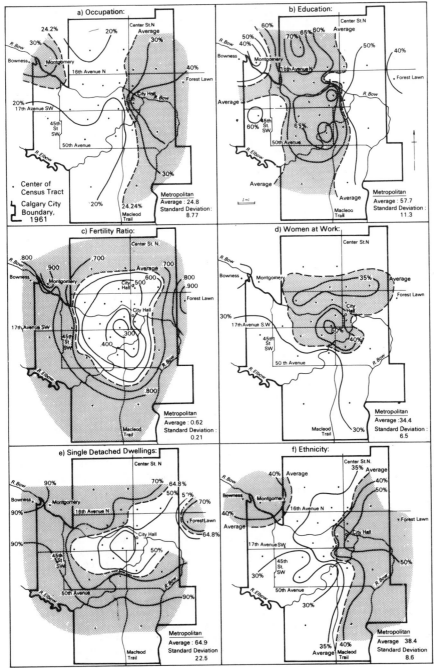

Figure 1.1   Distribution of six variables:
Calgary census tracts (1961)

13

arrive at a generalization, or summary statement of the basic patterns. The words 'frequently possible' are important ones, for the reader may not be able to provide a concise summary of the range of variations shown in Figure 1.1. Hence it must be remembered that the integrative procedure of the overlay process is still very much a subjective one, although the approach does represent some attempt at ordering the data.

Table 1.2
Correlation matrix of six variables: Calgary census tracts 1961

Variables		Correlations					
No.	Short Title	1	2	3	4	5	6
1.	Occupation	1.00					
2.	Education	-0.89	1.00				
3.	Fertility	+0.29	-0.41	1.00			
4.	Women at work	-0.11	0.22	-0.70	1.00		
5.	Single family dwellings	0.08	-0.16	0.82	-0.84	1.00	
6.	Ethnic	0.80	-0.77	0.16	0.11	-0.19	1.00

Source: Data from Calgary census tract information (Bureau of Statistics, Canada) 1961 Census. Pearson Product Moment correlations are shown.

(ii)   An alternative approach that can be used involves the use of measures of association between the individual variables. It is called the correlation bond method. The first stage consists of measuring the correlations between each of the variables over the same set of areas, in this case Pearson Product Correlation Coefficients (Table 1.2) since the indicators are measured in the interval scale of continuous values. In the second stage the variables most strongly linked to one another are then represented in diagrammatic form. Figure 1.2 shows the largest and second largest correlation links or bonds between each pair of variables. It is, of course, possible to translate this into an inferential problem and map all significant correlations, values that were significantly different from the chance (1%, 5%, etc.) of obtaining the value. But this would require that the data conforms to a whole range of rigorous assumptions (Poole and O'Farrell 1971). Given the purposes of this introductory description the correlation is treated as a single measure of the linear association between the variables. Two distinct clusters or correlations occur, one composed of the education, occupation and ethnic variables, a set that can be summarized or generalized as identifying socio-economic status, the other of the fertility, women at work and single family dwellings units variables, a set that can be described as family status. The coherence of these clusters of variables appears to be disturbed by the presence of negative signs. For example, the single family dwelling units

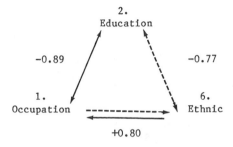

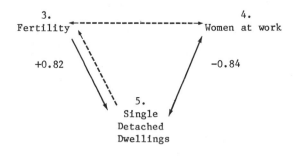

—————————→Directions of largest correlations

- - - - - - - -→Directions of second largest correlations

Figure 1.2 Correlation bonds between the six variables: Calgary 1961

variable is negatively linked with women at work. A moment's reflection will indicate that in the North American city it is likely that
since the single family dwelling units are occupied by families with
children, fewer women in these areas have a permanent job compared to
areas of apartments or duplexes.   Hence the variables will be
negatively related.

The basic advantage of the correlation bond method over the overlay
procedure is that the problem is shown to be clearly divided into two
distinct parts. The first part is one of defining the collectivities
of the variable set.   The second part is that of identifying the
regularity of the spatial patterns. Adoption of the correlation bond
procedure enormously reduces the size of the inferential leap needed to
describe the collectivities, but is only useful for the second problem
in so far as the appearance of two distinct clusters makes it possible
to justify the choice of one variable from each set as indicators of
these patterns.

In spite of these advantages the correlation bond approach still
deals with the problem one variable at a time. No measurement is made

of the degree to which there is a common variation, either in the variables or in the patterns. This is where factor analysis comes into its own.

(iii) A third approach uses factor analysis. The objective of the factorial approach is to deal with this basic problem by providing mathematical constructs called 'factors', one or more of which summarize the patterns of variation in a set of variables; the extent of common characteristics between each pair of variables having previously been determined by correlation or other similarity measures. A complete explanation of the way in which these factors are derived is provided in other chapters. Here it is enough to note that in the results produced from a Principal Axes Component analysis of the data using an orthogonal rotation (Table 1.3) the six variables are associated with two basic factors (often called axes, vectors or dimensions). Strictly speaking the factors are components; the component-common factor distinction is a basic one that is described in more detail in Chapter V. Each variable has a loading or weight with each axes which measures the amount of variation in that variable associated with the axis. These loadings look like correlations, in that they run on a scale from -1.0 (negative relation) through 0.0 (no relation) to +1.0 (complete positive relationship).

Table 1.3
Factor loadings of the six variable Calgary data (1961)

Variables	Components Loadings on 1	2	Communalities
1. Occupation*	-0.949	0.107	0.91
2. Education	0.932	-0.223	0.92
3. Fertility	-0.258	0.884	0.89
4. Women at work	0.003	-0.919	0.84
5. Single family dwellings	0.054	0.968	0.85
6. Ethnic	-0.929	-0.135	0.94
Variance of each component	45.05%	44.06%	

*Definitions provided in Table 1.1. The results were obtained from the Principal Axes technique using a component model followed by an orthogonal rotation.

Table 1.3 shows that the highest loadings of the occupation, education and ethnic variables are linked with Component I and the highest values or loadings of the other three variables are found on Component II. In a simple descriptive sense, therefore, the factorial approach has produced the same summary of the data set into two sources of variation as the correlation bond method. But this time the procedure has provided a measurement of the amount of common relationships for each source of variation by means of the loading. Indeed, the squares of the loadings tell us how much of the total variance of the variable is associated with each component. Taking the occupation variable as an example, 0.95^2 (or 90.25%) of the variability of this variable is associated with Component II, namely the second vector. Summing by column, instead of by row, it is possible to work out how much of the total variability of the similarity matrix is accounted for by each component. In the case of Table 1.3 the values are 45.1% and 43.9%, showing that the two axis solution accounts for 89% of the variability in the original correlation matrix. So by reducing the description of the data set from a 6x6 variable matrix to a 6x2 factor matrix only 11% of the variability has been lost by this linear representation of the data.

This discussion has described the way in which factorial procedures can be used to summarize the basic patterns of variation existing in a data set. However, a word of caution about the direction of the signs on the factor loadings is in order. Since the occupation and ethnic variables were originally calculated in terms of the manual/semi-skilled distinction and the non-British and French ethnic groups, so it must be expected that they would be inversely related to the education variable which measured adults with high levels of formal education. Similar reasoning may be applied to the other cluster.

Just as the factors, or strictly speaking components, each summarize the degree of common variation in a data matrix so it is possible to calculate component scores for each axis. These measure the importance of each area (census tracts in this case) upon each of the axes. In this example there will be two scores for each area and each score summarizes the variation in the area associated with each of the sources of variation. The six indicators for each tract are reduced to two new measures. When these scores are mapped a composite representation of the variation in each set of variables is obtained. Instead of mapping the scores in choropleth form, isopleths are used in Figure 1.3 to bring out the regularity of the spatial patterns. Component II is associated with the fertility, women at work and single detached dwelling unit variables and is concentrically distributed. Low values occur in the city centre where few single family dwellings are found, fertility levels are lower, and where there are higher numbers of women at work. Hence this axis, which has been summarized as measuring Family Status, has a specific spatial pattern. By contrast, Component I, described as Socio-Economic Status, displays a sector or zonal pattern, with negative values on the east of the city and a small wedge running along the Bow valley that expands in the Bowness-Montgomery region. High positive values are found in a wedge or sector extending from Mount Royal (M.R. in Figure 1.3) southwest along the Elbow valley to Glenmore, with an incipient sector on the northwest edge of the city. These positive scores pick out areas in 1961 that had a low percentage of manual and semi-skilled workers and ethnic groups, and have high percentages of people with formal education.

So far the factorial procedures have been treated purely as a black box mechanism, with the emphasis placed completely upon the results of the techniques, rather than on the way these results can be derived. The latter is dealt with in the next chapter. But even without understanding the basis of factorial procedures the results of Table 1.3 and Figure 1.3 are sufficiently clear to demonstrate that the

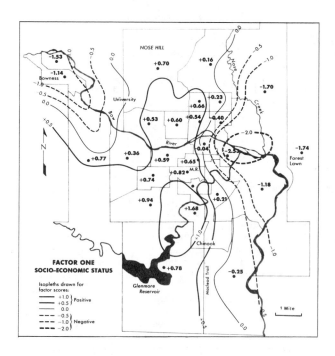

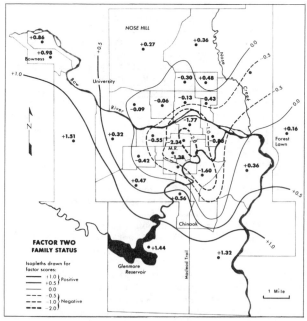

Figure 1.3   Factor scores of the Calgary six variable data (1961)

method provides a very valuable way of summarizing the latent pattern of relationships in a data set. Of course, the objection can be made that the correlation bond method, or even the intuitive approach, produced similar results. Hence there is little point in using such sophisticated methods. This may be partially true for the simple data set used in this example. But as the size of the variable set increases, so the advantages over alternative approaches become larger and larger. In any case the objection can only be accepted if the discussion relates solely to the problem of defining the collectivities in the data set. The factorial method also measures the importance of the observations in these factors or collectivities, and shows how much of the variance of the individual variables is associated with these generalizations or summary axes of differentiation. Other methods do not have these advantages. Moreover these results are derived systematically and depend upon a known data base, so that other researchers should arrive at the same conclusions. By enlarging the size of the data set the factors, or summary measurements of the collectivities in the data, may also be increased in number if there are additional sources of differentiation in the city. Hence the rest of the problems associated with the use of intuitively derived generalizations that have been discussed previously would seem to be solved. However the method is less successful in providing unequivocal evidence that these dimensions are the only ones that can be derived, or in demonstrating how the regularities or generalizations derived from one study can be related to other statistics. To successfully answer this question we must turn to another way in which factor analysis can be used and to the whole question of scientific explanation.

5. FACTOR ANALYSIS AND SCIENTIFIC EXPLANATION

The previous section has shown that factor analysis can be used as a systematic way of deriving generalizations, of reducing the complexity of the real world into a set of basic dimensional traits and their spatial patterns. Inevitably this means that some order is brought into those aspects of human geography and ecology suitable for this approach. But this systematization of the procedures for synthesis is not the only way of producing order. Instead of leaving the results as a generalization, or a systematic regularity an attempt could first be made to demonstrate that the pattern is an acceptable one (in the sense that other workers agree that it represents a suitable result) and, second, to investigate the extent to which the dimensions and patterns are typical of other cities. These are the purposes of all scientific explanations and are the basic objectives for adherents to any discipline claiming scientific respectability. In other words, studies are designed to connect together knowledge on separately known events by organizing the knowledge in a testable form so as to derive generalizations or laws from these relationships. To use Nagel's succinct summary, the scientific aim is to:

'provide systematic and responsibly supported explanations ... for individual occurrences, for recurring processes ...' (Nagel 1961 p.15)

Two rather different approaches have been adopted to attain this general scientific aim, the inductive and deductive methods. But before examples of these methods can be provided to derive the social dimensions of cities, two quite vital issues must be considered. First, it is usual to adopt the philosophical framework

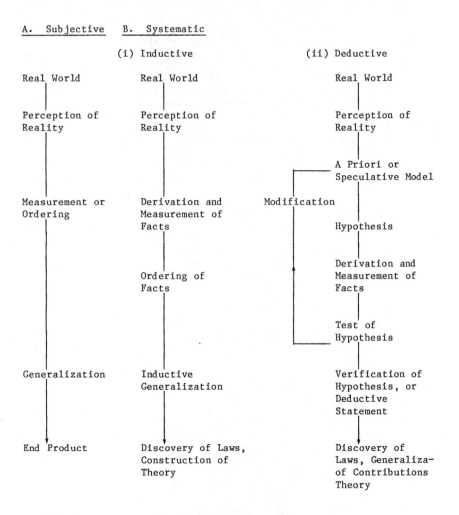

Figure 1.4   Inductive and deductive reasoning

known as epistemological dualism. In a sense this means that the notion of absolutes is rejected. Instead it is accepted that the systematic generalizations, or perceptions of the real world, cannot ever be identical to the hypothetical real world; complete understanding can never be attained. The importance of this idea in relation to human geography and ecology needs emphasis, given the remarks in the introduction to this chapter about alternative philosophical explanations. Second, it must be clear that the final objective of the inductive and deductive methods is basically the same. They differ in the route chosen to achieve this basic objective: in the way the initial generalizations or organization of knowledge is developed, and in their utility with respect to the verification process.

The principal stages in the inductive approach to the creation and verification of scientific generalizations are shown in Figure 1.4. In a geographical or ecological situation the investigator defines his problem and may begin by measuring the areal variation in a set of variables that were derived from his perception of the real world. The variables would then be ordered – perhaps by one of the methods described in the previous section – to produce an empirical regularity of the variation in the variable set or their areal patterns. This regularity could be regarded as an end product. But exponents of the inductive approach usually wait until a set of these regularities have been produced from several repetitive studies, and then attempt to integrate them into a general set of systematic generalizations or laws about the social variations in cities. In other words the investigator normally attempts to go beyond the singular instance of the studies; the repetitive findings would be used as the basis for the construction of a general statement applicable to all cases.

In urban ecology a good example of the inductive approach is provided by Hoyt's (1939) sector theory. He provided evidence for the sectoral variation in the residential structure of 142 American cities, and it is from these maps that he drew his conclusions about the importance of sectoral patterns in land values. Obviously the derivation of ecological conclusions from his maps does involve a major inferential leap. But at least he provides evidence of the spatial pattern that was identified as a key element in residential differentiation. So whether evidence of the patterns is presented, or whether the pattern is derived subjectively, the problem is the same, namely that the intuition of the observer is the vital ingredient in the production of the generalization. This is where the factorial approach is so useful. It provides a systematic method for deriving the generality from a data set, in other words it can be used as an ordering method to produce the generalization. Thus, in the Calgary social area example the six variables were shown to be differentiated in two basic dimensions of variation, described as Socio-Economic Status and Family Status, each associated with a spatial pattern. By repeating the calculations for other cities it becomes possible to measure the extent to which other areas can be described in terms of two social dimensions and two spatial patterns. In practice once more variables are used in any study it is likely that many more axes of differentiation are required to summarize the relationships.

Unfortunately, the utility of the inductive approach is more apparent than real as three basic problems show.

(i) First, the ordering process, whether factorial or not, does not tell us anything about why the variables are linked together, or why the spatial patterns emerge. In other words the method has provided us with traits and the patterns but not explained the process by which the pattern has been created. If the objective of the exercise is

21

classification for classification's sake, or is simply a summary description, this need not matter. Most human ecologists want to go beyond this stage.

(ii) Second, it is difficult to see why one should expect to be able to jump from the particular, in this case, Calgary, to the general case of other cities. On purely logical grounds no evidence has been put forward to suggest that the relationships discovered for Calgary have any relevance for other cities. Nevertheless, the factorial techniques do have the advantage of summarizing the patterns of variation in other cities in the same way, so they provide a systematic comparative procedure. The point here is that the reasons for the repetitive patterns have not been proposed.

(iii) Third, the question of why the six variables were chosen must be answered. Critics of the quantitative approach will point to the limited factual basis of the generalizations. Yet this limitation is clearly identified by the investigator - unlike the situation in the intuitive or subjective approach where it is difficult to find out how the investigator obtained the generalizations. Indeed the intuitive approach provides no objective measurement of each generalization (i.e., the trait or factor) and it is difficult to see they can be identified systematically. By contrast the factorial investigator can always expand his knowledge of the dimensionality of cities by incorporating new information. Although the inductively derived generalization does have a factual base, questions must be asked about its utility. The inductive approach apparently assumes that the derivation of variables and their measurement come directly out of our perceptual knowledge. A moment's reflection will indicate that we do not perceive in a vacuum; there is always some reason, some basis for the choice of variables. The inductive approach, in its simplest form, does not help to resolve this dilemma.

In many ways the deductive approach represents the inverse of the inductive method. It gives a greater role to the part played by the initial choice of speculation and makes a serious effort to control and channel this speculative activity. Figure 1.4 shows that the first stage in the deductive approach begins with our knowledge or understanding of the real world, but this time the knowledge is used as the basis for speculation on the relationships between the phenomena of interest. From this speculation, a hypothesis, or structured expected relationship, is postulated. Appropriate facts relevant to this hypothesis are collected and measured, and an empirical test of whether the facts bear out the hypothesis is made. If the test rejects the hypothesis, the investigator could either stop at this stage or modify the expectations and start again. If the hypothesis is confirmed it could either be related to other hypotheses and used to build a more general theoretical structure, or it could be put forward as the basis for a general law of the relationships between the phenomena.

One of the few examples of the application of the deductive approach in human ecology is represented by the work of Shevky and Bell (1955) who claimed to have used this type of methodology to explain the variations in urban social dimensions. They began by pointing out that the city is the most appropriate place to study changes in modern industrial society and maintained that the major societal differences from traditional society could be summarized as an 'increase in scale'. This was described as an increase in the scope and intensity of the social interactions which led to three primary changes.

(i) The increase in the division of labour led to a specialization in production by process, rather than by product. The result was a ranking of occupation according to skill, education and prestige.

(ii) The overall differentiation of function meant that the family ceased to be the productive unit. This led to a specialization of roles, eventually leading to the entry of women into the paid work force.

(iii) The widening horizons of society and the increasing mobility of social groups broke up the pattern of local associations. A consequence was the greater diversity of social groups within areas, leading to the isolation of some individual sub-groups from the mainstream of life and their localization within distinctive localities.

From this speculative model of the changes in society derived from the work of two anthropologists (Wilson and Wilson 1945), Shevky and Bell (1955) hypothesized that increasing scale could be identified by three basic constructs, or dimensions. These constructs could be indexed by a set of indicators or variables derived from the census. In fact they are similar to the six variables we have already used in the study of Calgary, namely the occupation, education, fertility, women at work, single family dwelling unit and ethnicity variables. In the original Shevky-Bell scheme these six variables were standardized within each city and converted to a ratio measured along a scale running from 0 to 100. Two composite social area indices were constructed by averaging the occupation and education ratios to produce a social status index, and averaging the fertility, women at work, and single family dwelling unit ratios to form a family status index. A four-fold division on each index was used to classify social areas into 16 basic types. In this way a hypothesis about the differentiation of urban society could be linked to the classification of social areas in a city.

Unfortunately, the Shevky-Bell approach contains a series of flaws, both in its technique and in its utility as a deductive methodology. For example, there is no empirical proof that the variables comprising the two indices always vary together in the same way - even if the ideas are restricted to western cities. Van Arsdol et al (1958) has shown that the variables do not co-vary in the southern cities of the United States whilst Herbert (1967) showed that the variable women at work is not related to the fertility and single family housing unit variables in the United Kingdom. To construct indices in this way, by adding together what could be unrelated variables, merely blurs the real pattern of relationships. Similarly, the between-city differences are obscured by reducing the variation within cities to the same 0-100 ratio scale.

It should be clear that both of these problems can be solved by the replacement of the social area measurement index by the factorial method, for the factor matrix will provide evidence as to whether the

1 Postulates concerning industrial society (aspects of increasing scale)	2 Statistics of trends	3 Changes in the structure of a given social system	4 Theoretical concepts or constructs	5 Sample indicators or statistics to measure the constructs	6 Derived measures	Specific indices
Change in the range and intensity of relations	Changing distribution of skills: Lessening of manual productive operations. Growing importance of clerical, supervisory, management operations	Changes in the arrangement of occupations based on function	Social Rank (economic status)	Years of schooling, Employment status, Class of worker, Major occupation group, Value of home, Rent by dwelling unit, Plumbing and repair, Persons per room, Heating and refrigeration	Occupation, Schooling, Rent	Index I
Differentiation of function	Changing structure of productive activity: Lessening importance of primary production – growing importance of relations centred in cities – lessening importance of the household as economic unit	Changes in the ways of living – movement of women into urban occupations – spread of alternative family patterns	Urbanization (family status)	Age and sex, Owner or tenant, House structure, Persons in household	Fertility, Women at work, Single family dwelling units	Index II
Complexity of organization	Changing composition of population: Increasing movement – alterations in age and sex distribution – Increasing diversity	Redistribution in space – changes in the proportion of supporting and dependent population – Isolation and segregation of groups	Segregation (ethnic status)	Race and nativity, Country of birth, Citizenship	Racial and national groups in relative isolation	Index III

Societal Change

Source: revised from Shevky and Bell (1955)

Figure 1.5 Social area analysis: construct formulation and index construction

24

variables are associated with three axes. In the Calgary case (Table 1.2) it has been shown that ethnicity is intimately linked to the occupation and education variables. So in this particular case a two axis model of urban social differentiation in 1961 appears to be a sufficient description of the data set. The key words here are 'sufficient description' and we need to return to this idea in later chapters. But to restrict the explanation to this level means that factorial techniques are only being used in a synthesizing role. They can be used in another way, in a rigorous hypothesis testing context. In this case the experiment would begin with a hypothesis that has been proposed – that the differentiation in the data set is associated with only two axes or dimensions – and once these relationships are abstracted from the data set only error variance remains in the data set. A variety of factorial techniques (Chapter V) have been invented to deal with this type of problem, but the vital part of the hypothesis is the proposition about the number of axes that are to be expected, and the significance test that would be used to accept or reject the hypothesis at the derived probability level. In other words the study tests whether the result could have occurred by chance at the 1% level (1 in 100 times) or the 5% level (5 in 100 times), etc. Since the 6x6 matrix of correlations has been reduced to a 6x2 factor matrix it is possible that not all the variation in the data set has been accounted for by the model, so some residual variance would remain. If the model fits the data exactly, the residuals left after abstracting the two axes would disappear, but with empirical data errors of measurement and random error are to be expected. Hence many factor significance tests actually measure the magnitude of the residuals, with the hypothesis of two factors being accepted if they are too small to provide another meaningful factor. The problem here is that 'meaning' for the content area of the investigator may be different to that of the statistician. Indeed significance tests often underestimate the number of factors that are statistically significant (Harman 1976 p.213). Now it is worth returning to the point made earlier about sufficient evidence – sufficient in the sense that other investigators would agree that the results are significantly different from chance. The normal statistical approach is to assume that the data is derived from a sample and the results depend on how this sample matches the complete population. Unfortunately for our study the data consists of all the census tracts in the urban area of Calgary in 1961, so significance tests are not really relevant for this example. In probability terms there is no way out of this dilemma, but we can demonstrate to other workers that alternative procedures and techniques produce substantially the same results, providing some degree of confidence in the stability of the results. Moreover, if similar results are obtained from other cities, greater confidence in the generality of the findings is achieved.

Although the application of the common factor model solves one of the major technical problems of the Shevky-Bell approach, other flaws still exist and these illustrate the dangers of the deductive methodology. The next obvious problem is that the results of the hypothesis test, or for that matter the inductive generalizations, only relate to the set of variables chosen for analysis. So to find two or three basic axes of differentiation in the social structure of a city does not mean that these are the only axes of differentiation along which cities can be structured. In our case study only a very limited data set was used. To search for others in a deductive context would mean that the initial data set would have to be adjusted. In any case there is always the possibility that the variables used in the control study may not be appropriate measurements. This is why most investigators in psychology require at least three or four indicators before any axis of differentiation is accepted. In the Calgary study it could be argued that ethnicity is not measured very clearly. Rather than using all people in a census tract not drawn from either of the two founding ethnic stocks in Canada (British and French) the proportion of Ukrainians, Germans, etc., could be used. A much clearer separation of ethnic groups may then be produced. This reinforces the fact that the choice of variables is always crucial for any analysis. Yet this is always true in any investigation. A researcher trying to confirm his intuition that a certain source of variation is as important will choose an indicator to measure this source. The problem in the intuitive approach is that the indicator has to be accepted as an independent measure of dimensionality. The utility of the factorial approach with a large enough data set is that each source of differentiation should be identified by several indicators. If these indicators are shown as clustering on individual axes the separation of the sources of differentiation is confirmed.

Most modern scientists favour the deductive route, or rather the hypothetico-deductive route, since it is supposed to begin with a reasonable hypothesis. Yet its difference from the inductive approach is not as clear cut in practice as most scientific papers would have one believe. Cattell (1966, 1978) has described the process of discovery as really being an inductive-hypothetico-deductive approach. He points out that:

'scientific work begins with data observation and returns to data observation ... It is usual, when playing up the elegance of science, to put more emphasis upon the deductive reasonings from a theoretical model to an expected result in a well designed experiment. But in the actual history of ultimate fruitfulness, to the initial inductive synthesis, whether from casual observation or more systematic data gathering, which gave both to the theory. Darwin's ruminations during the voyage of the Beagle to Franklin's watching of the thunderclouds ... suffice to remind us of the importance of the I in IHD.' (Cattell 1978 p.10)

Reference to Shevky and Bell's (1955) work demonstrates there are intuitive leaps at every stage, and it is not always easy to see how the constructs were derived from the principles of differentiation or the indicators from the constructs. Indeed the quotation demonstrates the vital importance of the initial inductive synthesis. In terms of the case study used in this study the component analysis of the 28 area x six variable data set can be thought of as being produced inductively by statistical means. However it will be shown in the next chapter that there are a series of decisions taken at various stages in the technical procedures. Naturally the results could have been produced in many different ways, but the factorial approach has the advantage of controlling the way the generalizations are produced. Obviously the vital decision in the investigation relates to the choice of variables and these could have been produced from an a priori theory of residential patterning, such as the one proposed by Shevky and Bell. If this latter route was followed it does make more sense to adopt the inferential basis of the hypothesis testing approach if all the requirements can be satisfied. This would ensure that the hypothesis that there are two or three factors can be accepted or rejected. But there is nothing to prevent the investigator using the component or data exploratory approach to arrive at the data set suitable for testing. In other words the hypothesis test may be the last stage of a lot of experimental work, in which component methods were extensively used in the early stages of the study.

But even if this route was followed the final difficulty of the deductive methodology shown in this example remains, namely the interpretation of the results. In the context of the social area theory of Shevky and Bell how can the classification of social areas be related to societal differentiation? Neither in the original monograph by Shevky and Bell (1955), nor even in the reply of Bell and Moskos (1964) to Udry's query (1964) on this issue, is there any satisfactory reason to assume that simply because urban society is differentiated so urban sub-areas must show variations. In practice there must be two separate questions involved, one involving a theory of societal differentiation, the other a theory of ecological or spatial variation. In the original Shevky-Bell formulations the two issues seem to be confused. Although the factorial appoach does not answer this problem, at least it clearly separates the two issues involved, namely the measurement of the collectivities of the data (the factors) and the measurement of the data on these collectivities (the factor scores).

6. BASIC OBJECTIVES OF THE FACTORIAL APPROACH

This introduction has demonstrated some of the ways factor analysis can help researchers resolve many of the problems identified in the beginning of this chapter, issues which provide difficulties for human geographers and ecologists. The result is that several major purposes can be seen to be fulfilled by factor analysis procedures.

(i) Factor analysis provides a means of identifying and measuring the relationships or basic patterns in a data set. As such it can be used to resolve the problem described in the first section, the frequent inability of human geographers or ecologists to identify key traits from the mosaic of overlapping relationships. Factor analysis can be used to screen data sets, to sift out, or lay bare the inter-relationships, and these are described in terms of factors which represent weighted summaries of the original variables. Alternatively investigators can pick out variables that best identify the various structures. For example, the occupation and single detached dwelling variables provide the best indicators of the Socio-Economic and Family Status axes identified in the results shown in Table 1.3.

(ii) Factor analysis provides a means of rewriting the original data into another more parsimonious form that can be considered to be a scaling of the original data, or a transformation into a new form more suitable for further analysis. In the example in Table 1.1 the 28x6 data matrix has been compressed into a 6x2 factor matrix and a 28x2 factor score matrix. In part, therefore, the approach makes up for the lack of experimental control experienced by human geographers, since all the essential information of the original set is contained in a smaller, more concise form. Hence standardized descriptions of the data are provided that identify the structures - the Family or Social Status axes - that lie behind the superficiality of the indicators. Obviously if investigators are interested in the original indicators, education, fertility, etc., this generalizing approach may be of little value. But by measuring the co-variation in the variables and reducing the data to more concise descriptions, the analysis enables investigators to illuminate the conceptual structures lying behind the individual facts: the structures sought after by the structuralist school of thought. Nevertheless, acceptable proof of the existence of these structures does require a well designed experiment on the available data - not just the collection of a wide set of indicators followed by factorial reduction. It must be admitted that most studies in the factorial ecology fall into the latter category.

(iii) Factor analysis has the flexibility of being used in an exploratory context, in data description or pattern recognition, as well as in a hypothesis testing mode. By the former route two axes of differentiation were produced from the 28 area x six variable data set. In the latter context the hypothesis that two axes of differentiation for this data set can be tested, and the attempt was made to separate the general or common patterns of variation from the specific or effects associated with individual variables. As yet most studies are in the exploratory mode, the approach that has proved so productive in psychology and is the method that is the explicit basis of Harman's (1976 p.6) overview of factorial methods.

The fulfillment of these purposes, however, does not answer the question of what relationship factors have to the various categories of concepts distinguished by scientists. In empirical terms it is apparent that the factors, whether Socio-Economic Status, Family Status, Ethnicity or others, can be regarded as typological devices for classifying phenomena according to their interrelationships. As such they are empirically derived concepts since they define existential categories of objects, in this case variables, although Chapter III will demonstrate their application to other things. In mathematical terms they are really formal or analytical concepts identifying a set of mathematical relationships among symbols. For example each row in Table 1.3 can be regarded as an equation, in which the variable is expressed in terms of weights attached to each factor. This point will be dealt with in more detail in Chapter II. In theoretical terms the factors can be regarded as constructs, or theoretical concepts, since they identify the structures lying behind the indicators, the constructs merely postulated by Shevky and Bell (1955). It is normally assumed that these theoretical concepts cannot be directly measured and they have relevance only to each other and to some general scheme, such as those sought after by the structuralist approach. In this sense a belief exists that there is a causal nexus behind the patterns that are observed. Hence the factors delineate or portray the structures that these causal relationships produce. Which of these concepts the factors are supposed to represent, depends principally upon the purpose of the analysis. So it is important that investigators should be clear about which construct they are attempting to identify in any study. As with all sophisticated multivariate methods, however, it is always worth checking the results back against the raw data and particularly the correlation matrix to determine the utility of the results. In this way investigators will not fall into the trap of accepting the factorial results as some kind of 'black box' manipulation of the data divorced from the original data.

2 BASIC PRINCIPLES OF FACTOR ANALYSIS

The objective of this chapter is to explain, in as simple a way as possible, how the basic patterns of variation existing in a data set can be summarized, or rewritten in a more concise <u>linear</u> form. In Chapter I it has already been pointed out that two rather separate problems are involved: first, the derivation of factors which measure the collectivities of the variables - or the way in which the variables 'hang together'; second, the calculation of factor scores, which measure the importance of the observations on the various factors and which can be used to study the geographical variations of the collectivities in the data set. This chapter provides an overview of many of the basic features of factor analytical techniques. It must be noted that the discussion follows a component rather than the strict common factor approach, because the latter depends upon statistical inference and hypothesis testing. The result is that the term 'factor' is used in its broadest possible sense here, as is usually the case in the human ecological literature. To restrict the term only to the inferential procedures of the common factor method seems unnecessarily pedantic at this time. Where the common factor approach is required the term 'common factor' is used to avoid any confusion. The chapter is divided into five sections, dealing successively with: the basis of factor procedures as described by elementary geometry; an example of the simplest form of algebraic factor solutions; the principles of rotation; the calculation of factor scores; and a conclusion summarizing the major decisions that have to be made in factor analyses. This elementary survey of the basic procedures means that no attempt is made to describe the variety of alternative factor solutions; these are discussed in Chapters V and VI.

1. THE GEOMETRICAL BASIS OF FACTOR ANALYSIS

The simplest way of explaining the basis of factor analytical procedures is to make use of the geometrical concepts of vectors and planes and the idea of correlation. A vector is a straight line running from a given starting point that has length and direction. Any two intersecting vectors form a plane. A correlation, on the other hand, is a measure of the relationship between any two variables that have been measured over the same set of observations, places or cases. Figure 2.1 shows that the relationship between any two variables (in this case two of the indicators described in Table 1.1) can be expressed in terms of a scatter diagram. The diagram shows that there is a positive relationship between the two sets of values. It is more precise to be able to replace this visual association with a single figure measuring the actual degree of relationship involved. Pearson's Product Moment Correlation Coefficient is one such correlation measurement and it runs on a scale from +1.0, measuring a perfect positive relationship between the two variables, through 0.0, indicating no relationship, to -1.0, which measures a perfect negative relationship. In Figure 2.1 the correlation is calculated at +0.82.

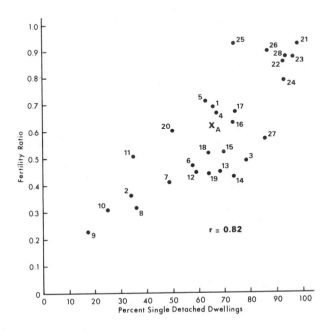

Figure 2.1   The correlation between two variables

The importance of the correlation in this case is that it can be expressed as a test vector. So instead of having to place the co-ordinates of the scatter diagram at 90° to one another, they can be rotated until the cosine of the angle between them is numerically equivalent to the correlation coefficient. Taking the variables V_1 and V_2 this means that the angle $(\emptyset_{12})$ between the vector representative of the variables can be written symbolically as:

$$r_{12} = V_1 \, V_2 \, . \, \cos \emptyset_{12} \qquad\qquad \text{Eq. 2.1}$$

Descriptively, this states that the correlation is equal to the product of the length of the two vectors multiplied by the cosine of the angle between them. If, as is usually the case, the lengths of the vectors are standardized at unit length, as for instance when the

31

variables are measured in standard deviation units or are standardized in this way, the correlation between any two variables can be represented simply by the cosine of the angle:

$$r_{12} = \cos \emptyset_{12} \qquad \text{Eq. 2.2}$$

Some examples of the way in which the correlation between any two vectors can be shown geometrically are provided in Figure 2.2a. The importance of this reformulation of the correlation as a vector is that it is now possible to portray the relationships between several variables as vectors on one diagram, instead of keeping them as separate scatter diagram plots. This can be illustrated by Figure 2.2b which shows the relationships between four variables as vectors V_1, V_2, V_3 and V_4. Hypothetical variables are portrayed, rather than those used in the Calgary example of Chapter I, to keep the argument as simple as possible.

Another elementary relationship that will subsequently be of use relates to the effect of dropping a perpendicular from one end of any vector A to another vector B. The perpendicular from A will cut this axis B at a distance along its unit length that is the same as cosine of the correlation between the vectors. The proof is easily provided, for the perpendicular creates a right angled triangle between the two vectors. For example:

$$\cos \emptyset = \frac{b}{c} \qquad \text{i.e.} \quad \frac{\text{side adjacent}}{\text{hypotenuse}} \qquad \text{Eq. 2.3a}$$

Since the hypotenuse is the standardized vector with length 1.0 the rewriting of the Equation 2.3a becomes:

$$b = \cos \emptyset \cdot c \text{ or } b = \cos \emptyset \cdot 1.0 \qquad \text{Eq. 2.3b}$$

(a) Representation of factors in two dimensions

From these relationships any matrix of correlation can be converted into an equivalent one that shows the angles of the vectors. In Table 2.1b the simple 4x4 matrix of Table 2.1a is converted into one that shows the angular relationships between the vector representations of variables 1 to 4.

Table 2.1
A hypothetical 4x4 matrix

(a) Intercorrelations

Variables	1	2	3	4
1	1.00	0.866	0.500	0.00
2		1.00	0.866	0.500
3			1.00	0.866
4				1.00

(b) Matrix of angles

Variables	1	2	3	4
1	0°	30°	60°	90°
2		0°	30°	60°
3			0°	30°
4				0°

This matrix is symmetrical so only one set of angles is shown.

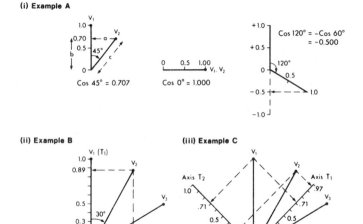

(i) Example A

(ii) Example B   **(iii) Example C**

Figure 2.2 Geometrical relationships of correlations

Potentially, there are as many vector planes as there are pairs of variables. But the geometry of the angles in Table 2.1b is such that the relationships in Table 2.1a can be expressed in one plane, or in the two dimensions coincident with the orthogonal axes V_1 and V_4. In Figure 2.2b it can be seen that these two axes are sufficient to describe all the angles between the four vectors. A moment's reflection will indicate that since all the angular relationships can be summarized in two dimensions there is no real need to use the 4x4 matrix to describe the relative position of the vectors. They can be

33

described equally as precisely in terms of the two vectors V_1 and V_4 as:

Table 2.2
Summary of the 4x4 matrix by two axes

Variable	Angles	
	(V_1)	(V_4)
V_1	0°	90°
V_2	30°	60°
V_3	60°	30°
V_4	90°	0°

By translating or converting these angles back into correlations the tabular representation of Table 2.3 is obtained. This summarizes all the relationships contained shown in the 4x4 table of Table 2.1 by a 4x2 matrix. Obviously this is a more parsimonious description of the pattern in Table 2.1, for the addition of two columns describing vectors V_2 and V_3, together with the appropriate entries for their correlations with V_1, V_2, V_3 and V_4, is redundant. In other words the pattern of points and relationships shown in Table 2.1 has been summarized by the 4x2 table.

Table 2.3
A 4x2 factor loading matrix

Variable	Axes	
	(V_1)	(V_4)
V_1	1.000	0.000
V_2	0.866	0.500
V_3	0.500	0.866
V_4	0.000	1.000

In algebraic terms Table 2.3 may be regarded as a rewriting of the original matrix of Table 2.1. The variability of each variable within the data set that was measured in terms of four vectors is now described as a function of two other vectors. The resultant equations, which are, in essence, factor equations, allow us to express, or to predict, the values of each vector from the two new reference vectors. These equations, predicting the values of the variables V_1 to V_4, are derived directly from the 4x2 matrix and consist of the following expressions:

$$V_1 = 1.000 \ V_1 + 0.000 \ V_4$$

$$V_2 = 0.866 \ V_1 + 0.500 \ V_4$$

$$V_3 = 0.500 \ V_1 + 0.866 \ V_4 \qquad \text{Eq. 2.4}$$

$$V_4 = 0.000 \ V_1 + 1.000 \ V_4$$

In this example the new reference vectors that describe the position of the original vectors are two of the original variables that happen to be at right angles to one another. There is, of course, no need to continue to use these vectors as reference axes or vectors. It is more convenient to assume that we have two new reference vectors, called T_1 and T_2, that are coincident with $\overline{V_1}$ and V_4. The only effect on the factor equations would be to replace V_1 and V_4 with T_1 and T_2.

Table 2.4
A 4x2 factor matrix: loadings, communalities and eigenvalues

	Variables	Factor Axes		Communality*
		T_1	T_2	
	V_1	1.000	0.000	1.000
	V_2	0.866	0.500	1.000
	V_3	0.500	0.866	1.000
	V_4	0.000	1.000	4.000
Sum of squared loadings by columns	Variance explanation	2.000	2.000	4.000
	Percentage explanation	50%	50%	

* Communality = sum of the squared loadings by rows

35

The new reference vectors used to describe the pattern of relationships in Table 2.2 may be regarded as hypothetical vectors, mathematical 'constructs', or more generally as 'factors' that geometrically
summarize the original variability of the variable. These factors,
axes, or vectors – the terms are used interchangeably – can also be
considered as surrogate variables, in the sense that the variability of
each variable is rewritten as a linear function of these new reference
vectors. The linear functions involved are expressed in precise
algebraic form in the factor equation 2.4. Each variable is represented as the sum of a weighted value of the two reference vectors.
When the vectors are standardized, the weights alone (the cell entries
in the matrix) provide a sufficient description of the relationship.
These weights are called <u>saturations</u> or <u>factor loadings,</u> and the type
of matrix shown in Table 2.4 is called a factor matrix, or more
accurately a <u>factor loading</u> matrix. Since these weights are analogous
to the regression coefficients in the regression model it may be
apparent that one can measure the amount of variability involved in
each loading by taking the square of each value. By summing the
squares of the loading across the rows a measure known as <u>communality</u>
is obtained. This is the amount of variability accounted for by the
factor solution, in this case as expressed in the two reference
vectors. In Table 2.4 the sums across the rows all add up to 1.00. In
the same way, but this time from summation by columns, it is possible
to obtain the amount of variance accounted for by the reference vector.
The square root of this value is usually called an <u>eigenvalue.</u> Each of
the vectors is standardized to unit length, so in theory the total
amount of variability to be measured will be 4 x 1.00, i.e., 4.00. In
Table 2.4 the sum of the columns both come to 2.00. By dividing these
amounts by the total amount of variance, which is the same as the
number of variables (in this case four) and multiplying by 100, the
percentage variance accounted for by that axis is obtained. Rather
than expressing these relationships as individual cases it is possible
to describe them in a more general symbolic form as:

$$Z_j = a_{jp} F_1 + a_{jq} F_2 \hspace{3cm} Eq. \ 2.5$$

Where:

Z_j equals the estimate of variable j; $a_{jp(q)}$ represents
the factor loading or weight of variable j on vector (or factor)
$p(q)$; $F1_{(2)}$ represents Factor $1_{(2)}$.

This example was deliberately chosen to portray the basis of the
factor model in as simple a way as possible. In any real world example
there is no reason why the hypothetical reference axes should be
coincident with the existing vectors. In practice, therefore, the
factor problem becomes one of searching for these hypothetical axes;
axes that are simply the descriptive tools used to summarize the
pattern of interrelationships in a data set. An important feature of
this search procedure is that there are usually many <u>alternative</u>
descriptions of a set of variables, thereby ensuring that factor
results are not necessarily unique. The point can be illustrated using
the same example as in Table 2.1.

In Figure 2.2c a hypothetical reference axis, TT_1, has been drawn at the midpoint or the centre of the splay of vectors. Another axis, TT_2, has been drawn at right angles to TT_1. The resulting angular relationships, and correlations between the original vectors and the new descriptive vectors called TT_1 and TT_2, are shown in Table 2.5.

Table 2.5

An alternative description of the 4x2 matrix

Variable	Angles		Correlations	
	TT_1	TT_2	TT_1	TT_2
V_1	45°	45°	0.7071	0.7071
V_2	15°	75°	0.9660	0.2588
V_3	15°	105°*	0.9660	-0.2588
V_4	45°	135°	0.7071	-0.7071

* The cosine of an obtuse angle is equal to the negative cosine of the angle taken away from 180°, i.e., Cos 105° = -Cos 75°

Therefore, the specific factor equations derived from this table are:

$$V_1 = 0.7071\ TT_1 + 0.7071\ TT_2$$

$$V_2 = 0.9660\ TT_1 + 0.2588\ TT_2$$

$$V_3 = 0.9660\ TT_1 + (-0.2588)\ TT_2$$

$$V_4 = 0.7071\ TT_1 + (-0.7071)\ TT_2 \qquad \text{Eq. 2.6}$$

Mathematically there is no difference in the results obtained from Equations 2.4 or 2.6. Both provide the same results for the variables. The variability of all vectors V_1, V_2, V_3 and V_4 are summarized equally efficiently, since the equations completely reproduce the original correlation matrix. But it is apparent that there is a difference between the two sets of reference vectors. Table 2.6 shows that the TT_1 axis accounts for almost three times as much variance as the TT_2 axis, whereas the T_1T_2 axes are of equivalent descriptive value in terms of the original vectors. This illustrates a vital point. A correlation matrix may be summarized equally effectively by alternative factor equations if the measure of effectiveness is the amount of overall explanation of the solution. Yet it is quite possible that the individual descriptive axes, or factors, each account for very different amounts of explanation as shown in Table 2.6. Hence the choice between these two solutions can only depend on some external criteria, such as maximizing the amount of variance on one of the vectors.

Table 2.6

Comparison between the factor loading matrices

Variables	From Figure 2.2b		Commun-ality (h^2)	From Figure 2.2c		Commun-ality (h^2)
	T_1	T_2		TT_1	TT_2	
V_1	1.000	0.000	1.00	0.7071	0.7071	1.00
V_2	0.866	0.500	1.00	0.9660	0.2588	1.00
V_3	0.500	0.866	1.00	0.9660	-0.2588	1.00
V_4	0.000	1.000	1.00	0.7071	-0.7071	1.00
Vector variance	2.000	2.000	$h^2=4.00$	2.864	1.636	$h^2=4.00$
Percent-age variance	50.0%	50.0%	100%	71.6%	28.4%	100%

The two sets of factor equations in Table 2.6 can be used to illustrate two other important relationships involved in factorial solutions. The first is that the cell values of the original correlation matrix can be derived from the factor loading matrix by taking the cross-products of the rows. For example, if one wished to derive r_{23}, the correlation (r) between row 2 and column 3, the cross product of the second and third rows of the factor loading matrix would be calculated as follows from the two examples:

From Equation 2.4:

$r_{23} = (0.866 \times 0.500) + (0.500 \times 0.866)$

$= 0.866$

From Equation 2.6:

$r_{23} = (0.9660 \times 0.7071) + (-0.2588 \times 00.7071)$

$= 0.6830 + 0.1830$

$= 0.866$

The second relationship is a geometrical estimation of the size of the factor loadings. The cosine of the angle between two vectors is the same as a measurement of the position where the perpendicular from one axis meets the other ensuring that the factor loadings can be measured directly from the diagrams. Figure 2.2c shows that the factor loading of vector V_2 on T_1 (T_2) is equal to the length of its projection on the reference axis T_1 (T_2). These values are equal to 0.50 and 0.89 in Figure 2.2b and to 0.26 and 0.97 in Figure 2.2c. Ideally, the exact values of the factor loadings can be obtained in this way, but this would always involve the construction of very detailed geometrical diagrams. In passing it can be noted that this illustrates one of the disadvantages of the geometrical as opposed to the algebraic representation of the factors.

(b) Representation of factors in three dimensions

In Table 2.7a the elementary 4x4 matrix is used as the basis for the construction of a more complicated 6x6 matrix in which two additional vectors (called 5 and 6) are added. The resultant matrix or angular relationships is shown in Table 2.7b.

Table 2.7
A 6x6 matrix

(a) Correlations

Variables	1	2	3	4	5	6
1	1.00	0.866	0.500	0.000	0.000	0.0000
2		1.000	0.866	0.500	0.000	0.2924
3			1.000	0.866	0.000	0.6018
4				1.000	0.000	0.6018
5					1.000	0.7986
6						1.0000

(b) Angles

Variables	1	2	3	4	5	6
1	0	30°	60°	90°	90°	90°
2		0	30°	60°	90°	73°
3			0	30°	90°	37°
4				0	90°	53°
5					0	37°
6						0

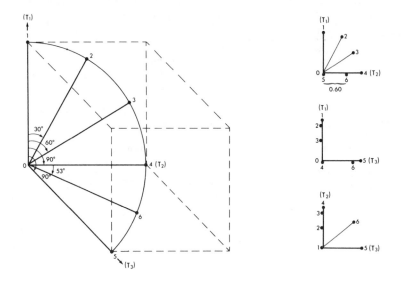

a) Position of the vectors in three dimensions.

b) Position of the vectors in three separate planes

Figure 2.3   Geometrical representation of the 6x6 matrix

If the same descriptive procedure used in the previous section is adopted to summarize the interrelationships in the data, a three dimensional set of reference axes has to be used. However, Figure 2.3 may be difficult to understand since the three axes lie in the two dimensions represented by the plane of the paper. Perhaps they can best be understood by imagining vector 5 as an axis standing out vertically from the page. Hence the reference axes T_1, T_2 and T_3 will be coincident with vectors 1, 4 and 5. These represent the sides of a cube as shown in Figure 2.3b. Measuring each vector from the three reference axes produces the summary description of the relationships in the factor loading matrix shown in Table 2.8:

Table 2.8
A three axis factor loading matrix

| Variable | Descriptive Axes | | |
	T_1	T_2	T_3
1	1.000	0.000	0.0000
2	0.866	0.500	0.0000
3	0.500	0.866	0.0000
4	0.000	1.000	0.5000
5	0.000	0.000	1.0000
6	0.000	0.6018	0.7986

40

From this table it can be seen that if only two of the reference axes are used then some of the information contained in the 6x6 matrix would be lost. For example, if it is assumed that only reference axes T_1 and T_2 are used, the revised factor loading matrix would be the one described in Table 2.9.

Table 2.9
Two axis description of the 6x6 matrix

Variables	T_1	T_2	Communality (h^2)
1	1.000	0.000	1.000
2	0.866	0.500	1.000
3	0.500	0.866	1.000
4	0.000	1.000	1.000
5	0.000	0.000	0.000
6	0.000	0.6018	0.3632
Variance (E)*	2.000	2.3632	Check h^2=4.3632
% Variance	33.33%	39.39%	E =4.3632

* These are the sum of the squared column entries

Two additional columns dealing with communalities and eigenvalues (variance contributions) have been added to this table. The previous section explained how the communality of a variable is constructed from the sum of the squares of each row. In this example it represents the amount of explanation of the two axis model. Table 2.9 shows that none of the variance of vector 5 is accommodated by the two axis model, and it only accounts for 36.32% (or .3622 out of 1.00) of vector 6. The last but one row in Table 2.8, the variance explanation, is obtained from the sum of the squares of the factor loadings as measured by columns. These values measure the amount of variance of the reference vectors T_1 and T_2, in this case 2.000 and 2.3632. Since it has previously been assumed that each of the vectors is of unit length, and that the total can be expressed as 1.00 (otherwise one could not use the cosine of the angle of the vectors to express the degree of correlation), the total amount of explanation of the six variable example must add up to 6.00 (six times one). If all the variance of each variable were accommodated on one of the reference vectors this value would be 6.00. It could not be greater than 6.00 unless rounding errors occurred. In practice, each solution (in this case the two vector or axis resolution) usually contains less than 6.00. The percentage amount of variance for each vector or axis can be obtained by dividing the variance explanation of the column by the total amount of possible explanation - in this case 6.00. So the amount of variance explained by vector T_1 will be 2.00 out of 6.00 (or 33.33%), whilst that of T_2 will be 2.3632 (or 39.39%). A check of these calculations is easily provided. The sum of the communalities will total the sum of the variance explained by the axes, subject to any rounding errors.

In Figure 2.3b the attempt has been made to show how these relationships can be obtained geometrically. Each surface of the cube is shown separately by two axes, and the relationship of each of the six vectors to each two dimensional surface is indicated by dots. Using the plane or surface on which vectors are located to represent the vertical and horizontal axes, it is possible to completely describe vectors 1, 2, 3 and 4 on T_1 and T_2 since they lie in the plane of these two dimensions. Vectors 5 and 6, however, cannot be described so easily. Vector 5 is represented by a point at the origin, and vector 6 by a point along vector 4. These points are only, of course, the approximate positions of vectors 5 and 6. They actually occur in another plane. Their position on the axes T_1 and T_2 can be determined by dropping perpendiculars from the actual position of the points on to the axes T_1 and T_2. In the case of vector 5 the perpendicular lands on the origin. In vector 6 it cuts axis T_2 at position 0.6, i.e., along the line running from 0.0 at the origin to 1.00 at the end. Perhaps an easy way of reconstructing this would be to build a three dimensional model, using pins to represent the vectors, separating each by the angular relationships shown in Table 2.8. If this model is placed on the T_1, T_2 surface and a light is held above it, the shadow of vector 6 will be seen to cut axis T_2 at position 0.60.

The same type of procedure could be used to obtain the position of all the other vectors on the surfaces T_1/T_3 and T_2/T_3, enabling the table of factor loading relationships found in Table 2.7 to be constructed. Obviously the procedure is excessively clumsy and restricts the analysis to the description of three axes. Moreover, it should be apparent that once we go beyond three dimensions there is no way of extending the idea in any geometrical sense. In hyperspace the researcher would be dealing with the multidimensional analysis of a myriad of contorted shapes. As a result the factor procedures for obtaining factors from 'n' dimensional examples must depend upon algebraic rather than geometrical methods.

2. THE ALGEBRAIC DERIVATION OF FACTORS

Until the advent of high speed computers with big memory stores, factor analysts were restricted in the size of matrix they could solve, simply because of the time taken to calculate the solutions. For instance, Harman (1967 p.156) pointed out that one of his 24 variable matrix examples would take 70 hours to obtain the first factor weights from hand calculations, instead of several seconds by computer.

At present there are a wide variety of alternative procedures that can be used to derive the basic factor solutions. These are reviewed in Chapter V, although it must be noted that the description follows the principles initially established in this text in dealing with the concepts behind the procedure, rather than dealing with the algebraic calculations. To undertake the latter is beyond the scope of this introductory text, and in any case would only repeat the valuable summaries already produced by Harman (1967, 1976), Rummel (1970), and

Cattell (1978). Similarly, given the objectives of this chapter, there is no space to go beyond an elementary description of the deviation of factors.

One of the simplest approaches to the problem, the centroid solution first proposed by Burt (1917), can be used to explain how a single factor can be derived from any matrix. Table 2.10 is a stage-by-stage description of the calculations involved in obtaining the position of the first centroid vector. The procedure is basically that used in all calculations of the centroid method. In such simple cases the second vector can be obtained by drawing one in at right angles to the first. Where more variables are involved, the derivation of second and subsequent centroid vectors is slightly more complicated, and depends upon rules for 'reflecting' axes that were derived by Thurstone to produce several common factors. These rules are described in detail in Fruchter (1951) and Harman (1967). Such methods have now been superceded by more sophisticated procedures and need not concern us here.

Table 2.10
The centroid method of deriving factors

(a) Angles between the vectors

Vectors	1	2	3
1	0	10°	115°
2	10°	0	105°
3	115°	105°	0

(Cos 115 = Cos 65)

(b) Cosines of angles between vectors

Vectors	1	2	3
1	1.000	0.9848	−0.4226
2	0.9848	1.000	−0.2588
3	−0.4226	−0.2588	1.000
Column Totals	1.5622	1.7260	0.3186

(c) Calculations of centroid axes

　　1) Sum the column totals 3.61

　　2) Take the square root of this sum 1.90

　　3) Divide each column total by the square root to calculate the first centroid vector loadings, namely:

43

Variables Factors	1	2	3
I	0.8221	0.9084	0.1677

4) Translate these cosines into angles, namely:

Variables Factors	1	2	3
I	34°40'	24°40'	81°20'

5) Assume the second vector (II) is at right angles to the first, by taking the angles in (4) from their nearest 90° or 180°

Variables Factors	1	2	3
II	55°20'	65°20'	171°40'

6) Translate into cosines or correlations

Variables Factors	1	2	3
II	0.5688	0.4173	−0.9894

(d) Loadings on the new axes or factors

Variables	Factors I	II	Sum of Squares
1	0.8221	0.5688	1.00
2	0.9084	0.4173	1.00
3	0.1677	−0.9894	1.00
Sum of squares	1.53	1.47	
% Explanation	51.0%*	49.0%	

* $\dfrac{1.53}{3.00} \times 100 = 51.0\%$

If this simple example is drawn geometrically, and then related to the discussion in the previous section, it will be apparent that the basis for the procedure consists of the search for the <u>centre</u> of a cluster of vectors. Once this is found, and the point is joined to the origin, a 'central' vector is created in 'n' dimensional space from which other points or vectors can be measured. Hence the algebraic solution answers the problem posed by Figure 2.2c, namely where can the descriptive vector T_1 be placed? In Figure 2.2c, the position of vector T_1 was simply assumed to be the mid-point of vectors V_1, V_2, V_3 and V_4. The centroid algebraic method provides an answer to the problem, one that can be seen as being analogous to finding the centre of gravity in 'n' dimensional space. The term 'centroid' represents a useful description of the approach. Today, the Centroid method of factor analysis is rarely used by factor analysts, since more sophisticated procedures (such as the Principal Axes method) are available for finding the summary factors or axes of differentiation. Matrix algebra is needed to understand the details of these various procedures, and such approaches are beyond the scope of this chapter. Nevertheless, Appendix A describes the Principal Axes technique by a workbook procedure designed to demonstrate the principles that are involved, using only basic arithmetic and elementary algebra.

3. PRINCIPLES OF ROTATION

So far this chapter has discussed ways of deriving factors from a number of matrices. However, there is no need to assume that these initial solutions represent the only problem of factorial procedures. Most of the factors of the initial solutions can be rotated to alternative positions from which the other vectors or variables can be described. So the question of rotation is a basic one in the field. The principles of rotation can be illustrated by referring back to Table 2.6 and Figure 2.2c. They showed that two very different solutions to the isolation of the basic summary vectors could be obtained, using either T_1 and T_2 or TT_1 and TT_2. But a moment's reflection will indicate that the two solutions are not the only alternatives that can be derived. By rotating the T_1 vector to position TT_1, and the T_2 vector to position TT_2, one of the solutions can be derived from the other. Yet by placing the descriptive vectors at any point between these extremes, other positions for the summary axes or or factors can be obtained. From this finding we can generalize and say that the description of any point, or set of points in any space, depends upon the orientation of the axes used as a frame of reference to describe the points. Although the axes may be arbitrarily assigned, just as with grid lines on any map, the description of the point is only made by reference to these axes. As any geographer knows from basic map interpretation work, the change in the position of the survey grid, or axes, does not change the relative position of any two points with respect to one another; it simply changes the way in which we know, or rather describe, these points. A simple example of this idea is shown in Figure 2.4 which shows point V located in a two axis space. The descriptions of V according to axes A_1 and B_1 is 0.2 on A_1 and 0.4 on B_2. So even if the axes are only moved around the origin 0, to three positions the following descriptions of V on the axes can be obtained:

0.35 on A_2 and 0.275 on B_2

0.425 on A_3 and 0.125 on B_3

0.477 on A_4 and 0.0 on B_4

Since this type of problem is going to affect the position of the factors, as well as any individual variable, the question that must be answered is how to choose between the alternative descriptions of V. Each description is perfectly satisfactory from a mathematical view-point. Researchers must make the choice on other grounds, and in most cases it involves the principle of parsimony.

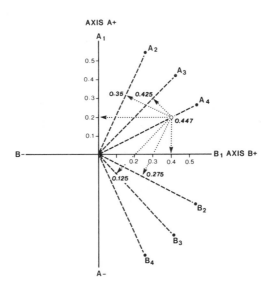

Figure 2.4   The rotation of axes

All these values can be measured off diagrams such as Figure 2.4 by scaling the various axes. Many other sets of axes can be constructed each with different measurements.

Parsimony invokes the need to produce as simple a description as possible of any set of factor loadings. If we assume the axes A_1 and B_1 in Figure 2.4 represent factors, and V is the position of a variable in the two-axis space, then the loading of V will be 0.2 and 0.4 on A_1 and B_1 respectively. In this case the description of variable V may be considered to be comparatively complicated, since two axes are needed to locate the point. The question arises as to whether it is possible to describe the point V more simply, perhaps with a zero or near zero value on one of the axes. So the first task is to introduce a measure of parsimony, or simplicity of description, of the point V. One measure can be produced from the <u>product</u> of the two co-ordinates on the axes OA and OB

i.e., 0.2 x 0.4 = 0.08).

Since the possibility always exists of negative values in any Cartesian space, it is more convenient to use the product of the <u>square</u> of the co-ordinates as the measure of parsimony

i.e., 0.2^2 x 0.4^2 = 0.0064

Symbolically, this procedure may be written as:

$$a^2_{jp} \cdot a^2_{jq} \qquad \text{Eq. 2.7}$$

Where: a_{jp} represents the co-ordinate of variable j on factor p;

a_{jq} represents the co-ordinate of variable j on factor q.

Generalizing to an 'n' axis or 'n' factor space this expression becomes:

$$\sum_{p=1}^{n} a^2_{jp} \cdot a^2_{jq} \qquad \text{Eq. 2.8}$$

It has already been shown that there are a multitude of possible descriptions of V in a simple two axis space. So by moving or rotating the axes the description of V in terms of these axes is altered. The simplest description of V will be produced when the co-ordinates of one of the axes is reduced to zero. In graphical terms this can be achieved by rotating the frame of the axes A_1 and B_1, through A_2B_2 and A_3B_3, to positions A_4 and B_4. Figure 2.4 shows that the co-ordinate of V on B_4 becomes zero. Algebraically the same process is achieved by successively calculating the product of the squares of the co-ordinates and searching for the minimum value. In Figure 2.4 the product of the squares of the co-ordinates become:

0.009 for axes A_2B_2

0.003 for axes A_3B_3

0.000 for axes A_4B_4 (because the co-ordinate value on axis A_4 is 0.447 and on B_4 it is zero).

47

In symbolic terms once again, the simplest description of any point in 'n' factor space will be found by the minimization of Expression 2.7 above. The addition of more than one variable does not affect the basic procedure, it merely produces a more complicated formula:

$$\sum_{j=1}^{m} \sum_{p=1}^{n} a^2_{jp} \cdot a^2_{jq} = \text{Minimum} \qquad \text{Eq. 2.9}$$

Where: $\sum_{j=1}^{m}$ represents the sum of the j variables, from 1 to m.

At this point it is necessary to recap on the definition of commun-ality. The communality of a variable is the total amount of variance (usually expressed as a ratio of 1:0) that is explained by the chosen or initial factor solution. It is derived by summing the squares of the factor loading of a variable, and is clearly fixed once the number of factors to be abstracted in the initial solution has been deter-mined. Hence, if position V represented the position of variable j in the two-factor space of Figure 2.4, the communality for the variable in the two-factor case would be: $0.2^2 + 0.4^2 = 0.20$. This may be expressed symbolically in algebraic terms for variable j and factors p and q as:

$$a^2_{jp} + a^2_{jq} \qquad \text{Eq. 2.10a}$$

or more generally as:

$$\sum_{p=1}^{q} a^2_{jp} \qquad \text{Eq. 2.10b}$$

where: 'a' represents the loading; 'j' represents the variable; and p and q represent the factors.

One important geometrical derivative of Figure 2.4 is that the length of the vector OV of variable j represents the square root of the communality, i.e., $\sqrt{0.20} = 0.447$. This can be measured from Figure 2.4. The communality of any variable is fixed by the decision taken in the preliminary factor solution about the number of axes or factors to be abstracted. In other words the rotation of the axes used to locate each variable point will have no effect upon the total communality, and the square of the communality must also remain a constant. By squaring the expression in Equation 2.11 and rewriting it it is possible to obtain the following equation:

$$\left(\sum_{p=1}^{q} a^2_{jp} \right)^2 = \sum_{p=1}^{q} a^4_{jp} + 2 \sum_{p=1}^{q} a^2_{jp} \cdot a^2_{jq} = \text{Constant}$$

$$\text{Eq. 2.11}$$

Generalizing this finding to j variables the following expression is obtained:

Eq. 2.12

$$\sum_{j=1}^{n} \sum_{p=1}^{m} a^4_{jp} + 2 \sum_{j=1}^{n} \sum_{p=1}^{n} a^2_{jp} \cdot a^2_{jq} = \text{Constant}$$

or $Q + 2M$ = Constant

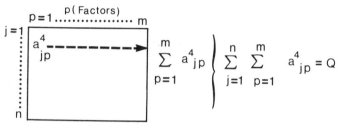

QUARTIMAX (Q)

Figure 2.5   Visual interpretation of the orthogonal rotation formula

M in Equation 2.12 is the same as the expression used as a measure of parsimony in Equation 2.9. As this measure depended upon the minimization of the term M, and Q + 2M equals a constant, it is apparent that the minimization of M and the maximization of Q must have the same effect. So two alternative approaches to the measurement of parsimony are suggested by this discussion. In practice it is Q, the maximization of the value of the fourth power of the factor loadings that is the easier to calculate, and Figure 2.5 provides a visual interpretation of the basis of this calculation for any matrix. It can be seen that in the first row (j=1) the fourth power of each cell is calculated and summed by row from Factor p to m. Then the values in the next row (j=2) are summed. All subsequent rows j=3,n are all summed and the total of all the row entries is Q. This provides the semantic as well as the computational basis for the orthogonal and analytical rotation technique known as quartimax (maximization of the fourth power or quart). Starting with the initial factor solution a series of iterations involving the calculation of the fourth power of the loadings is carried out. Once a maximum value is obtained the iterations are completed.

To help reinforce the general understanding of these procedures a specific example is provided. In a study of the social areas of Calgary described in Chapter I the distribution of six variables (Table 1.1) over 28 census tract areas was reduced to a 6x6 correlation matrix (Table 1.2) with a set of factor analysis results (Table 1.3) using an orthogonal rotation. If this rotation assumption is disregarded for a moment and the results are investigated in more detail, as in Table 2.11, it can be seen that a Principal Axes component analysis of the matrix produced two components explaining 89.1% of the total variance, the individual loadings being shown in Table 2.11.

Table 2.11
Principal Axes solutions for the 6x6 Calgary matrix

Variables	Principal Axes Component Solution		Quartimax Rotation	
	Factor 1	Factor 2	Factor 1	Factor 2
1. Occupation*	−0.771	−0.564	−0.949	0.107
2. Education	0.837	0.468	0.932	−0.223
3. Fertility	−0.788	0.476	−0.258	0.884
4. Women at work	0.625	−0.674	0.003	−0.919
5. Single detached dwellings	−0.616	0.748	0.054	0.968
6. Ethnic	−0.592	−0.728	−0.929	−0.135
Eigenvalues as % variance	50.60%	38.46%	45.05%	44.06%

* Detailed definitions are provided in Table 1.1

It is difficult to produce any substantive conclusion from this solution because most variables have high loadings on both components, producing a scatter of the individual loadings among the four quadrants of a two-axis diagram (Figure 2.6). Since the objective of the Principal Axes Component analysis is, as its name implies, to derive a solution in which the major axis is associated with the highest amount of variance, and the next axis abstracted is placed at right angles to the first one, one might expect to find these axes located between the scatter variables as shown in Figure 2.6. However, a close scrutiny of Figure 2.6 will reveal that it is possible to rotate both axes to the position shown in the diagram. Once this is carried out each variable becomes primarily associated with one of the axes; for instance, Vector I is associated positively with the education variable, and negatively with the ethnicity and occupation variables; whereas, Vector II is associated positively with single family dwellings and fertility but negatively with women at work. It is apparent that this graphical rotation has simplified the description of the six variables, producing the same two clusters of collectivities in the data set that were described by the correlation bonds in Figure 1.2. Algebraically, the same result can be obtained by the application of the Quartimax procedure. The results are shown in Table 2.11. Moreover, in this simple two axis example the loadings or weights can be read off the axes shown in Figure 2.6 by dropping perpendiculars from the variables to the various axes, and measuring the distance to the origin.

In Figure 2.6 the loadings for the fertility variable change from +0.48 and−0.79 on the original axes, to +0.88 and −0.26 in the rotated solution. The sum of the square of these values, the communalities,

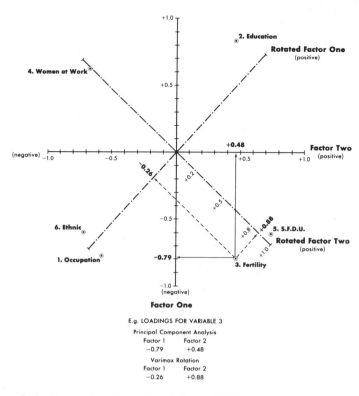

E.g. LOADINGS FOR VARIABLE 3

Principal Component Analysis

Factor 1	Factor 2
−0.79	+0.48

Varimax Rotation

Factor 1	Factor 2
−0.26	+0.88

Figure 2.6   Factor loadings for Calgary 1961:
the effect of rotation

remain a constant, since they are represented by the length of the
vector from variable 3 to the origin.   Again it demonstrates that
communality is not affected by rotation.   Of more importance is the
fact that if parsimony is an external criterion of assessment, then the
rotated loadings are more satisfactory descriptions of the relationships, since they provide a simple description of the position of each
variable in terms of high and low values.   Similar reasoning applies to
the other variables.   It must be emphasized, however, that since the
investigator has the choice of using the unrotated or rotated solutions
the choice must be justified on substantive grounds.

4.   THE DERIVATION OF FACTOR SCORES

The preceding sections have shown that the factor loading matrix,
whether in rotated, or unrotated form, summarized the variability of
the variable set according to a new of axes or factors.   This means
that the first problem described in the introduction to the chapter,
that of defining the collectivities of the variables in a data set, has
been dealt with.   In many geographical or ecological studies the
investigator may be more interested in summarizing the variations in a
set of observation units.   This is the main objective involved in the

calculation and use of the factor scores, for they measure the importance of each observation on every factor, or axis of differentiation.

The basis of the factor score calculations can best be appreciated by
remembering that the characteristic factor equations of the factor
loading matrix (Equation 2.5) measure, or more appropriately, estimate,
each variable in terms of the factors; they represent a row-by-row
description of the variables. In the same way each column represents a
description of each particular factor in terms of all the variables.
However, it is as well to recall that there is an important difference
between the totals of the row and column values. The sum of the
squared loadings by the rows is the communality. This can never be
greater than 1.00 because it is the amount of variance contributed by
one variable. The sum of the squared loadings by the columns is, of
course, the square of the eigenvalue, and these values vary according
to the importance of the factors. Hence any attempt to compare the
within-column distribution of values from factor to factor must take
this into account. The easiest way to standardize the vector of column
entries is to divide them by their eigenvalues. These values are
called factor coefficients to distinguish them from the factor loadings. A composite score summarizing the importance of each observation
can now be obtained for each factor (p) in turn. The factor coefficient for each variable (j) is multiplied by the original value for the
variable found in the observation unit (O_j) and the totals are
summed. This is repeated for the next factor. For any observation
unit (m) the procedure may be written symbolically as:

$$FS_{pm} = \frac{a_{1p}}{\lambda p} \cdot z_{1m} + \frac{a_{2p}}{\lambda p} \cdot z_{2m} + \frac{a_{3p}}{\lambda p} \cdot z_{3m} + \frac{a_{np}}{\lambda p} \cdot z_{nm}$$

Eq. 2.13

Or more concisely as:

$$FS_{pm} = \sum_{j=1}^{n} \frac{a_{jp}}{\lambda p} z_{jm}$$

Eq. 2.14

Where: FS_{pm} is the composite value for the observation m, or the
factor score of factor p on observation unit m.

a_{jp} is the factor loading of variable j on factor p

z_{jm} is the original value of j in observation unit m

λp is the eigenvalue of factor p

$\dfrac{a_{jp}}{\lambda p}$ is the factor coefficient of variable j on factor p

There is, however, a problem about these equations. They assume that
each variable is measured in the same units. This is rarely the case.
So it is usually necessary to standardize the raw data or the original
observation values before it is introduced into the factor coefficient
equations. Normally this is carried out by replacing each value by its
standard score (z), where:

52

$$z = \frac{x - \bar{x}}{S_x}$$ x is the original value; $\bar{x}$ is the mean of x; S_x is the standard deviation of x.

An example may help clarify the calculations. Table 2.12 shows the correlations and descriptive statistics produced from an analysis of the six variable data set used by Booth (1896) to describe the social condition of areas in London. Booth simply averaged the rank order values of the individual variables to produce a rank order scale of the variations in 'social condition'. Pearson Product Moment correlations were calculated between the six variables, followed by a Principal Axes factor analysis (component model) in which one axis accounted for 79.9% of the original variance. The calculation of factor, strictly speaking component scores, is shown in Table 2.12.

Table 2.12
Worksheets for the calculation of factor scores

(a)   Correlation matrix of Booth's data

Variables	Variables					
	1	2	3	4	5	6
1. Poverty	1.00	0.79	0.84	0.76	0.82	-0.76
2. Domestic crowding	0.79	1.00	0.77	0.50	0.75	-0.63
3. Young married females	0.84	0.77	1.00	0.86	0.79	-0.83
4. Birth rate	0.76	0.50	0.86	1.00	0.71	-0.76
5. Death rate	0.82	0.75	0.79	0.71	1.00	-0.78
6. Surplus of unmarried (males to females)	-0.76	-0.63	-0.83	-0.76	-0.78	1.00

(b)   Component loadings

Variables	Component I		Raw Data	
	Loadings	Commun-ality	Mean	Standard Deviation
1. Poverty	0.93	0.86	31.15	9.46
2. Crowding	0.83	0.68	32.19	13.32
3. Young married females	0.95	0.90	177.44	42.13
4. Birth rate	0.86	0.74	33.16	4.94
5. Death rate	0.90	0.82	20.33	3.90
6. Surplus of unmarried (males to females)	-0.88	0.79	38.99	22.71
Eigenvalue	4.79 (79.7%)			

(c) Factor coefficient equations

Component I =
$+0.193 \ z_1 + 0.172 \ z_2 + 0.198 \ z_3 + 0.179 \ z_4 + 0.189 \ z_5 + (-0.185)$

(N.B. Coefficient for variable 1 is $0.927 \div 4.791$ (eigenvalue) = 0.194)

(d) Raw data for areas 27 and 1

Areas	1	2	Variables 3	4	5	6
27 Hampstead	13.50	16.00	104.0	23.40	12.70	99.99
1 St. George's East	48.90	57.00	254.0	40.20	30.20	13.70

(e) Descriptive statistics

	1	2	3	4	5	6
Mean $(\overline{x})$	31.5	32.19	177.44	33.16	20.33	38.99
Standard deviation (S)	9.46	13.32	42.13	4.94	3.90	22.71

(f) Deviations from mean and standardized deviations

For Area 27	1	2	3	4	5	6
$(x - \overline{x})$	-18.0	-16.19	-73.44	-9.66	-7.66	+38.99
$\left(\dfrac{x - \overline{x}}{S}\right)$	-1.90	-1.22	-1.74	-1.95	-1.96	+2.69

(g) Factor score coefficients (fc)

	1	2	3	4	5	6
1. fc	0.193	0.172	0.198	0.179	0.189	−0.185
2. fc . $\dfrac{x - \bar{x}}{s}$	−0.367	−0.210	−0.345	−0.349	−0.370	−0.498

Total of row (2) above = −2.138.
This is the factor component score for Area 27 (Hampstead)

(h) Rest of the factor component scores

Areas	Score	(Rank)	†	Areas	Score	(Rank)	†
1. St. George's East	1.98	(1)	1	15. Islington	−0.14	(15)	15
2. Holborn	1.24	(2)	2	16. Lambeth	−0.15	(16)	16
3. Whitechapel	1.25	(3)	3	17. Fulham	−0.27	(17)	17
4. Bethnal Gr.	1.17	(4)	4	18. Woolwich	−0.41	(18)	18
5. Shoreditch	0.99	(6)	5	19. Camberwell	−0.43	(19)	19
6. St. Saviour	1.02	(5)	6	20. Wandsworth	−0.64	(20)	20
7. St. Olave	0.99	(7)	7	21. Marylebone	−0.69	(21)	21
8. Stepney	0.98	(8)	8	22. Hackney	−0.76	(22)	22
9. Poplar	0.55	(9)	9	23. St. George H.Sq.	−0.83	(23)	23
10. Mile End	0.47	(10)	10	24. Paddington	−1.21	(24)	24
11. St. Pancras	0.15	(11)	11	25. Kensington	−1.31	(25)	25
12. Greenwich	0.01	(12)	12	26. Lewisham	−1.64	(26)	26
13. Strand	−0.07	(13)	13	27. Hampstead	−2.14	(27)	27
14. Chelsea	−0.13	(14)	14				

† Booth's Combined Orders

In this particular example the results from Booth's simple rank order approach and the rank of the factor scores are very similar. This is really a product of the fact that Booth's assumption that the data measured a simple source of variation which he called 'social condition' turned out to be a reasonable one. After all the only factor extracted, the Principal Component in this case, accounted for 79.9% of the variance. If additional sources of variation were present in the data set, as in the two axis case in the Calgary example, then Booth's rank order approach would not be sensible since addition of the rank orders would be measuring unlike entities. What the factor scores have produced is a composite measurement of the variability of each area as measured by the variables, or rather the common or general part of these variables.

5. SUMMARY

This discussion of the basis of factor analysis procedures has concentrated upon the problem of rewriting some elementary data matrices in a more concise form. Using the two case studies as examples of the way in which parsimony of description can be produced, it has been shown that the Calgary 6 variable x 28 observation unit matrix could be reduced to two new matrices: first, a 2 factor x 6 variable factor loading matrix which summarized the interrelationships existing between the variables with a loss of only 10.95% of the original variance; second, a 2x28 factor score matrix which measured the importance of each observation unit on the factors. In the London example (from Booth 1896) the 6 variable x 27 area example was reduced to a single 6x1 loading matrix and a 27x1 score matrix. But this chapter has not been primarily concerned with the derivation of the factor procedures which were used in Chapter I to illustrate the utility of the factorial approach. Rather the objective has been to demonstrate the fundamental principles of the factorial approach, particularly the features that contribute to the basic invariance of the procedure. For the sake of completeness it is worth summarizing these principles.

(i) Potentially, there are as many factors as there are variables in a data set. But if there is any underlying order in the data a limited number of factors account for most of the variance. It is these axes that are used to describe the variable set, with the result that the very minor axes are rejected.

(ii) It is incorrect to conceive of only one factor solution for a data set. For example, the alternative results of the 4x4 matrix in Table 2.5 showed that different factorial descriptions of the relationships between the variables can be obtained. Each of these are mathematically satisfactory but may not be intuitively meaningful in terms of the substance of the descriptions.

(iii) The derivation of any initial set of factors from the similarity matrix is not the end of the factorial procedure. Figure 2.5 has shown that it is possible to rotate the factor axes to new positions, producing a new factor loading matrix. However, it must be remembered that the rotation cannot affect the communalities or the total explanation of the solution; these are fixed by the initial solution. In the example only the case of Orthogonal rotation was considered.

These features mean that the basic problems of factor analytical procedures consist of resolving three basic issues.

(i) The first is the number of axes to be used in defining the factor space. For example, if an explanation level of 80% of the variance is considered to be a satisfactory result by a researcher the two axis solution of the 6x6 matrix shown in Table 2.9 could be used in place of the three axis solution.

(ii) The second is the amount of explanation of each variable that is to be accounted for by the solution. For example, if the 'cut-off' decision for Table 2.9 was set at a level in which over half of the variance of all the variables had to be explained the three axis solution would be chosen in place of the two axis solution.

(iii) Finally, the position of the factor vectors in the factor space needs to be identified. This involves decisions such as whether the vectors are to be located directly by the initial factoring method, or whether rotations are allowed.

To a very great extent these decisions about the number of axes, their position in the factor space, and the size of the communalities remain the prerogative of the investigator. In the component model, where it is assumed that all of the variance can be analyzed, the investigator attempts to justify one or more of these decisions by reference to external criteria. In the common factor model, where it is assumed that only part of the variance of the data set is described as variance, these decisions represent vital parts of the hypothesis that is to be tested, and also depend upon outside considerations. All these decisions are, however, outside the scope of this chapter. It is enough to have isolated the basic issues in factor procedures, and demonstrated that different criteria can produce a variety of factor solutions. This, combined with the fact that there are a series of different ways of factoring and rotating data sets, means that the choice between alternative procedures becomes a very real problem in factorial ecology. But before recommendations for particular courses of action are made (Chapter VI), much more basic problems involved in any investigation must be covered. These issues form the basis of the next three chapters, which deal successively with: the ecological content and range of relationships that can be investigated by factorial methods (Chapter III); the data problems that have to be faced in any study (Chapter IV); and the range of alternative factorial techniques and models that are available (Chapter V).

3 CONTENT AND RELATIONSHIPS IN FACTORIAL ECOLOGY

1. CONTENT

The content and range of relationships dealt with by factorial ecology
are essentially the same as those in human geography. They can best be
appreciated by Figure 3.1, a simple conceptual model which extends and
modifies previous work in allied fields (Foley 1962; Duncan 1965;
Davies 1970). Like all such models it is a tentative framework, or
arrangement of working ideas, that is useful only insofar as it is
fruitful in synthesizing and clarifying the range of relationships to
be found in the field. For descriptive purposes four basic parts, or
conceptual components of this model can be identified: elements,
scale, perspectives, and society.   Although these parts are normally
separated for descriptive purposes, in practice they are interrelated.
The elements describe the types of data that are of interest to the
investigator, whilst the perspectives represent the range of ways in
which this data can be viewed or interpreted.   By contrast, scale
refers to the degree of resolution of the data, the size of the areal
unit upon which the data is collected, whilst society represents the
vital human or social organizational background from which the data has
been abstracted.   Since most of these component parts comprise a
variety of different relationships they need to be looked at in more
detail.

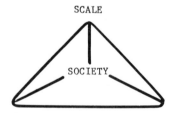

SCALE

SOCIETY

ELEMENTS

1. Population
2. Functional Activities
3. Morphology
4. Environment

PERSPECTIVES

1. Formal (Structural)
2. Functional (Relational)
3. Perceptual
4. Temporal

Figure 3.1   The scope of the field

(a) Ecological elements

The ecological elements represent the data base of interest to the investigator, and may be summarized under the four headings of: population; functional activities; morphology; environment.

(i) Population summarizes the whole range of characteristics of the population, from age and ethnicity to beliefs and perceptions.

(ii) The term functional activities relates to the results of the activities engaged in by the human groups, whether social, economic, or political. In total the patterns of these activities may appear to be chaotic, but they are usually highly organized. This element comprises not only the activities themselves, but also the effect of these activities upon areas, which enable the identification of specialized land use zones, etc.

(iii) Morphology describes the man-made skeleton of a town or region and is much more static in nature, dealing with such issues as the type and character of buildings.

(iv) Environment refers to the whole range of physical environmental features. To the more obvious divisions such as climate, relief and natural resources must be added areal size and locational attributes, as well as those intermittent changes in natural conditions such as floods, earthquakes that have such a profound effect on individual places.

(b) Geographical scale

Although it is possible for ecologists and geographers to deal with individuals if these form part of the characteristics of area, most spatially minded investigators work with some type of areal or spatial aggregate. The aggregation is not based on the type of group characteristics familiar to the sociologist, for instance a group identity derived from some behavioural trait, or from membership of a club or society. Rather, it is based on an areal or spatial grouping by means of which the individuals are grouped into areas. Hence their characteristics are used to describe the areas. It is the totality of the area, rather than the variation within the area, that is of primary interest at that scale of analysis - although these internal variations could be studied if larger scale units were available.

In theory, information can be organized in a continuum of different sized areas, from the smallest parcel of land up to areas the size of the earth. Haggett (1965) has attempted to introduce some order into our measurement of this continuum. He showed that if the diameter of objects, or areas to be studied, are measured in centimeters and are arranged on a logarithmic scale, a G scale can be derived which provides an immediate point of reference for all studies. Haggett's study also illustrates how the scale can be used to limit the content of geography or ecology to that of earth the home of man. It would only deal with scales running from 10^2, the approximate height of an adult man, to a point between 10^9 to 10^{10}, the circumference of the earth. Atomic radii at a scale of approximately 10^{-10}, or objects the size of the Milky Way at 10^{23}, are clearly outside the scale of human geography or human ecology.

Haggett's demonstration of a continuum of geographical scales does not mean that a range of alternative scales can be attained in practice. Investigators often aggregate information into groups in order to make the data easy to handle before carrying out any analysis. In the same way census authorities usually collect information upon an enumeration area (or unit) basis, that is the area or district to which one enumerator is allocated. These are then aggregated together to obtain census divisions. With the growing use of computers in data analysis, as well as the practice of most census authorities to release information at an enumeration area level, it becomes easier to aggregate data at a variety of scales. Hagerstrand (1969) was one of the first to show how the adoption of grid co-ordinate reference systems will speed up this process even further, subject, of course, to the restrictions used by census authorities to preserve confidentiality.

(c) Perspectives

Four very different approaches can be used to summarize the variety of possible relationships existing between the ecological elements.

(i) Static, structural (or formal). Most studies are still carried at the static or structure type of perspective, in which the formal characteristics of an object or an area represent the primary focus of attention. The study of age distributions, or land use patterns provide examples of this type of a structural type of study. Davies (1983) has described how structure in this context has a very different meaning to its use in structuralism, the search for deep-seated causal mechanisms.

(ii) Functional (or connectivity). Most objects or areas have connections with, or interact with, other objects or areas by reason of the functioning of activities in these areas. For example, in any study of the commercial structures of towns it would soon be apparent that various shops and offices have different patterns of connectivity, not only in the inflow of customers, or the journey to work of the employees, but also the organizational patterns of these activities. All of these relationships involve the functional perspective.

(iii) Perceptual. The current acceptance of the need to look at human distributions, in what Sprout and Sprout (1965) called the environmental milieu, may be summarized by the term perceptual. Originally, the study of perceptual characteristics - as demonstrated in Firey's (1945) use of 'sentiment' and 'symbolism' in a study of Boston - represented a critique of the whole human ecological approach of the Chicago school, since the land use generalizations were primarily linked with visible or 'economic' structures. Within the last two decades Lynch (1960), Gould and White (1968) and Downs and Stea (1977) have pioneered investigations which investigate the perceptual character of areas in an analytical fashion. The result is that study of spatial perception now occupies a firm place in human ecology.

(iv) Temporal. Study of the existing structural and functional character of an area can only deal with the system of relationships at this one point in time. The study of any element, or any system, can be expanded by dealing with the process of change, either by looking at the past relationships, or by projecting these relationships into the future. In this way the process of change can be investigated by providing a fourth or temporal perspective to the study of the ecological elements.

A word of warning must be introduced. Not all of these perspectives can be applied to all the ecological elements. For example, it is difficult to conceive of a study dealing with the connectivity of the morphological element at any single time period. Certainly the activities of the buildings could be investigated, or the changing morphology could be looked at over a succession of time periods. But at any point in time the building is fixed. In the same way, many environmental elements may be considered as static at any one time period, although the dangers of regarding the environment as immutable can be seen by the changes or damage caused by floods, earthquakes, and storms.

(d) Society

The relationship of any ecological study to the wider society of which it is a part was all too frequently taken for granted by human ecologists in the early days of the field. So for many years the societal framework was rarely the subject of explicit comment. Given the increasing popularity of cross-cultural and temporal studies this situation has now been remedied. Many of our measures of the ecological elements have been shown to be society-specific and their meaning can only be understood if viewed against the values and technology of the individual society. For example, the difference between the distance of sixty miles and four miles may appear to be large, but they can represent the same time-distance concept of one hour travel: in the first example by a car used on a motorway in the western world; in the second to an Indian, or medieval peasant reliant upon his feet for transport. Examples of the variations in value placed upon educational attainment, or the effect of segregation laws in different societies, all demonstrate the difficulty of equating a measure derived from one society directly into another. Also, distinctions can be drawn between the meaning of a term, or the evaluation of an element, in the society of the investigator, as opposed to its meaning in the society being investigated. The result, in academic terms, could be a misunderstanding of the basic principles of different societies. In practical terms, a good example is provided by the effect of western missionaries and explorers upon native societies in the nineteenth century, which led to eradication of much of the cultural diversity among groups. More general than these specific examples has been the growth of interest in structuralist arguments in geography. Harvey's (1973) pioneering work searching for the underlying causes in urban variations has been extended by Castell's (1977) study. Together, with Dear and Scott's (1981) recent survey, the study of societal background and the search for the underlying constructs of change, is now a major research frontier in urban and human geography, one linked to the wider concerns of social science in general and political economy in particular (Forrest et al 1982).

2. RELATIONSHIPS

The analytical divisions that have been used to summarize the content of human geography and human ecology illustrate the enormous scope of the field. In practice, it is impossible to simultaneously handle all the possible components within the confines of a single study. So it is usual to recognize three approaches to the study of this data, namely: the individual, aggregative, and contextual levels. Although the factorial ecologist is primarily interested in the aggregative level and also the contextual level, a brief discussion of the other levels of analysis will make it easier to appreciate the scope of the field.

(a) Individual level

In this type of study the data is usually drawn from samples of individuals, and the results relate to the characteristics of these persons. For example, a social survey could identify a set of individuals, determine the race and literacy level of each person, and calculate the correlation between these two variables for the individuals. In this type of study both the dependent variables (those to be predicted) as well as the independent (or explanation) variables, relate to a number of individuals.

(b) Aggregate level

Frequently, an analyst is not be interested in individuals per se, but in the characteristics of aggregations or groups of individuals. These groupings must be carefully defined in advance of the analysis. Two different types of aggregations can be distinguished.

(i) Ecological. The first relates to an aggregation of the population into areal units, the so-called ecological or geographical approach.

(ii) Structural. The second relates to an aggregation of the population into some structural group, such as an age group, or social club. In both cases the correlations will be calculated on the basis of the defined aggregations not the individuals, so the relationships discovered will relate only to these defined aggregations. Naturally, the dependent and independent variables are both aggregative figures. For example, an investigator could calculate the percentage of the population (in the age group 15-30 years) of an area who have low income, and relate this to the percentage of the population in an area (in the age group 15-30 years) who are black. In this one-variable example the calculation of the percentage of blacks, as opposed to the percentage of whites, could also be carried out on the basis of membership in organized social clubs and then extended to spatial communities, producing a social as opposed to an ecological type of study.

(c) Contextual level

Studies of individual or aggregative characteristics at the ecological or structural levels do not exhaust the range of possibilities for study. It has frequently been observed that the explanation of any individual phenomena or ecological phenomena demands a consideration of both individual and spatial characteristics. For example, the behaviour of people in 'Skid Row', or the characteristics of Skid Row as an area, can only be explained by looking at both behavioural and areal traits. It is clear from the writings of the classical school of urban ecology in Chicago in the 1920s that the members were aware of these interlinkages (Park 1925). Studies of particular groups were set within the geographical area of the city in which they were found, and studies of the areas were related to the groups. Only rarely, however, was there any analytical attempt to differentiate the individual and ecological approaches; they were combined. This led to the use of the term 'contextual' to identify the approach that attempts to look at both levels. In other words the purpose of the contextual approach is to explain an individual behaviour pattern in terms of the social context (or milieu in Sprouts's words (1965)). This context is the one lived in by the individual so when several of the social and other personal attributes of the individual are held constant the effect of this context can be estimated.

Hence, the contextual approach takes as its dependent variable some aspect of individual behaviour (such as voting in a study by Cox (1969)) but uses as the independent variable, the characteristics of the social context (a social structural approach), or the environment or milieu (an ecological approach), as well as individual attributes. Factor analysis can be used to define and measure these characteristics. For example in the Calgary study in Chapter I the dimension of Socio-Economic Status replaces the occupation, education and ethnic variables. Usually the latter are held constant so that the impact of the context in which the behaviour occurs can be precisely measured. There is, however, no reason why the dependent variable cannot be the territorial aggregation, the characteristics of the ecological context, with the explanatory variables comprising more global features, such as ecological concentrations or geographical areas, as well as the individual features. In this type of study any investigator would have to exercise great care in moving back and forth betwen the aggregative and individual levels. Limited attention is paid to contextual problems in this study given its focus on the deviation of structural character.

3. DATA ARRANGEMENTS AND MODES

(a) Geographical data cubes

This discussion of the relationships between the three levels of study demonstrates that the ecologist is primarily interested in the aggregative level, specifically with aggregations that are based on area. Yet for a complete understanding of the individual problems of any area it may be necessary to extend the study to a contextual level. But even with this restriction the scope of the study described in Figure 3.1 still has an enormous range, and it may be difficult to appreciate the way in which various components are linked together in any actual study. In practice a single study cannot simultaneously handle all the possible arrangements, so only a limited number of relationships can usually be explored at the same time. A useful insight into the ways in which the data may be handled is provided by

Figure 3.2, a modification of the traditional geographical data cube first proposed by Berry (1964).

For purposes of illustration, Figure 3.2 takes it for granted that a study is carried out within the framework of a single society. The horizontal divisions of the face of the cube, the rows of the cube, represent examples of the series of scales that are used by factorial ecologists, from the enumeration level of intra-city studies to the national level. The vertical divisions on the face of the cube have been subdivided according to three basic components described as: static or structural, functional and perceptual. The third dimension of the cube, the one outside the plane of the paper, represents the temporal perspective. If a study of the regional characteristics of urban places were to be carried out at one time it would be represented by a row which is one cell deep in the cube. If temporal differences in one variable for the towns were to be incorporated, a horizontal 'slice' would be needed. Figure 3.3 provides examples of the way in which information organized at the four different perspectives can be dealt with, and may demonstrate the range of possible approaches at the ecological level. These are best summarized as: the R and Q modes which deal with variables and areas at one time period; the P and O modes which deal with many variables and times but only one case; and the T and S modes dealing with one variable in many cases and times.

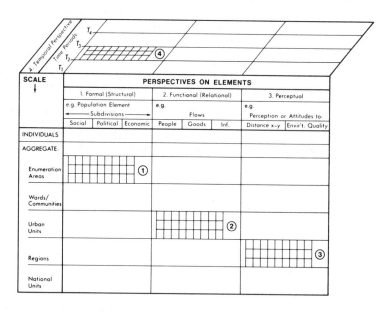

Figure 3.2 A geographical data matrix cube

(b) R and Q modes

The first example in Figure 3.3 is concerned with the study of population at the enumeration area scale. A detailed breakdown of the

64

population characteristics of a set of enumeration areas is seen in the columns of the matrix, whilst the areas form the rows of the matrix. Two very different ways of summarizing this data matrix can be conceived.

(i) The first is the R mode route, which is to relate each one of the population characteristics to the other in turn. The extent of similarity between each of columns of the matrix would be shown by a single similarity measure in the cells of the new matrix R.

(ii) The second is the Q mode approach. In this case each of the rows of the matrix would be related to one another and summarized in a similarity measure. The cells of the new matrix Q, therefore, would represent a row-by-row resolution of the original matrix. It would provide a measure of the similarity between the enumeration areas of the original matrix on the basis of the population characteristics.

This example indicates that the dimensionality of any variable in the part data matrix can be summarized either in terms of the variables (columns), in the attempt to define the collectivities of the variables, or in terms of the observation (rows), the collectivities of the areas in an ecological example.

The second example in Figure 3.3 shows one way of dealing with a data set consisting of the functional connectivity of a set of towns. In this case the flows from the towns are occupying the rows, and the flows to any town are shown in the columns. An R or Q mode resolution, or summary, of this matrix can be visualized, but as the example indicates, the biggest disadvantage is that only one type of flow at a time can be handled.

The third example shown in Figure 3.3 is a matrix dealing with the perceptual perspective at the regional scale. A representative sample of individuals in Region 1 could be asked about their attitude in moving from Region 1 and living in the whole range of Regions 1 to N. Their responses - perhaps in the former of rank order preferences - would be contained in the cells of the matrix B. From this data a similarity matrix can be derived - either by the Q mode case or row-by-row comparison or the R mode variable column-by-column comparison - with the factor analysis applied to the similarity matrix.

The final example shown in Figure 3.4 is a matrix of enumeration areas arranged against temporal changes in a particular characteristic of the population. Studies of change in which percentage or ratio variations are used are always dangerous to interpret because the absolute values of the size of the population is suppressed. Nevertheless the example demonstrates how investigators can use the factorial approach to identify phases of change by the R mode model or clusters of areas with similar sets of changes.

As with all R and Q mode analysis some care needs to be taken with the results. In the R mode approach the factors will identify clusters of the variables (the columns), in which the factor loading measures the strength or amount of variance explanation of the variable accounted by the factor (or eigenvector) within the context of the

Example 1. Population Characteristics at Time T_1.

Enumeration Areas:	Age				Ed.				Ethnicity					
1														
2														
3														
4														
5														
.														
N														

R / Resolution / Q

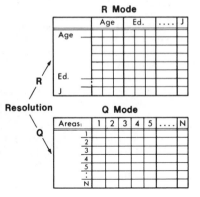

R Mode

Q Mode

Flows	To Towns:						
	1	2	3	4	5	...	N....
From Towns: 1							
2							
3							
4							
5							
.							
N							

Example 2. Functional Connections between Towns (Journey to Shop Flows) at T_1.

Example 3. Attitudes of Residents to Moving to Regions 1–N at Time T_1.

Attitudes	To Regions:						
	1	2	3	4	5	...	N
From Regions: 1							
2							
3							
4							
5							
.							
N							

Enumeration Areas:	Pop. Character at Time:			
	T_1	T_2	T_3	.........T_N
1				
2				
3				
4				
5				
.				
N				

Example 4. Temporal Changes in Population Characteristic T_1, T_2 in the Cells.

Figure 3.3   Examples of human ecological studies using aspects of the data cube

factor solution. The factor scores measure the importance of the rows (or cases) upon the factors or the composite clusters of parts of the variables. In the Q mode approach the factors measure the clusters of rows (or case), and the factor scores will demonstrate the importance of each variable to each factor. Britton (1972) has provided one of the few examples of this approach in structural geographical work and Davies (1972, 1979) has suggested the Q mode is preferable in studies of flow patterns. It is worth noting that the reversals in explanation involved in this description of the R and Q mode approaches, does not necessarily mean that identical results will be obtained from the use of both methods, so care should be taken over interpretation. To take one example, the scores are usually standardized for each component or factor to a zero mean and unit variance, and have to be estimated in common factor procedures. Hence differences in the relative scaling of the variables or the cases may result, particularly where different communality estimates are employed for the various variables – as in a common factor model. Care should be taken to determine whether the R or the Q mode procedure is the most appropriate for the problem of any single study. It could, of course, be argued that the R mode approach is more appropriate for the sociologist, since it identifies structures or dimensions in the variables of a data set. By contrast, the geographer or ecologist, may be more interested in the Q mode approach, since areal associations are derived directly, and sub-groups in the population being analyzed can be most easily identified. Yet no such disciplinary bias can be identified in the geographical or sociological literature, although geologists who are interested in areal classifications seem to prefer Q mode approaches (Klovan 1976).

(c) S, T, P and O modes

R and Q mode approaches are by far the most popular modes used by factor ecologists since the majority of factorial studies in all disciplines have been concerned with the cross-sectional relationship composed of cases and variables. One important issue in geographical or ecological study is the addition of temporal relationships. This allows the analysis of processes. Cattell's (1952) classic review of the 'cases- variable-time' problem is still one of the best summaries of the problems involved, in which a series of different research designs are needed to deal with the data cube problems in two dimensional terms: S and T modes are used for the time and cases relationships, whereas the P and O modes are applied to the time-variable linkages. These are shown in Figure 3.4 where an additional set of labels is proposed. This is a mnemonic based on which two of the C (cases), V (variable) or T (time) possibilities are being used. The order of the labels describes the modes being adopted, with the similarities calculated on the second letter. Hence CV represents the standard R mode of cases and variables, with the similarities calculated on the variables, whilst VC represents the Q mode, for the similarities are calculated on the pairs of cases. As it is frequently confusing to remember which sets of data appear in the factor loading and factor score matrices Figure 3.4 shows the appropriate entries for the loadings and scores for all modes of analysis.

Comparatively few of the alternative modes of analysis have been used by geographers, although the S mode has been used in several studies. For example, King and Jeffreys (1971), as well as Pigozzi (1975), have

ALTERNATIVE FACTOR MODES

a) *Two Dimensional Modes*

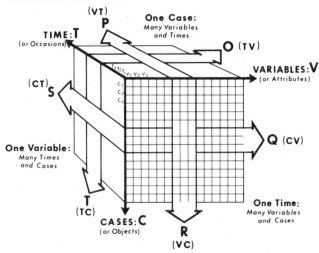

b) *Three Dimensional Modes*
(Two Geographical Axes)

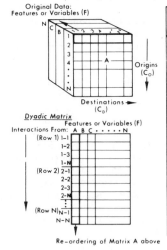

Re-ordering of Matrix A above

MODE (SYMBOLS)	SIMILARITY MATRIX BASED ON:	ENTRIES IN FACTOR MATRICES	
		LOADING MATRIX	SCORE MATRIX
(a) 2D			
One Time:			
R (VC)	Variables	Variables	Cases
Q (CV)	Cases	Cases	Variables
One Variable:			
S (VT)	Cases	Cases	Times
T (TC)	Times	Times	Cases
One Case:			
P (VT)	Variables	Variables	Times
O (TC)	Times	Times	Variables
(b) 3D			
Interactional Matrices			
C_oC_oV	Variables	Variables	Flows
C_oC_oV	Flows	Flows	Variables
C_oC_oT	Times	Times	Flows
C_oC_oT	Flows	Flows	Times

Figure 3.4 Alternative factor modes

used the S and T approaches to look at the cases – time relationship for a set of cities (cases), in which the single variable used is the 'rate of unemployment'. In both studies, the factors identified sets of towns with distinctive profiles in their rates of unemployment, demonstrating that the fluctuations in the general level of economic activity had an urban component, in the sense that particular groups of towns were affected by unemployment at different times. The results illustrated the spatial patterning of the cyclical impulses in the economy, although not all the town sets were associated with distinctive regions. For the T mode the correlation in the cases-time relation for a variable would be based on the time periods. Cattell (1952 p.503) referred to this as the 'social climate' thermometer, since the approach would identify the clusters of times at which major changes occurred. One example could be when attitudes to a regional devolution question are monitored, and are shown to peak at individual times; another would be the indexing of phases in the business cycle.

In the P and O modes the variable-time sequence can be investigated for a single case (usually a community or region is the major field of interest) and they are particularly suitable for longitudinal studies. Cattell and Abelson (1951) and Gibb (1956) provided early examples of the P approach by measuring the dimensions of social change in the USA and Australia; clusters of variables associated with particular social characteristics were shown to emerge at particular periods. The method, therefore, provided a precise calculus of social change. In the O approach the correlation is related to time, so it is particularly suitable for the analysis of stimulus situations. For example, the effect of a new administrative policy on a set of variables in one area could be monitored through time, and the phases of major changes would be identified by the clusters of the time periods represented by individual factors.

Although the six modes investigate pairs of relationship the investigator must be careful to specify which type of relationship has primacy in an individual study. After all the modes are not necessarily interchangeable, as Cattell (1952) has pointed out. For example, it is possible that different weightings can be produced from the loadings of the P approach, which identify the groups of variables by factors, as opposed to the scores of the O approach which scale the variables on the time periods. Few difficulties occur if the appropriate mode for the problem at hand is carefully thought out in advance of the data collection and analysis.

(d) Three mode analysis and the dyadic procedure

These alternatives do not complete the ways in which the relationships can be explored in geographical studies. It is apparent that Cattell's (1952) suite of modes represent ways of breaking down what is a three dimensional problem into a set of two dimensional sets. Recent developments in factor analysis procedures by applied statisticians have focussed on this problem. Indeed Tucker (1966) has produced a three-mode factor method which simultaneously deals with any three modes of data in the Principal Component case. More recently Bentles and Lee (1979) have provided a common factor approach, complete with decision rules for testing hypotheses. Cant (1971, 1975) was the first to use the three-mode approach in geography, although the method is becoming more popular in other social sciences.

An alternative way of dealing with three dimensional data sets is to adopt the dyadic approach, one which has proved to be particularly useful in the analysis of flows between towns. Black (1973) applied the idea in a study of commodity flows between states in the U.S.A., whilst Davies and Thompson (1980) showed how the method could unravel the different patterns of linkage structures in an analysis of 15 commodity types between the largest towns in the Canadian Prairies.

The basis of the dyadic approach is shown in Figure 3.5, where a set of commodities (Z) are traded between a series of towns which act as both origins (X) and destinations (Y). The data set can be visualized as a three dimensional cube that can be broken down into a set of two dimensional origin-destination data arrays (XY) for each commodity (Z). Instead of dealing with each commodity (Z) in turn, then factor analyzing the origin-destinations information, and trying to integrate the individual studies, the dyadic approach converts the three dimensional information cube into a single two dimensional array. This is achieved by taking the origin-destination matrix of <u>each</u> commodity type in turn, and rearranging it in a single array. For example, the second and subsequent columns of the Z_1 commodity matrix can be arranged underneath the first column. This single column or array obviously incorporates <u>all</u> the flows from each origin to every destination for an individual flow type. In the normal assymmetrical data matrices which characterize data sets of this type, the flow X_1Y_3 will be different from Y_3X_1 since the first flow is from origin 1 to destination 3, the other is from destination 3 to origin 1. By putting together all the individual flow columns the three dimensional data set is transformed into a two dimensional matrix. This can be factor analyzed in the usual way. By the R mode route the factor loadings will pick out the flow types that are most closely associated with one another, and the scores will scale the dyads or links between origins and destinations according to their contribution to this association of flow types. Since flows to and from a town involved intra-urban rather than inter-urban movement the diagonal cells in the original two dimensional matrices X_1Y_1 and X_2Y_2 are usually excluded.

(e) Direct factor analysis

So far this dicussion of the alternative modes of analysis has made one major assumption, namely that there is always a similarity matrix calculated between the original data matrix and the derivation of the factor loading matrix. Although most factor analyses base their extraction of factors upon the similarity matrix, there is no need to follow this accepted route. As its name implies the alternative approach involves the <u>direct</u> analysis of the data matrix with the factors obtained directly from this data. Factors, or more precisely eigenvectors, can be obtained from any square symmetrical matrix, but normally the approach is only used when the data runs on a scale from 1 to 0. Some of the earliest examples of the direct factor aproach can be seen in sociology, where investigators such as Macrae (1960) used a simple binary scaling device (1 for friendly, 0 for non friendly associations) to characterize the sociometric choices between each pair of individuals. The resultant factors, therefore, identified the

THREE DIMENSIONAL DATA ARRAY

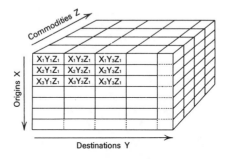

Commodities Z

Origins X

Destinations Y

$X_1Y_1Z_1$	$X_1Y_2Z_1$	$X_1Y_3Z_1$	
$X_2Y_1Z_1$	$X_2Y_2Z_1$	$X_2Y_3Z_1$	
$X_3Y_1Z_1$	$X_3Y_2Z_1$	$X_3Y_3Z_1$	

TWO DIMENSIONAL ARRAYS

(I) Multi - Analyses

Origins X

Destinations Y

(a) Commodity Z_1

$X_1Y_1Z_1$	$X_1Y_2Z_1$	....
$X_2Y_1Z_1$	$X_2Y_2Z_1$	....
$X_3Y_1Z_1$	$X_3Y_2Z_1$	

(b) Commodity Z_2

$X_1Y_1Z_2$	$X_1Y_2Z_2$	....
$X_2Y_1Z_2$	$X_2Y_2Z_2$	....
$X_3Y_1Z_2$	$X_3Y_2Z_2$	

(c) Commodity Z_3

$X_1Y_1Z_3$	$X_1Y_2Z_3$	....
$X_2Y_1Z_3$	$X_2Y_2Z_3$	....
$X_3Y_1Z_3$	$X_3Y_2Z_3$	

(II) Dyadic Analysis (2D summary of 3D Arrays)

Cases are dyads XY

	Z_1	Z_2	Z_3	
	$X_1Y_1Z_1$*	$X_1Y_1Z_2$*	$X_1Y_1Z_3$*	⎫ From
	$X_2Y_1Z_1$	$X_2Y_1Z_2$	$X_2Y_1Z_3$	⎬ Column 1
	$X_3Y_1Z_1$	$X_3Y_1Z_2$	$X_3Y_1Z_3$	⎭ of Z_3
	$X_1Y_2Z_1$	$X_1Y_2Z_2$	$X_1Y_2Z_3$	⎫ From
	$X_2Y_2Z_1$*	$X_2Y_2Z_2$*	$X_2Y_2Z_3$*	⎬ Column 2
	$X_3Y_2Z_1$	$X_3Y_2Z_2$	$X_3Y_2Z_3$	⎭ of Z_3
	$X_1Y_3Z_1$			
	$X_2Y_3Z_1$			
	$X_3Y_3Z_1$*			

—— Variables are commodities Z ——→

If Origin X_1 is the same as Destination Y_1 intra-urban flows would be incorporated so these dyads are usually removed. Directional differences are built into the analysis since X_1 to Y_3 is different to X_3 to Y_1.

Figure 3.5 The dyadic approach to the analysis of three dimensional arrays

clusters of friendship groups. In geography the most popular use of the direct factor procedure has been in the analysis of communication structures. Gould (1968), Taafe and Garrison (1973), and Tinkler (1972) coded the presence of a direct road link between towns with a 1 and used 0 for the absence of a connection. Garrison and Marble (1964) have gone one stage further in applying the same idea to a 59x59 interconnection matrix of airline services between Venezuelan towns. Separate factors associated with each of the major regional patterns in the patterns of connectivity were identified, with the factor loadings of each centre on the individual factors measuring the strength of the connections of a place to latent structure identified by factors.

There is little doubt that the direct factor approach provides another example of the flexibility of the factorial method but a few words of caution are in order. If the approach is applied to small data sets critics are likely to point out that the same results could have been obtained by less sophisticated methods; the only defence is that a measurement of the latest structures has been produced. For transportation networks it is probably only worth using the method where large data matrices are involved. In addition, it is worth remembering that the approach works <u>directly</u> on the data matrix so this must be symmetrical. An investigator must be sure that the problem requires a factor solution based on the initial data matrix – not one associated with the similarity of the rows or the columns, the Q or R mode approaches of the typical cases-variable arrangement. In this context it should be obvious that the absence of a similarity measure such as the correlation coefficient means that the size differences between the cells of the matrix are incorporated into an analysis. In the more typical 'data – correlation – factor procedure' the correlations are based on the standardized data. This means the influence of size differences between each row or column is removed. These points do not necessarily mean that the raw data has to be used. Different powers could be applied to the matrices, and the resulting data set can be analyzed by direct factor procedures. Hay (1975) has shown that a direct factor analysis of a connectivity matrix produces the same results as a powered coefficient of association based on common presences. Although he preferred this to a method based on absences, Tinkler (1975) disputed this view.

4. FALLACIES

This discussion of the content and range of relationships that could be explored by ecologists using factorial methods would not be complete unless the limitations of the various approaches are understood. In particular, it is vital to distinguish between those generalizations derived from studies of areas, and those obtained from studies of non-areal aggregations, and those from individual phenomena. Unfortunately, urban ecology researchers seem to have unconsciously linked the two approaches together. For example, in the Chicago school of urban ecology, descriptions of areas of the city, such as the ghetto, are found alongside studies of individual or group behaviour, such as the hobo, or the gang (Theodorson 1965). The interrelationships that were identified between behavioural and areal phenomena certainly contributed a great deal to the understanding of cities. But the failure to clarify the precise nature of the linkages – especially as regards inferential statements between the two levels – produced a great deal of confusion about the utility of the ecological approach. It eventually led to W.S. Robinson's (1953) classic paper on the ecological fallacy. Robinson showed that if any set of variables are correlated at an ecological level and also at an individual level, different results can be produced. Using data from the 1930s, Robinson produced a correlation of +0.20 between illiteracy rates and the percentage of blacks in the U.S. population when these variables were measured from sample survey data on individuals. However he found a correlation of +0.95 between the same variables when the data was measured on an areal basis. This led Robinson to conclude that individual behavioural relationships could <u>not</u> be inferred from ecological associations.

At first sight Robinson's paper appeared to destroy the whole credibility of the ecological approach, but the flaw in the argument was soon recognized by researchers in the field. The differences occurring between individual and ecological correlations is <u>only</u> vital if one is inferring individual characteristics from ecological data. Today, most ecologists would argue that there is no reason why this type of inference should be made. The appropriate attitude would be to assume that differences <u>will</u> occur in the individual and ecological relationships, so the conclusions of both studies will differ because they relate to quite different types of analysis: one dealing with individuals, the other dealing with areal aggregations. Both approaches, however, are still at the level of measuring casual associations; there is no necessity to impute causal relationships. If one seeks to explain the patterns, two separate explanations can be put forward. The first assumes that literacy variations between black and white people is a genetically determined difference, with the advantage with the latter. The other assumes that the differences are the result of the social environment. In the 1930s, for example, black people in the U.S.A. were concentrated in areas with low levels of urbanization, low standards of living and low modernity, so high illiteracy would be a function of both the poor quality and length of schooling.

This example illustrates the problems of inferring individual characteristics from ecological data and demonstrates the dangers and the limitation of the ecological approach. However, Alker (1969) has extended the argument by identifying a whole range of similar fallacies that confront the research worker. By modifying Alker's ideas, these fallacies can be reduced to the five primary types shown in Figure 3.6, namely individual, cross-level, contextual, universal, and cross-sectional fallacies. They all stem from a failure to clarify the inferential processes that are used to interpret the results of any study. As such they are important problems for factorial ecology.

(a) The individual fallacy

This fallacy stems from the assumption that the behaviour of individuals, or even the characteristics of individuals, can <u>only</u> be explained on the basis of individual features. Although some features may be the product of individual behaviour, taken to its extreme it represents a denial of the possibility of making generalizations about aggregations.

Levels of Relationships and Fallacies

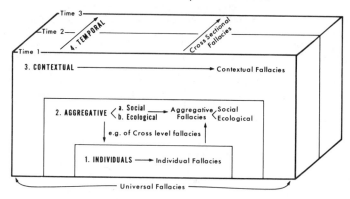

Figure 3.6   Levels of relationships and fallacies

(b) Cross-level fallacies

These fallacies consist of making inferences about the behaviour, or characteristics of one level of an analysis, from another level. The classic example would be one discussed in previous sections, the ecological fallacy, the assumption that the relationship found at the ecological level is true for individuals. Similar difficulties, perhaps to be called group fallacies, exist when one infers that the aggregative characteristics of social or other groups necessarily occur at the individual level. To avoid confusion it must be noted that some revision of Alker's (1969) terminology has been made. Alker actually used the term 'cross level' to describe the fallacy where groups, or aggregations, could be given the attributes of individuals. It would seem that since this is really the reverse process to the ecological fallacy, the term 'cross-level' is more appropriate as a summary for the embracing group and ecological fallacies. A good example of the group fallacy can be seen in the German school of geopolitik. Here, 'states' were endowed with the characteristics of individual organisms, and this was used as a justification for following certain policies. For instance, it was pointed out by analogy from the behaviour of organisms that if states did not expand territorially they could not survive. Useful descriptive generalizations may be obtained from these ideas, and might lead to the discovery of unknown relationships. Nevertheless they frequently obscure the fact that human actions chart the progress of the aggregate. As Harold and Margaret Sprout (1965 p.34) noted in their perceptive survey of human ecology:

'In the case of the state reification buttresses ethnocentric values and images which characterize much of the discussion of foreign policy, military defence and international politics in general.'

In some ways the 1970's fad for the application of general systems theory to ecological relationships, principally the argument that there is a vital similarity between processes operating at several levels or in many different fields, may be looked at in the same way. It may be construed as leading to cross-level fallacies if the causal processes turn out to be rather different. So far this discussion has concentrated only upon the problems of inference between the individual and aggregative levels. It would not seem difficult to extend the argument to cross-level fallacies with the contextual level.

(c) Contextual fallacy

To deny the fact that the 'context', whether social, or ecological, can alter the strength of a statistical relationship would represent a contextual fallacy. For example, it is likely that the relationship between illiteracy and any low status ethnic group would be stronger in a ghetto, not only because there are more members of the ethnic group in the area, but because of the poverty of schools in the area. The sense of 'hopelessness' present in some ghettos (Stokes 1962) would also make the attainment of an acceptable standard of literacy more difficult. Hence, the 'context', the area in this case, would probably increase the ecological relationship. It is difficult to measure the extent of this effect but examples are provided in Chapter VI Section 5 (Herbert 1976).

(d) Universal fallacy

All too frequently social scientists have used subjective local examples and relationships found in any one city or region have been assumed to have global relevance: one that occurs in many circumstances. This type of approach lays the ecologist open to the charge of committing a universal fallacy. To find a high correlation, either individually or ecologically, between negroes and illiteracy in one city does not necessarily mean it occurs in all cities. This is, of course, the problem of the Shevky-Bell (1955) set of constructs that forms the basis of much factorial work: Social Status; Family Status; Ethnicity. The indicators were originally based on a theory of society that accounted for the transition from traditional to industrial society, and were designed to index the latter. To assume the dimensions or constructs occur in all cities represents a universal fallacy.

(e) Cross-sectional or temporal fallacies

The final set of fallacies are rather similar to the idea of a spatially universal relationship. But this time it is the extension of the idea back in time that is important. Unlike many physical phenomena the changes that occur in society over a human life span are very considerable, and it is important to take these changes into account when identifying variables. In the classic Shevky-Bell (1955) schema, fertility was measured by the number of children, 0-4 years old, as a percentage of the fertile females, those aged 15-44 years. In western society today most females who have children are likely to finish childbearing by the early thirties, so females over 35 years of age have few children under four years of age. The result is that a curvilinear relationship may exist between the number of young children and the age group of fertile females, instead of the more linear association of 20 years previously (Johnson 1971). In the same way that temporal changes affect the measurement of individual variables, it should be expected that constructs relevant at one period may not be relevant at another. For example, the change from an industrial to post-industrial society has produced several changes. One of the changes identified in factorial ecologies (Davies and Lewis 1973) is the differentiation of a pre-family group of young adults. If an investigator continues to ignore such changes by keeping to one set of constructs, the research may be subject to cross-sectional or temporal fallacies. One other point is worth mentioning. Frequently, the changes in society occur in marked breaks or jumps, instead of a simple linear curve. For some time periods, therefore, the assumption of linear relationships may hold true. In others the presence of abrupt changes in society through time may lead to serious underestimation of the actual trends and relationships.

5. SUMMARY

This chapter has attempted to summarize the scope of the human ecological field to which factorial methods can be applied, and to identify the problems of a conceptual and methodological nature, not those of particular data and techniques. In human ecology and geography relatively few of the various data arrangements and modes have been subjected to factorial methods and only a limited part of the potential content has been explored. Similarly, factorial ecologists have only dealt with some of the possible sets of relationships. Indeed most of the examples in this book reflect the emphasis upon identifying the social structures, patterns and flows in the literature rather than dealing with behavioural issues. This has meant that relatively few of the ecological fallacies can be identified in the factorial ecology literature, so the examples described have often used single rather than multivariate explanations. Nevertheless such problems will undoubtedly surface as more and more human ecological studies are carried out, whether using factor analysis or other multivariate techniques, so it is important to be familiar with these issues.

4 DATA PROBLEMS AND CHOICE

Many of the criticisms made about the use of factorial methods in human geography and human ecology can be summarized by the acronym GIGO: garbage in; garbage out. This one 'word' summarizes the results produced by poor data selection and insufficient problem orientation of a factorial study. This demonstrates that the use of factor analysis in any study, by itself, does not guarantee any necessary utility or respectability of conclusion; the factorial procedures only produce worthwhile results if they are used with care and caution within the context of a clearly defined problem. This re-emphasis of the need for explicit problem definition, as well as the derivation of an appropriate data set to which the techniques used in the analysis can be applied, acts as a counterweight to some of the wilder 'flights of factorial fancy' that characterized some of the early studies. Unfortunately for those seeking a set of foolproof guidelines, the wide variety of purposes fulfilled by factorial methods (Chapter I, Section 4), the range of data to which it can be applied (Chapter III, Section 2), as well as the set of alternative relationships that can be explored (Chapter III, Section 1), mean that one cannot present a rigid or definitive set of instructions governing the choice and preparation of data. However, there are a set of recurring considerations and problems in the applied factorial literature that do need emphasis – particularly if the results of factor analyses are to be used to create a comparative body of rigorously related literature in regional and urban study. These issues for the basis of this chapter.

The first two sections deal with the separate parts of any geographical study: first, the study area and observation units; and second, the choice of variables. Although this survey is phrased in an R mode framework, it should be noted that the position will be reversed if a Q mode approach is adopted. This reversal frequently causes problems for the inferential basis of the study if samples are employed, since the theory of common factor analysis is based on sampling from homogeneous populations, only rarely the situation in human geography or ecology. The third section deals with distributional characteristics and assumptions of variables. The fourth section takes the discussion to the penultimate stage before factor analysis is applied; it describes the alternative measures of similarity that can be used in a factorial context.

1. STUDY AREAS AND OBSERVATION UNITS

Two closely associated, but rather different types of problems relating to area occur in geographical or ecological studies: the study areas chosen for analysis, and the observation units within these areas upon which data is aggregated.

(a) Study area

The choice of any study area obviously remains the prerogative of the researcher. If the investigation is only concerned with one area, then much of the following critical commentary is misplaced as it relates to areal generalizations. If the results of the study are used to derive generalizations that are thought to be applicable outside the confines of the area (normally the reason why the generalizing procedures of factor analysis were chosen) then it is vital to justify the choice of areas used as a case study. All too frequently factorial ecologists have failed to provide this justification. The result is that much of the literature resembles a collection of eclectic studies, rather than an integrated body of comparative literature. One of the best examples of this problem can be seen in the research work on urban social structure, where one of the earliest factorial studies compared the cities of Boston and Helsinki (Sweetser 1965). Although it was discovered that there was a remarkable degree of similarity between the most important dimensions in the two cities, no really coherent reason for the comparison of these two particular cities was provided. Unfortunately, many of the subsequent factorial studies of urban structure have tended to follow the same approach. Inevitably the results have not been very penetrating. But it is difficult to see what else this type of approach will produce, since the results are very close to 'chalk and cheese' comparisons. In their own study, for instance, Palm and Caruso (1972) compared the factor dimensions of ten U.S. cities. These cities were not only different in economic and ethnic composition, but consisted of examples drawn from most of the regions of the country. In addition, differences in annexation policies and the extent of suburbanization were introduced by their decision to deal only with political entities. Not surprisingly, only Life Cycle and Economic Status emerged as being typical components in all cities: it would have been remarkable had any other result been achieved! Given the existing knowledge on cities at the time of their study (Johnston 1971) it can be expected that the dimensions of urban social structure are likely to vary with at least size, economic type, or regional location, whilst the local peculiarities of individual cities will provide another source of variation. Hence it can be concluded that if the study of urban social structures is going to advance beyond the stage of describing empirical regularities in one centre, the various sources of social variation need to be systematically identified and explored (Davies 1983). To ignore these differences is to fall into the universalist fallacy (Chapter III). In this context it is intriguing to note that Cattell (1978) recently made an eloquent plea for psychologists to move away from the analysis of total populations, and to search for types, within which the extent of variation in factor dimensions, etc., can be analyzed. He reminded his readers that the theory of factor analysis assumed that the factors are measured from a homogeneous population. If different types of populations are thrown together in one study a great deal of confusion is produced, since

78

the factors in the population become blurred and reduced in importance. This type of advice is important in factorial ecology since investigators rarely know (with the necessary precision) how areas differ from one another. Davies (1978), however, in his set of analyses of the social structure of five Prairie metropolitan centres (Davies 1975, 1977) had previously carried out an inter-metropolitan analysis of Canadian centres, demonstrating that the case study cities were sufficiently similar in social pattern to be considered a separate type in Canada. This removed the charge that different Prairie city types were being considered together. Such an approach would seem to be the first stage in any comparative analysis of the cities of an area, if the identification of general patterns of variation are not to be blurred by specific local patterns.

Similar problems occur in other types of ecological study. In many studies of functional regions a common mistake is to impose a type of 'closed system' framework upon the results; for example, by tightly defining the study area and excluding places outside the specific area of interest. The result is that factors define functional regions that are biased by artificial boundary restraints. Davies (1972, 1980) has shown that one advantage of the factor analysis of flows is that the method has a self-defining property. Those places in a Q mode analysis that have less than half their variance explained by the general factor patterns are identified by communalities under 0.5, illustrating that their interaction patterns are not primarily associated with the general factors.

(b) Observation units

The problems associated with the observation units of a R mode study (the places in the typical geographical study) are rather different. In Chapter II the discussion of the differences between individual and ecological correlations concluded that all ecological relationships are best assumed to be 'scale specific'; they relate to the size of the unit used as the basis for aggregation. This means that the problems associated with the choice of observational unit are as important to the human ecologist as those relating to the variables. Unfortunately they are rarely given as much attention.

The most appropriate way of discussing the problems of the observation units is by making a distinction between what Rees (1972) called modifiable and non-modifiable observation units. In the former, the observation units are considered to be aggregations of area that can be altered; in the latter, the areas are fixed by the objectives of the analysis. For example, a factorial ecology of a set of political areas would deal with the areas as defined units, in the same way as a classification of towns deals with towns as defined points, even though the units could be built-up or administrative units. A study of intra-urban variations within a city would not necessarily have a deterministic set of areal units, so the study would be classified as one using modifiable units. Each of these problems, and some suggested solutions to the problems faced by factorial ecologists are discussed below. A third section deals with the related problems of the degree of homogeneity in observation units.

(i) <u>Non-modifiable units</u>. In studies where the units of aggregation are defined by the purpose of the analysis, comparatively few problems occur, once it is recognized that the results have relevance only to the areal set. Examples can be found, however, of inappropriate units of aggregation. For example, Moser and Scott's (1962) classic study of British towns was not really a study of towns per se, but of administrative areas over 100,000 population size. This does, of course, beg the question of what is a town or city, and in this context it is worth noting that very different definitions of urban centres can be obtained by using political, land use, or functional region criteria (Carter 1972). The result is that a definition of the units will vary according to boundary policies, annexation practices, and the extent of suburbanization. Usually, comparisons between administrative cities or units are likely to be seriously affected, since in many areas middle or upper class suburbs are excluded from the political city, thereby producing quite false interpretations of the differences in 'urban' as opposed to 'metropolitan' structure. Broady's (1965) factorial comparison of Southampton and Portsmouth illustrated this type of problem. In this example the very different social patterns of the two centres are more a consequence of the way the political boundaries delineate the built-up or functional regions of the cities, rather than any radically different dimensional structures, although the cities do contain rather different proportions of people at various levels of socio-economic rank. More careful studies of the effect of different definitions of city areas have been carried out by Rees and Berry (1970) in Chicago as well as by Hughes and Carey in New York (1972). In the latter case the results from a factor analysis of the New York Standard Metropolitan Statistical Area were compared with those of the city of New York and with one of the political subdivisions in the city, Bronx. At the metropolitan area scale five axes were identified, indexing: Social Status, Life Cycle, Resources, Ethnicity and Occupied Housing. Hughes and Carey (1972) showed that this last axis could not be picked out at the city level. In the study of Bronx, Social Status and Life Cycle were identified, although the latter was linked to ethnic groups whilst a Segregation axis was found amongst the other three axes. This had high associations with Sub-standardness and had some parallels with the Resources vector found at the metropolitan scale. In addition, factors describing Housing, and 'Female Clerical – Children under Five' were also picked out, but these had few parallels with the axes at the other scales. Hughes and Carey concluded that factor structures were <u>not</u> invariant with changes in the scale of observation, a not surprising conclusion given the very different social characteristics of the population found at the various observation scales. Nevertheless, the study provided a valuable cautionary note to the implicit assumption in the literature that some universal axes of differentiation could be found.

(ii) <u>Modifiable units</u>. In many study areas the question of the units to be used in an investigation is an open one, since data is available for a series of different units. Not only does this mean that a series of scales of analysis are possible, but the units of investigation are modifiable from one to another. This situation is more and more typical of intra-urban studies in many western countries, since information is often available on a street, or block front basis, as well as at the enumeration area or census tract scales.

One of the most rigorous case studies of this problem can be found in the work of Openshaw (1973). He looked at the distribution of dwelling units from six architectural periods in South Shields (England), aggregating his data into a variety of different cells from one hundred metres to one kilometre squares. The results showed that increases in cell size led to larger means and standard deviations, although skewness was less marked. Increasing cell size was associated with larger correlations, so a factor analysis of the data produced larger eigenvalues and higher loadings for each fator. Subsequent work by Openshaw (1978) stressed the fact that the results of spatially aggregated data are not independent of the choice of zoning system. He concluded that no simple or general purpose solution to the difficulty is in sight and went on to suggest that:

'there can be no longer any reasonable excuse for the researcher failing to investigate the empirical consequences of choosing a particular zoning system.' (Openshaw 1978 pp.781-794)

Although this conclusion was primarily linked to the use of different sized units in a particular situation the general thrust of the recommendation does not seem unreasonable, given the evidence, unless the investigator is only interested in describing his results at one scale and in one place.

As yet comparatively few exhaustive empirical studies of the question of invariance by geographical scale have been carried out in the factorial field despite the progress made by sociologists such as Blalock (1964) and particularly M.T. Hannan (1971), in the study of aggregation and disaggregation problems. Even where comparative studies of the same data set at different scales have been carried out, it must be admitted that there is little unanimity in the results. For example, Berry and Spodek (1971) in a three level study of Bombay using wards, circles and sectors, found factorial results that were described as 'remarkably stable' even though the stability was subjectively determined. By contrast, Romsa, Hoffman and Brozowski (1972) produced quite different results in a 30 variable study of Windsor (Ontario), using two areal breakdowns, 43 census tracts and 340 enumeration districts. Coefficients of congruence (Harman 1976) were used to measure the similarity between the axes, and only two of these coefficients were greater than 0.7 with four under 0.5. Perle's (1977) study of the neighbouring metropolitan area of Detroit using a 43 variable set represented a rather more comprehensive study of geographical scale differences, since an analysis of 62 sub-community areas was compared with the results obtained from 444 census tracts. The author concluded that:

'although there are some similarities between the two sets of results, major differences persist in factor interpretations ... (and) ... the different scales of areal aggregation lead to rather different perspectives on the structural configeration of the city.' (Perle 1977 p.557)

It is possible to be rather critical about the over-emphasis of housing and employment variables in Perle's study, since it leads to imbalance in the sources of variation. Nevertheless, Perle's conclusions about 'major differences' seem unduly pessimistic when one

Table 4.1
Loadings for first order axes in Detroit (1971):
444 census tracts (and 62 communities)

Census tracts (communities) explanation: 70.2% (86.9%)

1. Status: Housing/ Possessions [Housing Quality (Life Cycle)]	2. Status: Employment, Education [Socio-Economic Status (S.E.S.)]
-89 (92) Single detached	-85(-93) College
-81 (97) Two cars	-79(-71) Years education
-75 (76) Persons/house	-76(-88) Professional
-73 (82) One + baths	-61(-77) $25,000+ income
-73 (77) Owner occupation	-55(-44) Income
-72 (92) Auto to work	-58(-) Owner-occup. value
-60 (79) Rent	-52(-66) Service
-56 (48) Under 18 yrs.	-50(-) Renters rent
-55 (78) Medium income	-42(-41) 1 + baths
(65) Husband/wife families	(-46) Medium age
(57) Years education	(-46) Old age
(51) New housing	
	(47) Persons/house
(-52) Unemployed	(52) Manuf.
(-44) Old aged	40 (-) No car
(-50) Limited education	45 (47) Unemployed
41(-62) Old housing	52 (59) House workers
51(-76) Under $4,000/year	57 (78) Overcrowded
57(-59) Recent movers	79 (89) Foremen
68(-87) Widows, divorced	87 (82) Limited education
70(-90) No bath	
72(-91) No plumbing	
72(-87) No car	

3. Ethnicity [Ethnicity-Segregation (E/S)]	4. Females (Housing-Age) [Females Labour Force (Type of Housing)]
-83 (88) Foreign stock	
-68 (79) Polish	86(-) Married females,
-60 (63) Italian	labour force
-58 (70) Manufacturing	58(-50) Females, labour force
-53 (--) 20 year movers	54(-49) Husband/wife families
	(-67) Housing value
(-73) Females L.F.	
(-53) Houshold workers	
(-51) Median age	(45) Young family
(-59) Old age	(44) Unemployed
58(-59) Service	(74) Under 18 yrs.
74(-58) N.W. renters	(56) owner occupied
74(-56) N.W. owner-occup.	(83) renters
77(-56) Blacks	

82

Table 4.1
(continued)

5. Life Cycle [Age]	6. Housing (Non White and Housing) [Non White Housing (Blacks - Females)]
76 (57) Age	72 (-) Renter
68 (-) over 65	45 (-) Old housing
	42 (-) Owner occupied (%)
-41(-) Overcrowded	41 (55) Non white (%)
-44(-52) Recent movers	41 (56) Non white renters
-50(-) Pop. under 18 yrs.	(48) Blacks
-59(-71) Young families	(45) Married females L.F.
	-60(-) Value housing

Census Tract figures are given first; loadings for community scale are in brackets. Perle's titles are given in square brackets. Decimal points are excluded: -41 is -0.41.

Source: revised from Perle 1977

reviews the pattern of loadings, at least in a descriptive sense for the author did not use any quantitative measure of similarity. Table 4.1 presents the evidence for this revised opinion, where a short summary of the original variables is given to ease comprehension of the indicators. The comparison shows that the three largest axes dominate the study. (They are renamed as: Status - Housing and Possessions; Status - Employment and Education; and Ethnicity in Table 4.1.) But these are, in fact, very similar in substantive terms despite differences in the size of the loadings, and the greater complexity or generality of results at the coarser scale of the large community area. The 'Age' axis also has some similarity, for the highest loading variables at the two scales although the loss of some specific age variables to the Ethnicity and Type of Housing axes of the community scale must be noted. The other two minor axes are, of course, associated with rather different characteristics which are much more scale-dependent. For example, the Females in the Labour Force axis at the census tract level is only partially identified at the community scale for its most important loading is more closely linked to the ethnic axis at the community scale. In general, therefore, the detail found at the census tract level has been lost by the use of the larger community units. Not surprisingly, therefore, the inter-factor correlations vary at the two levels.

Corroboratory evidence that the larger or more dominant axes are often similar at different scales of analysis has been provided in Davies's (1983) comparison of 20 wards and 541 enumeration areas in a study of the built-up area of Cardiff (Wales), an area of a third of a million inhabitants. He showed that the use of larger collecting grids (such as wards) produced higher levels of explanation for the axes, as well as more general vectors. This is, incidentally, also a feature of Perle's Detroit study. Moreover, the quantitative comparison of the axes shown in Table 4.2 confirmed several features. The first is that the three largest axes were very similar as shown by the size of the congruence coefficients. Another axis, Urban Fringe, needed modifica-

Table 4.2
Results at two scales in Cardiff, Wales, 1971

Enumeration Districts	Con-gruence Co-effic-ients	20 Wards
1. Socio-Economic Status	+0.88	1. Socio-Economic Status
2. Young Adult – Non Family	+0.94	2. Young Adult – Non Family
3. Life Cycle and Tenure	-0.81	3. Life Cycle and Tenure
6. Fringe	+0.63	5. Fringe – Mobility
5. Late Family – Mobility	-0.56	4. Female
4. Ethnic – Housing		Not found (-0.69) with Life Cycle
Total Explanation: 67%		91.5%
Source: Davies 1983		

tion in its title, since it measured additional characteristics. The third feature is that the Ethnic-Housing axis of the enumeration level was not found at the ward scale. Finally, the Late Family-Mobility axis was replaced by one identifying Female variables. Given the very drastic differences in the scale of the two studies such similarity in the results are really rather encouraging. After all, it cannot be expected that such a localized characteristic as the 'Late Family Areas' in the middle city, or the ethnic concentrations in the inner city of Cardiff, will be identified in an analysis restricted to a mere 20 wards, unless of course, the areas were deliberately drawn up to isolate such characteristics. It must be concluded from this evidence that intra-city scale differences for the major axes may not be as important as originally thought, although the minor axes can be expected to vary. In the last resort it must be remembered that the degree of similarity between the axes is a function of the size and homogeneity of the collecting units, a rather obvious conclusion that has frequently been forgotten in fascination with the sophistication

84

of the factorial techniques. Nevertheless, Openshaw's (1978) recommendations for investigations at different scales should be endorsed, if the study objectives are sufficiently flexible to allow this type of digression.

The fact that investigators should be cautious about the effect of using different sized collecting units upon a set of results apply with equal force to areas that possess very different shapes. The danger of the ecological fallacy (of assuming that relationships hold at different levels) is always present. Yet it must be emphasized that investigators are not bound to accept the initial set of units upon which to measure their variables. The units used in any study can always be modified. One approach is by amalgamation to remove size or shape differences, thereby producing a more effective analysis. Another is to weight the data to take the different sizes into account. Other strategies include: measuring the internal homogeneity of the units; or avoiding the problem of dealing with complete populations by sampling within the data set. Each of these issues must be considered in turn.

1) Amalgamation. The simplest way of resolving the problem of different sized and shaped small areas is to amalgamate them to produce greater homogeneity. Such an approach has a long history. Even Charles Booth at the end of the 19th century (Davies 1978) recognized the problem of distortion in his comparison of social conditions between London areas and used districts of approximately similar population size so as to ease the process of comparison (Booth 1893). Jones (1969) grouped the 2105 collectors districts of Melbourne into 611 areas which he maintained were as internally homogeneous as possible for his social structure analysis. Yet the mere process of amalgamation alone does not really resolve the difficulty, unless adequate justification for the grouping is provided. In most cases this justification must be quite specific. The objective of achieving equality between the areas does not explain whether it is equality of population, or of area that was desired. Another objective could be to define internally homogeneous areas, or meaningful social units, but again this begs the question of how this is to be carried out. Ideally, grouping procedures could be applied to the initial data set to cut down the number of original units. But most investigators have settled for areas that they consider to be meaningful for their problem. For example, in many North American cities community areas have been defined by planning or local community associations, and these may be appropriate units for study as Davies (1975) has shown in Calgary. Yet is must be recognized that once this type of decision is made the definition of area ceases to be modifiable. The analysis has been restricted to a certain scale, and in new areas of study it may be important to test the effect of areal aggregations. Bourne's study of Toronto (1971) provided one of the few rigorous factorial examples of this problem. He amalgamated 301 census tracts of Toronto into two sets, one of 59 equal areas, the other of 62 areas of equal population. Although he found that the major relationships and patterns remained stable between the two studies, some minor factors proved to be particular to individual scales and were described as being indicative of localized processes. Again, therefore, these results show that there is a vital need to justify the use of one scale over another, if an

individual study seeks to contribute to a rigorously comparative body of literature.

2) Weighting. The alternative approach to the problem is to adjust the correlation measures that are input to the factor analysis by taking into account the relative size of the areas upon which data is aggregated. This was the procedure followed by Robson (1969) in his factorial study of Sunderland by correcting the correlation coefficients according to the square root of the population size of the unit. Unfortunately, the utility of this procedure cannot be fully evaluated for the results were not compared with those obtained from the unadjusted values. Indeed, in the more general cases of correlation and regression, the utility of weighting is in some doubt. Robinson (1950) demonstrated the dependence of ecological regression and correlation measures upon area size, and illustrated the need to weight the values by the area of the counties. Thomas and Anderson (1965) showed that Robinson's solution was only a special case, and that the variations in correlation and regression parameters obtained from different groupings of counties were not significantly different if the values were considered to be samples from all possible systems. This demonstration of relative invariance does, of course, conflict with Openshaw's (1978) evidence referred to earlier. Part of the difficulty may be resolved by viewing the ecological regression issue as referring to different samples of areas at approximately similar scales, rather than to the differential geographical scale problem investigated by analyzing enumeration areas and census tracts. For example in Calgary in 1971 there were 778 enumeration areas and 77 census tracts. It is hard to envisage that the results of such different scales can be reconciled by differential weighting. Nevertheless, at any scale the question of how different units can be equated remains. In a theoretical context many of those issues have been investigated by Blalock (1964) and his followers, particularly Hannan (1970). Unfortunately, few of the recommendations of these studies have been incorporated into the factorial literature to date.

At this time it is difficult to provide any unequivocal advice for investigators in the field. Certainly the necessity of weighting has not been conclusively proven. By contrast, it does appear to make intuitive sense to amalgamate areas for the appropriate scale of investigation and then explore the effect of data aggregation. It is possible that a series of relationships at different scales may be identified, although the work of both Davies (1980) and Bourne (1971) would indicate that the modification is likely to be primarily associated with the minor axes of differentiation. In some ways factorial investigations may be in the classical 'chicken and egg' situation. Until the analysis is completed, the types of relationships involved at different scales may not be known, so it may be a matter of initial guesswork to identify the appropriate scale for a study. This might be compared to the situation in which a physicist was confronted with a blindfold choice between working at the fundamental particle, electron, or atomic level. Perhaps we can only hope that as our knowledge of the spatial system improves, the appropriate scales of analysis for different problems and processes may be more easily definable. Alternatively, certain ecological relationships may be stable at a variety of different scales. At this stage in our

knowledge it is better to be cautious and assume that there are
differences - unless the study shows otherwise.

(iii) <u>Homogeneity of units</u>. The problems discussed to date relate
to the choice of area and to the effect of distortions such as size and
shape on the results of any analysis. A related issue (which has not
been given much attention in the factorial literature) is the question
of the degree of homogeneity <u>within</u> the areas being studied. This is a
particularly important problem in studies of regions or sub-areas of
towns, where the units are abstracted from a spatial continuum,
especially when the study is being carried out to indicate the general
environment of the area. By focussing upon the average values of the
units of investigation only the intensity or mean importance of the
characteristics is measured; whilst the relative degree of heterogeneity (or homogeneity within the areas) is ignored.

Newton and Johnston (1976) have addressed themselves to this question
of areal homogeneity in the factorial case, using a previously
constructed index of qualitative variation (IQUI). This index, called
a homogeneity score by the authors, measures the social variations in a
population across a set of categories that are mutually exclusive.

Homogeneity Score (IQV) =

$$\frac{\sum\limits_{i=1}^{k} \sum\limits_{j=1+1}^{k} n_i \, n_j}{\frac{k(k-1)}{2} \left(\frac{N}{K}\right)^2} \quad \text{x } 100 \qquad \text{Eq. 4.1}$$

The formula is calculated as the ratio between the sum of the
inter-category products, from k categories, and the maximum value of
these products if there were an even distribution of populations over
these k categories. The score ranges from 0, complete concentration of
the populations into one category, to 100, where there is an equal
number in each category.

Homogeneity scores were calculated for 320 census districts in
Christchurch (N.Z.) on 11 sets of variables that indexed the social
dimensions produced by previous factor analyses. Some of the scores or
variables (such as age structure and income) displayed little variation. The vast majority of the scores were in the 70-100 range,
indicating that the census districts were very heterogeneous in
composition. Others, particularly the size of dwelling and the value
of property variables, showed distributions that were much more evenly
spread across the homogeneity scale, indicating more variable degrees
of homogeneity. A Principal Axes component analysis of the data,
followed by Varimax rotation, produced the following three axes that
accounted for 70% of the variation in the 11x11 correlation matrix
based on the 11 variables: (1) Socio-Economic Status; (2) Life
Style/Age; (3) Housing Type. Since two of these axes were very similar
in type to the classic social area dimensions (Shevky and Bell 1955)
the authors concluded that:

'the different indicants of residential area characteristics spatially co-vary in their homogeneity as well as their intensity.' (Newton and Johnston 1976 p.548)

These homogeneity results were compared with the sources of variation produced by two other factorial studies in the same area, one using census data, the other housing information. The input variables were the three components identified in all three studies. Three quite distinct axes were produced, and identified dimensions called: Age - Life Style; Status; and Housing Type dimensions. Low quality housing, however, was linked to the Age and Life Style axes. The authors noted that the homogeneity axes were associated with similar content area axes derived from the intensity studies. This study demonstrated that the social dimensionality of cities identified in the intensity studies was parallelled in the homogeneity domain. Although the index used in this particular study was sensitive to the numbers of categories and observations, Newton and Johnston's (1976) approach represents a promising start to investigations of the range of variation associated with intra-urban dimensionality.

(iv) Sampling. It has repeatedly been shown that reliable estimates of the parameters of any homogeneous population can be obtained by taking a sample of the population, rather than by measuring the characteristics of the complete population. Much of the inferential theory of statistics, therefore, is based on the concept of estimating the extent to which a result obtained from a sample of cases - such as correlation between two variables - differs from the value derived from complete population. The common factor analysis model follows the same procedure. Many different types of sampling procedures (Conway 1967) have been proposed and are described in basic statistical texts in geography (Gregory 1978). A systematic sample would involve choosing every 100th or 50th case in a population, or some other systematic sequence; in a random sample the population may be numbered, and tables of random numbers can be used to pick out the cases to sample. This criterion of randomness maximizes the probability of the sample reflecting the general properties of the total population, by dealing with independent, unrelated individuals or observations. Obviously, the size of the sample is of crucial importance, since sufficient cases or observations are required before the population estimate becomes reliable, although there is always some range in the size of the population estimate.

In most factorial studies in human geography samples have not been employed except in perception studies, where individuals, not areas, are being analyzed. Investigators have usually tried to explore the variations in the complete population of urban enumeration areas or cities in a nation. It must also be noted that the mere fact of spatial variations in the characteristics of area means that the individual areas cannot be combined to form a homogeneous population; different cases are likely to have varying character. In the purely statistical models of factor analysis (Harman 1976) these two issues cause limitations for the applied analyst, since the tests of signif-icance applied to the factor results assume that samples from a homogeneous, independent population are used to obtain the data. The result is that the traditional significance tests of common factor models can rarely be employed in any strict sense in factorial ecology studies, at least not in the early stages of any investigation. One way around the problem is to carry out a series of exploratory data analyses to establish the subgroups of the population, then to produce a final study within these groups, using the complete battery of the inferential test procedures. Obviously the complete population will not be studied in these tests. But in testing the bases of differentiation by means of the statistical model the results could be expressed in inferential terms. This can be followed by a complete analysis. The problem here is that many of the significance tests are designed for observations in the R mode, with hetergeneous areas as the cases or observation units, whereas persons or individuals are being dealt with by psychologists. In the geographical context the individuals are often areal aggregations, hardly homogeneous or independent cases. Moreover, it is dubious whether the current sampling distributions used for R mode common factor cases hold true for the alternative models described in Chapter III Section 3, in which time or variables are sampled. Cattell (1978) has provided an excellent discussion of the issues involved for applied factor analysts.

2. CHOICE OF VARIABLES

The second set of problems facing any investigator are those associated with the choice of variables. For convenience, the problems are discussed under six headings: variable domains, variable relevance; scale of measurement; complementary and overlap; comprehensiveness and redundancy; with a final section discussing the issues posed by missing data. In all studies the variables chosen for analysis must be carefully chosen and should relate to the assumptions, and the particular research problems being investigated if maximum utility is to be obtained from the individual pieces of research.

(a) Variable domains

Most factorial studies in geography and ecology have used information drawn exclusively from a particular domain or content area, for example, socio-economic indicators from census sources, or preference data from perceptual surveys. The consequence is that our overall understanding of the sources of variation in areas is rather limited. This restriction on the comprehensiveness of studies, by an over-emphasis upon particular domains or types of data, can be criticized in terms of the long term goal of understanding the extent or spatial variation in human distributions. But in the short term the emphasis may be useful, because investigators can concentrate upon identifying the structure of variation in particular domains. Once these structures are well known, and marker or key variables identifying these structures are defined, the variables from these different domains can be mixed together for a more comprehensive study. To attempt this before the structures are clearly established is to invite confusion. Once these domains are looked at in conjunction with one another it is likely that some structures that have been given a title in one domain, will also be found in another one. It is much less likely that different structures will be found, each exclusively related to these individual data domains. A good illustration of this conclusion is provided by the literature on inter-urban or urban system analyses of the USA, where the so-called 'quality of life' indicators (although measured by objective interval scale variables) seem to have given different results (Smith 1972) to those investigations based on census based indicators (Berry 1972). (Incidentally B.C. Liu's (1977) results are not components because individual indicators are added together without benefit of factorial methods.) A factorial study by Davies and Tapper (1979) of the USA urban system (which is discussed at greater length in Chapter VIII) combining census or socio-economic and 'quality of life' indicators integrated the two sets of studies. This study

Table 4.3
Dimensions of inter-urban variation in the USA

Census-based Variables	Quality-of-Life Variables
Berry (1972)	Smith, D.M. (1973)
1. Size	1. Socio-Economic Status
2. Socio-Economic Status	2. Social Deprivation
3-7 Economic Base axes (5)	
8. Female Economic Participation	
9. Elderly Males; Working-Commuting	
10. Recent Employment Experience	
11. Family/Age	
12&13 Minority Groups (2)	3. Racial Segregation
14. Growth/Mobility	4. Pollution
	5. Social Problems

demonstrated that the two sets of indicators overlapped the dimensions, and showed that the socio-economic indicators were probably the most useful comprehensive source of data, since they defined most of the sources of variation. Nevertheless, Davies and Tapper's study was still primarily in the realm of so-called 'objective' indicators. Whether the variation is reflected in the behavioural or subjective domains remains a moot point. Nevertheless, this tendency for individual structures to be recognized in several different content areas of variables does seem to occur in the inter-urban field where Knox and MacLaran (1978) looked at subjective and census based

indicators of 'quality of life' measurements in Dundee and observed
that:

'in view of the positive correlation we have found between object-
ively measured circumstances and both values and perceptions of
most life domains, we are forced to conclude that for the purposes
of generally describing or evaluation ecological disparities in
well-being conventional 'hard data' are as good a surrogate as
any.' (Knox and MacLaran 1978 p.241)

This type of discussion soon runs into the problem of non-homogeneous
populations discussed in the previous section. A classic illustration
of this point can be seen in Boal and Orr's (1978) study of the
residential preferences of N. Ireland university students, where
Catholics displayed greater preference for parts of the Irish Republic,
whilst Protestants gave higher scores to Northern Ireland and parts of
Scotland. These obviously complicate the consistency of the local
peaks of 'hometown' or 'home region' effects in the general preference
surfaces identified by Gould (1968, 1972) in many countries. Such
issues, however, go beyond the major concern of this section, the
problem of variations in different domains. When one moves away from
strictly classificatory studies more problems are found. Studies of
urban sub-areas at the neighbourhood level (Smith 1978, 1980) demonstrated the great complexity of the relationships involved, because of
variations in the social, physical environmental and behavioural space.
Although the social and physical environment may have an important
effect on behaviour, they are in Wirth's words:

'at best conditioning factors offering the possibilities and
setting limits for social and psychological existence and development.' (Wirth 1945 p.177)

Yet, despite this note of caution, there seems little doubt that the
contextual effects described in Chapter III are often influential in
affecting behaviour. Smith (1980), for example, expanding on previous
work on communities, showed that economic and demographical variables
are likely to have stronger effects on certain urban problems - such as
child neglect - than attitudinal or neighbourhood assessment
variables.

Nevertheless, it is vital not to assume that particular social
structural or physical structural indicators will directly lead to a
particular behavioural response. Palm and Caruso (1973) demonstrated
the way individual behavioural patterns varied in areas, even though
the areas were similar on a range of human ecological indicators. The
results showed that cities or city areas appearing to be similar on,
for example, socio-economic indicators (whether ghetto or high class
residential) can display quite different behavioural associations.
This, of course, should be expected, given the discussion about
contextual fallacies in Chapter III. It is always dangerous to assume
that similarity in one context, such as comparable socio-economic
structure, will produce similar responses in different contexts, such
as territorial behaviour. They are, after all, different parts of the
spatial mosaic! Nevertheless, the study of the degree of consistency
between behavioural and structural indicators can be helped by factor
analysis procedures, since the approach can be used to define the

sources of variation in individual data sets. Alternatively, the canonical correlation method (Briggs 1977; Clark 1973; Johnston 1978), can measure the degree of association between the two data sets. The R mode factorial approach produces factor scores which measure the importance of each area on the dimensions of variation. Hence the application of this approach to a structural data set provides behavioural investigators with at least one means of establishing their sample of areas within which interviews can be carried out on a rigorous quantitative basis.

(b) Variable choice and relevance

If the use of factor analysis in human geography and human ecology is to become increasingly oriented towards comparative analyses, rather than to the study of single city or regions, questions about the relevance of individual variables become more and more pressing. This is especially true if the comparison is at the cross-cultural or cross-temporal level, for in these cases the fallacies described in the previous chapters have constantly to be avoided. Three rather different types of problem can be recognized, and these are described as questions of: indicator derivation; ethnocity; and particularism.

(i) <u>Indicator derivation</u>. Most of the theories in the social sciences relate to conceptual structures that cannot be directly observed – in the sense that they are entities that manifest themselves in terms of certain effects that are only partially identified in individuals and observable variables. For example, in the Shevky-Bell (1955) schema discussed in Chapter I, the 'occupation' and 'education' variables are both indicators of some overall dimension called Social Status. This relationship between the structural dimension – either assumed or identified by factor analysis – and the indicators used in any study cannot really be measured so directly. Several stages are usually involved in the move from the conceptual context to the precise indicator used in the study. This procedure is usually followed intuitively, but the options available at each stage are worth outlining in order to demonstrate the variety of choice involved.

Problem	Stages	Examples
Problem	Constituents	Housing, employment, etc.
	Aspects	Quality, overcrowding
	Indicants	Price, age, structure, etc.
	Variables	% Housing > 30,000 pounds

Figure 4.1 Decision stages in deriving indicators

Examples of the way this process works in factorial ecology have been shown by Knox (1974), as well as by Knox and MacLaren (1978). Suppose that the basic problem of interest was at the 'level' of 'quality of life'. Figure 4.1 shows that one of the constituents of 'quality of life' is the condition of housing, so certain aspects of housing, such as the quality of accommodation or overcrowding, can be measured. A variety of indicators can be used to define the characteristic of quality, for example, price, age, structure. The precise variable can be chosen on the basis of the percentage of housing in an area over some limit perhaps over 30,000 pounds in price. For ease of future interpretation, however, it is advisable to define the variables as simply as possible. Complex variables should be avoided, since they are likely to confuse the precise sources of variation involved.

Whether the final factor dimensions confirm that the original constituents of the problem really are separate entities can only be determined in the final analysis. The point here is that the various analytical stages in Figure 4.1 demonstrate the gap between the original concept - either the theoretical idea or the range of data being explored - and the final choice of variable. This type of specification problem is really a data analysis problem (Davies 1972) and demonstrates that we cannot assume simple one-to-one relationships between variables and theoretical ideas; there are a range of alternative possibilities. Unless a great deal of care is taken in the specification of variables - such that they are linked to the original problem - much of the utility of factor analysis in a theoretical sense will be lost. Fortunately, the factorial method can be used to 'screen' data sets to pick out the various sources of variation. It is recommended, therefore, that an investigator carry out a number of preliminary studies to carefully define the variable sets that are used in the final analysis.

(ii) Ethnocity. The term 'ethnocity' seems to have been first used by Wheatley (1963) in human geography to emphasize that social systems of cultural groups, both in time and space, are often fundamentally different. The importance of this idea for factorial investigations is that it demonstrates the function performed by any indicator, variable or measure of association, must be related to the social system of which it is a part. To use the variable outside this system may result in a complete misunderstanding of the function of the indicator in a particular society. In other words investigators cannot be sure they are measuring the same thing in different societies - even if the same indicator is used. The classic example is provided by suicides. Since such acts are condemned by the Roman Catholic Church, in studies of western cities sociologists have used suicides as indicators of social breakdown or anomie. Hence the indicator has proved useful for identifying areas of social disorganization. But the same function would not be fulfilled by the use of the variable in Japan to measure social breakdown. Suicides are still socially respected in Japan, and indicate honour, commitment and strength of character. Another example may be taken from the very different roles of social or cultural associations. In medieval western cities, social associations and organizations found within the city are specifically urban institutions. In medieval Muslim societies, Lapidus (1966) has shown that membership in some socially defined groups, such as a school of law, have united urban residents with village people. These social linkages created major division within the Muslim urban entity ensuring, in Susan Landay's (1971 p.303) words, 'the map of social space does not match the map of physical space'. Similarly, within western society today, the work of Melvin Webber (1968), in particular his 'non-place urban realm' concept, reminds us that many of the functions or interactions performed by the leaders of society may not be spatially determined on the basis of a particular location. Many of the linkages would still occur if the person moved, since they are not determined by that location. Only a time-budget investigation could disentangle the local and world scale interactions. Hence a traditional view of the process of functional interchange in which the concept of space is viewed as a rigid container (i.e., absolute rather than relative space) can lead to many misinterpretations.

Similar problems will occur as factorial ecologists move away from the use of census based variables and begin to explore such aspects of urban and regional character as the 'quality of life'. It seems inevitable that 'quality' will mean various things in different societies. In western industrial society, an increase in the 'quality of life' may mean more leisure time and more leisure alternatives. To unemployed residents of underdeveloped countries leisure may be an attribute they have had enough of! They may want accelerated economic progress to provide the material benefits of the average western society. Similar contrasts can be made with the planned economies of the communist world, which have a very different set of objectives to those in the capitalist west. The result is that unless we are to fall into the trap of the universal fallacy (Chapter III), a series of different types of society should be recognized as the background for factorial studies. Berry (1972) suggested the typology as a threefold one; between western, underdeveloped and communist countries. Timms (1971) identified several different types, each of which have their own set of dimensions and relationships. (These are described in Chapter IX.) Only future investigations will enable us to resolve the precise differences between these and other proposals. At this stage it is enough to have demonstrated the ethnocity problem, and illustrated that investigators are searching for ways to circumvent the problem.

(iii) Particularism. Once the problems posed by ethnocity have been resolved, comparative researchers should use a standardized set of variables to ensure that the results of two independently derived research studies can be directly compared. Unfortunately, the attempts at standardization that have been made in the factorial field have been far from satisfactory. For example, a standard set of variables proposed by the Centre of Urban Studies in London in 1961 contained indicators such as the percentage of Maltese and Cypriots. These may be useful variables in parts of inner London but were not appropriate in studies of British provincial towns such as Cardiff (Herbert 1970) for the proportions of people of these ethnic stocks are very low. So even where some agreement upon a key set of variables can be made, the utility of a blanket application of the set to all cases must be questioned. A good example of the same problem occurs in the work of Palm and Caruso (1972) where the identical set of variables used by Rees in Chicago (1970) were applied to a study of ten US cities. on intuitive grounds it is most unlikely that the complete set of variables useful in Chicago will be appropriate for such towns as Portland (Maine), San Diego (California), or Tampa (Florida). From our descriptive knowledge of these places we know that there are regional variations in the ethnic and economic character of the cities. So, as has been pointed out in the previous section, it is not really surprising to find that Palm and Caruso only found two general axes, for many of the other axes will be linked to 'place-particular' associations given the variable set and techniques used.

The deceptively simple decision of adopting a single standardized data set has further ramifications of note. If Product Moment correlations are used as measures of similarity the minor percentage variations of particular variables are also magnified, since the sizes of the original values are suppressed in favour of the variations, once the data is standardized to zero mean and unit variance. This is the normal procedure in most factorial studies. In other words, unless an investigator is very careful relatively small percentage values could have a very influential effect in any analysis. One way of removing the effect of exaggeration of those minor values is to exclude variables that do not achieve a sufficiently high mean value for the study areas as a whole. For example, Davies and Barrow (1973), adopted a cut-off value of 5% for any variable in their comparative study of three Canadian prairie cities. Unless a variable achieved a mean value of 5% in the city as a whole the variable was excluded from the analysis. Quite obviously the danger with this procedure is that some of the individual characteristics of cities may be lost – although one can take the precaution of describing the place-particular associations before the factor analysis results are presented. These approaches, however, are only subjective approaches to the problem. They are far from satisfactory when a variable attains a very high value in one city and is small in another. However some recent factorial work has shown at least one way out of the dilemma.

Table 4.4

Comparison of factor studies in Toronto and Montreal

(a) Coefficients of congruence with French
ethnicity included (excluded)

Components		Toronto			
		1	2	3	4
Montreal	1	76(77)	22(19)	40(37)	05(06)
	2	49(43)	79(82)	19(13)	01(02)
	3	04(01)	10(00)	08(13)	82(81)
	4	30(33)	39(33)	64(71)	03(19)

(b) Coefficients of congruence using partial
     correlation coefficients

Components		Toronto			
		1	2	3	4
Montreal	1	89	23	19	17
	2	04	84	13	25
	3	06	04	36	70
	4	27	24	77	29

Decimal points are excluded: 76 is 0.76.
Source: Wiltshire, Murdie, Greer-Wootten (1973)

Wiltshire, Murdie and Greer-Wootten (1973) faced this type of problem
with respect to the French ethnicity variable in their comparison of
Toronto and Montreal. Rather than excluding the variable (which would
ignore one of the major sources of differentiation between the cities)
they used partial correlation coefficients to control for the ethnicity
variable. In its simplest case, as in a three variable situation, the
partial correlation coefficient measures the correlation between two
variables in such a way that the effect of the third variable is
controlled. Table 4.4 shows the results obtained by Murdie et al using
coefficients of congruence (Chapter VI, Section 4) to compare the
comonents abstracted by the various analysis. In the first part of the
table the sizes of the congruence coefficients (which run from 0.00 to
1.00), show that the effect of including and excluding the French
ethnicity variable is rather minimal. In the second part of the table,
however, the effect of using partial correlation coefficients increases
the size of the congruence measure for the two largest components.
This led the authors to conclude that there was a small, but nevertheless perceptible movement towards what was considered to be the
theoretical structure of perfect congruence. Unfortunately, experiments with a different variable set for the same cities produced the
reverse effect, so that the general utility of the partial correlation
approach in factor analysis studies remains to be proven.

The partial correlation method may resolve the difficulties when a
small number of variables are involved. However, it does not help us
when doubts arise about the whole basis of comparison between two
cities. In some ways this takes us back to the point made in the first
section of this chapter about 'chalk and cheese' comparisons. Using,
once again, the example of the internal social structure of a set of
cities, it is apparent that one can depend primarily upon intuition, or
previous knowledge of urban areas, to tell whether the cities are
similar enough to compare in any meaningful way. However, there is no
need for the comparison to remain at the intuitive level. One could
begin by carrying out a preliminary analysis of the city system to

define the groups of cities suitable for rigorous comparison. Alternatively, analysis of variance techniques can be applied to each variable in the data sets within each city or region, and the difference between 'within city' and 'between city' variations can be calculated. If the range of variation between the cities is greater than that within the cities, it may be suggested that the attempt is being made to compare different structural types of centres. The widespread adoption of this, or similar types of approach, may help factor analysts to place their comparisons on a much sounder basis.

(c) Measurement scales

Once suitable decisions have been made about the variable set to be used in an analysis, the next question concerns the scale of measurement of the variables to be adopted. In many cases the information is derived from national censuses, and consists of a count of the number of individuals or objects with a certain characteristic in a defined area; for example, the number of people who are 65 years of age, or the number of houses with bathrooms. Usually these values are divided by an appropriate aggregate such as the total population or the total number of houses. The variable is then treated as if it were measured in an interval scale - as a measurement that has the possibility of continuous variation. Unfortunately, it is frequently forgotten that in the calculation of percentages, as opposed to ratios, a closed number system is imposed since they can only run on a scale from 0 to 100. Other issues associated with percentage distributions are dealt with in the next section. Here the primary point is that there is no reason why the interval scale data on Table 4.5 cannot be rearranged into different scales of measurement in order to resolve the closed number problem. An example of changing measurement scales is given in column 4 of Table 4.5 where the data are arranged in order of their importance, with the highest numbers given a rank of 1, the lowest a rank of 10. In column 5, all values greater than the average are coded one, the smaller values are coded zero. These procedures alter the original measurements to rank order (or ordinal) scale, and nominal scales respectively.

The simple example of Table 4.5 shows that a great deal of original information content of the data set is lost by this alteration to rank order and nominal scales. This must be a matter of concern. But in many cases factorial ecologists have found that the only description of an appropriate variable consists of data measured in rank order or nominal scales. This is the case in studies of networks, such as roads or streams, where symmetrical connectivity matrices (destinations as the columns, and origins as the rows) are used to organize the data. If direct connections exist between an origin and a destination the cell is coded 1. If no connection exists it is coded with a 0. This matrix can be factor analyzed to pick out the major nodes on the network (Gould 1968). One of the earliest examples of the use of rank order scales in factorial ecology came from Gould and White's study (1968) of the desirability of living in the different counties of Britain. Students were provided with a list of counties and asked to rank them in order of their preference for living in the areas. The result was a rank order matrix of rows (students) and columns (rank order preferences), which was converted to a rank order correlation

Table 4.5
Alternative measurement scales

1 Area	2 Amounts	3 Interval Scale (%)	4 Rank Scale (Ordinal)	5 Nominal Scale (1 = value above average)
1	16	8	9	0
2	8	4	8	0
3	16	8	5	0
4	14	7	6	0
5	28	14	4	1
6	10	5	7	0
7	4	2	10	0
8	36	18	3	1
9	40	20	1	1
10	38	19	2	1
Totals	200	100%		

matrix and then factor analyzed to identify the basic dimensions or structure of the data set.

These examples demonstrate the flexibility of the factorial approach in measurement terms, but one must remember that there is no substitute for the careful collection and measurement of data. This point is especially important when one is dealing with measurement levels with very little range of tolerance, as for example, at the nominal scale. Certainly one must not try to overmeasure, to force the data into categories that are really too sophisticated for the problem. One of the first perception studies in geography using factorial methods (Gould and White 1968) provides a good example of this type of problem. It is most unlikely that all students would be able to discriminate between the relative desirability of, for instance, Merionethshire, Clackmananshire and Huntingdonshire in Britain, and put them into a single rank order scale with any real degree of meaning. In this case, a simpler nominal scale approach would have been more appropriate for the problem; an approach in which students were asked to note their attitude to the counties according to a five point, or seven point scale. For example, the question could be phrased as follows:

If you had to move from your present county of residence to any of the following places would you be:

	Aberdeenshire	Berkshire	Other Counties, etc.
Very Pleased	V.P.	V.P.	V.P.
Pleased	P.	P.	P.
Indifferent	I.	I.	I.
Displeased	D.	D.	D.
Very Displeased	V.D.	V.D.	V.D.

Respondents would be requested to circle the appropriate answer for each county.

This type of approach makes use of a scale called the semantic differential scale. This procedure, developed by Osgood (1957) and fellow workers in the classic study on the 'Measurement of Meaning', is discussed at length in such excellent texts as Oppenheim (1966) or Snider and Osgood (1969) and takes one into the area of interview technique and questionnaire design. This study is not the place to discuss such issues at length. However, the basis of the approach needs to be discussed by means of an example. Normally, the concept to be measured is represented by a series of bi-polar adjectival scales, with an equal or indifferent category in the middle, and an equal number (one, two, or usually three categories) on either side. This produces a three, five or seven point scale. For example, Palmer, Robinson and Thomas (1977) used 35 different indicators in their factorial investigations of countryside images. These were grouped into seven hypothetical 'dimensions' composed of a set of indicators. Each indicator was measured on a seven point scale.

For example: Is the centre 'accessible'

Polar Term _____ Convenient

1. Very
2. Quite ‾‾‾‾‾
3. Slightly ‾‾‾‾‾
4. Neither/nor/equally ‾‾‾‾‾
5. Slightly ‾‾‾‾‾
6. Quite ‾‾‾‾‾
7. Very ‾‾‾‾‾

Polar Term _____ Inconvenient

To avoid bias in the answers given to each question, the ordering of the scales and the location of the favourable side of the term were randomized. These scales were applied to a homogeneous group of 60 higher income individuals using the point biserial coefficient as the measure of similarity.

The results of a Principal Axes analysis followed by Promax (oblique) rotation showed that 65% of the variance could be accounted for by seven axes. The hypothesized Evaluation, Crowding, Accessibility and Emotion/Reflection structures were basically confirmed, even though the occasional hypothesized indicator did not load in the expected way and additional indicators were linked to the axis. Two other axes were labelled as Interest-Activity and Scenery, whilst the third was a difficult one to label. It was linked only with one major variable, and was tentatively called Contrast. All three smaller axes were found on one of the higher order axes. By contrast the first order evaluation axis was associated with another higher order vector, and the three other marginally defined structures were linked to a third higher order structure. This study comes closer than most factorial ecology investigations to validating several of the originally postulated scales. Yet the study does appear to have over-factored the data set (an issue discussed in Chapter V). Nevertheless, the fact that two-thirds of the variation is accounted for by this factor model does lead to one conclusion. It is that people like different places for reasons which are essentially the same. This provides added confidence

Table 4.6

Factor analysis of semantic differential scales

Hypothesized Dimensions	Concepts: Polar Terms	Axes						
		7	6	5	4	3	1	2
A. Accessibility	1. Convenient – Inconvenient				0.59			
	2. Near – Distant		-0.38	-0.55	0.95		-0.46	0.41
	3. Fast roads – Slow roads				0.61			
	4. Accessibility – Inaccessible				0.79			
	5. Short Journey – Long Journey							
B. Activity	6. Lots to do – Nothing to do					0.41		
	7. Suitable for picknicking – Unsuitable		-0.78					0.38
	8. Open – Enclosed			0.69			0.58	
	9. Unorganized – Organized							
	10. Good for walking – Bad							
C. Crowding	11. Few people – Many						0.65	
	12. Quiet – Noisy						0.49	
	13. Empty – Full	0.38					0.59	
	14. Deserted – Crowded						0.72	
	15. Few cars – Many cars						0.79	
D. Facilities- Settlement	16. Few buildings – Many						0.69	
	17. Old – Modern							
	18. Well catered for – Poorly			0.39			-0.42	
	19. Uncommercialized – Commercialized						0.69	
	20. Unspoilt – Spoilt					0.43		

Table 4.6
(continued)

Hypothesized Dimensions	Concepts: Polar Terms	Axes						
		2	1	3	4	5	6	7
E. Scenery	21. Varied scenery – Monotonous	-0.55		0.47		0.49		
	22. Sparse – Lush							
	23. Wild – Tame						-0.70	
	24. Mountainous – Flat	-0.68					-0.48	
	25. Hard – Soft							
F. Evaluation	26. Warm – Cold	0.71						
	27. Relaxing – Disturbing			0.57				
	28. Secure – Insecure	0.57						
	29. Calm – Turbulent	0.70						
	30. Peaceful – Dramatic	0.52						
G. Emotion-Reflections	31. Meaningful – Meaningless			0.76				
	32. Contrasting – Similar			0.49				
	33. Inspiring – Uninspiring			0.72				
	34. Pleasing – Annoying			0.57				
	35. Refreshing – Depressing			0.57				
First Order Factor Titles		Evalua-tion	Crowding	Reflec-tion and Emotion	Access-ibility	Interest and Activity	Scenery	Contrast
Second Order: Axes		I		II			III	
Loadings		0.94	0.77	0.71	0.63	0.71	-0.56	-0.75

Source: Revised from Palmer, Robinson, Thomas (1977)

to the assumption that there are basically similar structures in people's opinions of places; images are not completely unique. In the author's words:

'composite dimensions seem to represent an underlying continuum in the perceived variation amongst countryside locations. At one extreme of this continuum are the convenient, planned and organized sites, which are assessed functionally in terms of their accessibility and their facilities. At the other extreme are the relatively wild, remote and aesthetically pleasing places which seem to induce more profoundly emotional responses. In between are places which blend emotional and functional assessment more or less equally.' (Palmer, Robinson and Thomas 1977 p.748)

The choices involved in measurement scales (especially when they relate to the measurement of subjective responses) are of critical importance for any investigator using factor analysis. Nevertheless, it is apparent that these problems are not restricted to factor ecology; the issues have wider application in the field of human geography and ecology. Three points relevent to factor analysis studies do, however, need emphasis.

1) If some scale of measurement other than the interval scale is chosen for a data set, the investigators should not blindly apply Pearson's Product Moment Coefficients to the data set to derive the similarity matrix. Appropriate correlations for rank order data, such as the Phi Correlation, should be substituted. Cattell (1978 p.469) provides a summary. Unfortunately, most existing package computer programs dealing with factor analysis do not allow this flexibility.

2) If investigators are worried about not being able to satisfy the normality, and sample base assumptions of interval scale correlation measures (particularly important considerations for significance testing) the re-organization of the data into a non-parametic scale – such as a rank order – may be appropriate. Siegal (1956) provides one of the best summaries of the utility of non-parametric methods. If a pattern or structure in the data set is particularly strong, an investigator may obtain very similar results by factor analyzing the rank order correlation matrix. Moser and Scott (1962), in their now classic classification of British administrative areas, applied their factor analysis twice, once to the interval scale data and once on the rank order data. The results were similar in both cases. Yet a warning must be added. This particular finding does not give an investigator carte blanche to apply whatever scale seems to be easiest. It merely means that in this case the structural pattern of British towns was considered to be independent of the scales of measurement used. In other words the loss of information involved in moving from an interval to a rank order scale (where the marginal difference between two variables varies from any number in the interval scale to one in the rank order case) apparently did not affect the dimensionality of the data set.

3) The third point is the question of whether research workers should use mixed mode data sets (i.e., a mixture of percentage, rank order or nominal scale data) within the confines of a single analysis. In correlation analysis nominal or ordinal scale data can be related by

102

similarity measures, such as the point biserial correlation, etc., whilst Aitchison (1978) has shown the utility of Gower's general index of similarity in taxonomic studies. But few factorial ecologists have explored the advantages of using such measures. Even Rummel (1970), in his otherwise comprehensive survey on applied factor analysis, downplays the problem. Many of Rummel's examples consist of mixed mode data sets from the field of politics. The implication must be that he does not consider this problem to be important. Similarly, some of the earliest applications of factor analysis in geography, for example, King and Henshall (1966), also used mixed mode data sets. On theoretical grounds, it is difficult to justify this decision. Two points are particularly relevant. If the inferential method of common factor models is being used, normality and linearity assumptions need to be satisfied, and mixed mode data sets may not satisfy this requirement. Where factor scores are calculated, the original data set is multiplied by the factor loading matrix to produce scores, and these <u>must</u> be affected by the mode of data. It is true that in some cases <u>there</u> is no way that all the phenomena of interest can be measured in one scale. For example, Henshall's study of agricultural patterns in Caribbean Islands (1966) used a presence or absence scale to measure rock type, although the majority of variables used in the study are at the interval scale. It is difficult to see a way of completely resolving this problem. One way is to analyze data sets of similar scales separately, and subsequently relate them together via canonical correlation methods (Clark 1973; Johnston 1978). Another would be to calculate appropriate correlations (such as rank order, or point biserial correlations) between each pair of variables in the data set. Even these solutions do not really solve the problems involved in calculating factor scores with mixed mode data sets. Despite Horn's (1969) review this is another unresolved question for factorial ecologists. As yet nobody has given a convincing demonstration of the effect of using mixed mode data, so it must be admitted that the recommendation to try to avoid mixed mode data sets is intuitively based.

(d) Complementarity and overlap

Another major criticism of factorial ecology has been that rigorous data selection procedures have <u>not</u> been applied to the choice of variables. One of the simplest examples of this problem consists of using two variables that are the mirror images of one another; for example, using the percentage males and females in an area in a study. Statistically, these variables will have a correlation of −1.0; inferentially they only have one degree of freedom. In factor analyses they wil appear on opposite sides of a factor and, as Mabogunje's study of Nigerian towns (1968) shows, they will over-emphasize the importance of an axis by contributing two units of variance to the solution. More importantly the determinant of the matrix cannot be calculated since one of the rows of the similarity matrix is the mirror image of another. The same type of problem occurs with the use of any closed number sets, such as sequences of variables that add up to 100%. Many earliest factor analyses display this problem, especially in the categories of age variables such as: 0-14 years; 15-24 years; 25-44 years; 45-59 years; 60+ years (Davies and Lewis 1973). The use of all these variables in any analysis undoubtedly biases the correlations, and hence the position of the factor axes for any one of the variables can be predicted from the others. Thus, over-counting occurs. Where a large data set is being investigated, and the factors are identified by many variables, the bias in the position of the axes may be quite small – especially if there are many categories in the closed number variables. In small data sets the effect can be quite marked. Yet one advantage of using all the indicators is that it produces a more complete understanding of the interrelationships between various stages in the Life Cycle dimension. This must be set against the obvious statistical bias, and the fact that if variables do add up to 100% it may not be possible to obtain a factor solution, since the determinant of the matrix cannot be calculated, a necessary calculation for many factor routines. Fortunately, it has been gradually recognized that closed number sets should be avoided in factor studies and explicit attention has been drawn to this issue (Davies 1975, 1978).

In the multiple category case complementarity in the variable set can result in spurious relationships between the indicators. This general problem of closed number systems and complementarity in the variable set is a problem with a long history in the statistical literature, as studies from Pearson (1893), Kuh and Meyer (1955) to Schuessler (1973) show. Dent and Sakoda (1973) have drawn the attention of factor ecologists to the problem by showing how a spurious correlation can be produced. If two variables (V_1 and V_2) are divided by the same denominator (D), then the correlations between the two indexes or ratios (R_1 and R_2) is likely to be higher than the correlation between the original variables. Kuh and Meyer (1955) indicated that if the index or ratios, rather than the original variables, are the real object of attention, then this higher correlation cannot be regarded as spurious. However, if the original variables are the real concern of the analyst – for instance when a ratio was calculated simply to remove the influence of the size of the collecting areas – then the correlation between the ratios should not be trusted as it is probably spurious.

Table 4.7
Spurious index correlations

(a)   Correlations for 54 census tracts
      Syracuse (New York) 1960

Controlled Variable	Alternative Definitions	
Child–Adolescent (aged 0–19 years)	Young Adult (aged 20–39)	Mature Adult (aged 40–59)
Divided by total population	−0.131 (divided by total population)	−0.477 (divided by total population)
	+0.324 (divided by total over 20 years)	+0.563 (divided by total over 40 years)
Source: Dent and Sakoda (1973)		

104

Table 4.8
Derivation of measures for age distributions

	Denominator	Numerator (age)	Measure
1	Total population	0- 9 years	Children
2	Age 10 years and over	10-19 years	Adolescents
3	Age 20 years and over	20-30 years	Young Adults
4	Age 30 years and over	30-50 years	Mature Adults
N	Total population	65 years plus	Old Age

Table 4.7 shows one of the examples used by Dent and Sakoda (1973), in which the proportions of youths and children are negatively correlated with young adults and mature adults when one might expect positive relationships. If the overlap between the variables is removed – by changing the denominator values to make sure that the ratios are calculated from categories of the variable that are mutually exclusive – positive and higher correlations are produced. These values may be closer to the original expectations.

Another solution to this type of problem is to use partial correlation coefficients. This removes the effect of the overlapping variables. A far simpler approach is to readjust the variables, by successively eliminating categories from the denominator after they have been used as the numerator. Table 4.8 shows the way in which this could be carried out for a set of age distributions. It can be seen that in the last variable the chosen denominator is 'total population' because any other valuable is unlikely to represent a reasonable sub-set of the population age category. In addition, numbers in the old age category are usually very small, so that any spuriousness introduced by non-orthogonality is likely to be minimal.

A final word of caution is worth recording. The potential spuriousness of the index correlations due to overlap does not come from the coefficients themselves, but rather from the interpretations placed on the index. If the indexes per se are the object of attention we must reject the idea of spuriousness; it is only when the investigator is concerned with the numerators alone that the difficulty occurs. Dent and Sakoda (1973) maintained that in most factorial studies in human ecology there is an implicit assumption that the original numerator variable is the value that is basically of interest. If the denominator was a constant value in all cases – as it would be if the population of the observation units were the same – there would be no need to divided by the value. In general, their point is well taken, and provides yet another question to which investigators should address themselves in the planning of any study.

(e) Comprehensiveness and redundancy

The degree of comprehensiveness and redundancy in variable sets are related problems that have not been given much attention in the literature. The reason is that factorial ecologists have been concerned principally with data exploration or, perhaps testing whether the three axes of differentiation proposed by Shevky and Bell can be found in any data set. Three separate types of difficulty within this general area can be identified.

(i) <u>Over-representation</u>. If the data set used in any study is dominated by one type of variable then it is inevitable that the factor dimensions will reflect this domination, for the axes cannot transcend the input data. For example, 20 out of the 57 variables used by P.H. Rees (1970) in Chicago were ethnic and religious indicators. Not surprisingly five out of the ten factors abstracted were concerned with race, national origin or religion. In a study of Swedish cities, Jansen (1971) found three separate familial dimensions, but 19 out of 44 variables measured demographic or family status. Even the variable list recommended by the Centre for Urban Studies (London) in 1961 included seven ethnic variables out of a total set of 26 indicators. Investigators should be careful of repeating this problem by using too many variables from one domain – unless this emphasis is desired.

(ii) <u>Under-representation</u>. This is really the opposite type of problem to the one discussed above. Yet it does have two rather different aspects. First, possible axes of differentiation may be ignored because only one variable of possible relation to each axis has been included in the data set. Inevitably these are swamped in the analysis, and the potential axis can be missed. Second, a whole set of vital characteristics may be excluded, either by negligence, or because the necessary data is difficult to obtain. In studies of intra-urban structure relatively few researchers have included mobility variables. In cases where migration indicators have been included they emerge as a separate factor axis (Rees 1970; Davies and Lewis 1973). This provides a vital link between the literature of intra-urban mobility and ecological structure. Again a note of caution must be introduced. If a study is explicitly concerned with only part of the characteristics of areas, then the under-representation of certain variables may be a minor issue. But all too frequently it is implied that the 'basic' dimensions of urban or regional structure have been uncovered. This is particularly true of studies in urban ecology, where the set of variables consist <u>exclusively</u> of those associated with the three axis of differentiation proposed by Shevky and Bell (1955), namely: Ethnicity, Social Status and Family Status (urbanization). This research provides a test of whether the three axes are present, as for example in Hunter and Latif's study of Winnipeg (1973). But they fail to tell us anything about the completeness of the urban description. How basic are these axes? Can other axes be derived? The problem is really magnified when this type of variable set is standardized through time. The question of a temporal fallacy is soon raised, since societal changes are quite likely to produce variations in the axes of differentiation in cities through time. This ensures that the variables become less and less reliable as measures of the basic axes of differentiation.

(iii) Redundancy. The technological changes consequent upon high speed computers make it relatively easy to add more and more variables to a data set. But it is important to ask whether the addition of these variables add anything of substance to the results of the analysis. Apart from the problems caused by intercorrelation or multi-collinearity (which are discussed in the next section) certain variables may be redundant. If this is the case, much of the power of the factor method is being wasted. One way to resolve the problem is to re-run the analysis on successive reductions of the variable set. In this way Schmid and Tagashira (1964) reduced the number of variables in their study of Seattle (USA) from 42 to 21 variables without any loss of axes or dimensions. However, they showed that certain minor axes of differentiation, described as 'Down and Out' and 'Residential Stability' were eradicated when the analysis was reduced to ten variables. Davies (1978) in a study of Calgary using 60, 50 and 14 variables found substantially the same results. Axes measuring Family Status-Age, Socio-Economic Status, Migrant Status, and a bi-polar Young Adult Participation/Non-Affluent vector, remained very stable but two ethnicity axes were less clear. Perle's (1979) study of Detroit partially confirmed stability of the major dimensions in different variable sets, although he showed how minor axes merged. This led Perle to observe that if the information is in one domain, as, for example, in census information:

'ecological structure at one point in time is little affected by the number of variables used.' (Perle 1979 p.413)

The use of smaller variable sets in the final study than those used in the preliminary, perhaps exploratory studies does have much to recommend it. Yet it does beg the question of how the indicator variables are to be chosen. Usually the least complex, intuitively clear, and highest loading variables are abstracted, although some additional care may be taken to take distributional criteria such as near normality into account. Jollifee (1972) has provided a succinct summary of the 'variable-discarding' process and some of his points are relevant to the geographical and ecological fields. Yet one must not over-emphasize the advantage of simplicity provided by this data reduction process. After all, simplicity may not always be a good guide. For example, Davies and Lewis (1973) deliberately retained several migration variables in their study of Leicester, even though there was a risk of redundancy. The redundancy involved seemed a small price to pay for the increased clarity of the results. It must be admitted this is the view of empirical practitioners, rather than inferential statisticians, a distinction emphasized by Cattell (1978) in psychology.

Many of the problems raised in this section can be quite easily resolved by the adoption of careful data selection procedures. In particular, more attention should be paid to the relevance of particular variables to the problems dealt with by the study. For instance, it is relatively easy to continue to emphasize the importance of ethnicity among the white population in North American cities by constantly introducing ethnic variables. This could be exaggerating ethnicity as a major source of residential differentiation. Such a viewpoint does not mean that ethnicity should be ignored as a part of urban life. Rather it is likely its effect has diminished over the past 40 or 50 years, when it is compared to other sources of social variation. With this in mind it is worth concluding this section by making the following recommendation – even for component or exploratory studies. The ideal approach would be to identify variables on the basis of the expected axes of differentiation, taking care to have a balance between the categories of variation and good marker variables – variables previously used by other analyses that have been shown to have good value in defining different structures. Many applied factor analysts (Comrey 1973 p.209) argue for at least _five_ indicators per expected axis, but this is often difficult to achieve in the non-experimental situations found in factorial ecology which are usually dependent on secondary data sources such as the census. The hypothetical axes need not be subjected to common factor procedures – especially if there is doubt about the independence of the cases. Nevertheless the procedure helps to move ecological studies towards this more rigorous goal and the approach is recommended in organizational terms. After all it adds systematization to the research, provides a useful comparative basis for the study, and makes it easier to compare the results with other studies. If axes are difficult to predict in a particular situation, or if the study objective is to search for more comprehensive sources of variation in a content area than have been considered in the past, the indicators can be grouped into different categories to avoid the analysis being biased by one variable type. For example, Davies and Lewis' (1973) use of 60 variables in Leicester was designed to explore the general dimensionality of cities linked to census data, rather than only focussing upon the Shevky-Bell (1955) three axis model. The question of how many variables to choose is more difficult, but investigators should always try to have many more _variables_ than _cases_. Cattell (1978) repeated previous suggestions to have four times as many _cases_ as _variables_ in psychological studies. But if the objective of the study is descriptive, not inferential, such a high ratio may not be really necessary.

(f) Missing data

Most analysts are confronted with situations in which data for one or more variables cannot be obtained for certain observations, a familiar problem when information is obtained from several sources. Obviously, the processing of a data matrix with blank cells by package factor programmes will distort the results (particularly the factor scores) and must be avoided. A variety of procedures for treating the problem can be suggested. The most obvious is the exclusion of the variables if there are too many missing observations. Useful alternatives involve replacing the missing case value by the average for the variable, or estimating it by means of regression. If such procedures are not suitable the judgement of an informed observer may be appropriate. Rummel (1970) describes other alternatives. However, it must be emphasized that investigators should always bring this issue to their readers' attention, with a full explanation of the estimation procedures involved. In addition, it is recommended that the data should be analyzed without those variables having missing values and should be compared with the results obtained by using the estimated values. In this way a check on possible distortions is provided. If major discrepancies in the factorial dimensions found in the two studies are produced, doubt must be cast upon the utility of the estimated values.

3. DISTRIBUTIONAL ASSUMPTIONS AND CHARACTERISTICS

One of the most worrying set of problems in applied factor analysis stems from questions about the distributional assumptions of the variables used in any analysis. Critics such as Etrenburg (1962) have pointed out that factor structures may not be comparable if the distributions of variables are different between studies – although authorities in the factor analysis field (Rummel 1970; Cattell 1978) emphasize the essential robustness of the factor methods in identifying the general linear structures of data sets. Nevertheless, the importance of checking the distribution of variables used in any study cannot be under-estimated. Two basic issues are important.

(i) The simplest one is that since factorial studies deal with very large data sets it is difficult for an investigator to simultaneously recall all the details of the relationships involved. If he could there would be no need for the analysis! It is almost always the case that errors other than measurement ones creep into the data (perhaps during the data collection process or the transcription onto cards) so a reliable data screening procedure is essential. Once the data codes and cards have been checked for accuracy it is worth running the information through a simple standard score programme. The standard score reduces each variable to a zero mean and unit deviation such that the variables are expressed in deviation units. High standard scores are more easily identified, and can be checked for accuracy. If the values are incorrect they must be changed. But if the values are accurate it is apparent that this method has pinpointed the outliers, or extreme values in the data set. It is difficult to be precise about outliers since it is only a descriptive term. Normally, however, standard scores above 3.0 (three standard deviations from the mean) are treated as outliers. Their presence in a data set will cause problems in any linear analysis such as the factor procedures used here. If variable A has values of 1, 2, 3, 4 and 5 on five areas, and variable B scores 5, 4, 3, 2, 1 on the same areas, a perfect negative correlation of -1.0 will be produced. The addition of a sixth observation, with a score of 100 will produce a correlation of +0.67. To use data sets with such extreme outliers, will therefore, produce all sorts of problems for the investigator.

It may be better to exclude such variables from the analysis, and substitute other values, or transform the data to reduce the impact of such scores. One approach is to change the data to a rank order scale as described above to reduce the influence of the outliers, and then to use rank order correlations in the study. (Alternatively, examples of transformation procedures are given in Appendix B.)

(ii) A more complex issue relates to the fact that the factor model makes certain assumptions about the data – particularly linearity and normality – although the issues of homoscedacity, multi-collinearity and auto-correlation are also relevant. Before dealing with each of these problems in detail it is worth observing that they have added force when Pearson Product Moment Correlation Coefficients are used, and the hypothesis testing procedures of the common factor method are adopted in place of the component model.

(a) Linearity

In Chapter II it was shown that the observed variables of any data set are expressed as linear functions of the hypothetical variables, mathematical constructs, or factors. Although non-linear factor methods (Macdonald 1967) have been developed recently, as yet they do not seem to have been used in factorial ecology. Hence the methods that are being used assume the observed variables must be linearly related to one another - otherwise the factor model is not able to successfully reproduce the relationships of the similarity matrix. In dealing with this point Harman (1976 p.386) observed that this linearity requirement ensures that certain kinds of data - he refers to nationality, colour of eyes - cannot be studied in a factorial framework. Obviously, Rummel (1970) and other workers in applied factor analysis do not accept this type of constraint upon the range of data used by factorial studies; nevertheless they still emphasize that linearity is the basic requirement of the factor models used today.

Despite this general agreement, few factor analysts outside the mathematical and psychological literature have bothered to check that the linearity conditions are satisfied. In such circumstances it is fortunate that Cattell (1978 p.381) has pointed out that significant non-linearity is much less common that most critics suggest, providing evidence that only 5% of correlations between a series of objective personality and ability tests were significantly non-linear at the 0.05% probability level. As yet, it is not known whether such results are applicable to the ecological literature. What, therefore, should investigators do in these circumstances? The best advice is to screen the variables and search for non-linear relationships. Offending variables may be excluded - without loss to the analysis if the type of variation accounted for is found in another indicator. If the variable is one that seems essential, it may be possible to redefine the variable by working back through the constituants to the variables as shown in Figure 4.1. Frequently one does not have to go so far. For example, Johnston (1971 pp.216-319) has shown that the relationship between children under four years of age and the age of fertile females is likely to be curvilinear in modern society, because social change has produced a tightly defined average child-bearing period. The relationship would be more linear if the number of children under 14 was substituted for those under four years of age. Another approach would be to transform the variables by, for example, using the logarithm of one or more of the indicators. Transformation, however, is not the magic wand! Without careful scrutiny of the distributional characteristics of the variables, a transformation can do as much harm as good - as will be seen in the next section. Hence it is always important to test the degree of linearity in the relationships after as well as before the changes. Standard econometrics texts (Malinvaud 1980) provide examples of such tests. One of the simplest involves the use of two way contingency tables. Figure 4.2 shows how the curvilinear and linear distributions of a data set are reduced to contingency tables by taking the number of cases above and below the means of the two variables, and recording them in a 2x2 or four cell table. Even a subjective evaluation shows that linearity between the two variables is not satisfied in one of the cases, although it is obviously appropriate to test the relationship, perhaps by some of the methods proposed by Goodman (1966).

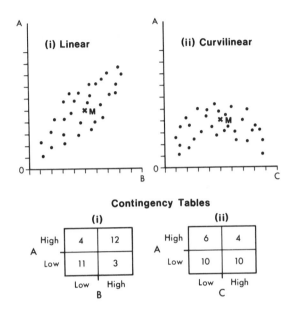

Figure 4.2 Curvilinear relationships and contingency tables

There is little doubt that the search for linearity can be a long one and this explains why investigators such as Cattell (1978) have proposed screening procedures to speed up the simplistic examination of scrutinizing single variables at a time. One of the most promising of these is the <u>relational simplex</u> method, which is based on the assumption that any transformation which improves the correlation between two variables is to be welcomed, since it reduces the error term. By applying a whole series of transformations to a variable and measuring its average correlation to all other variables, an improvement in fit could be said to have been produced when this average is maximized. The process is then repeated for all variables.

Such developments must be welcomed, because they are bound to improve the overall fit of the factor model to the data. By contrast, the use of untransformed data is bound to produce underestimates of the linear association between variables. Yet the investigator may be prepared to accept such interpretations if they still account for a considerable amount of the variance, simply in the interest of simplicity or comparability with other studies. In this context it may be comforting for applied factor analysts to be reminded of the view of one of the most experienced applied factor analysts that:

'empirical research results applying the normal extraction procedure and model to plasmodes with variables deliberately made non-linear to their relations demonstrate the factor model to be quite hardy.' (Cattell 1978 p.381)

Yet Cattell goes on to emphasize that this interpretation applies to the identification of general structures − not to the details of the loadings. Nevertheless the gap between the presence of non-linear relationships and the use of a linear model cannot be allowed to widen too far. Investigators should try to exclude the seriously non-linear relationships, unless either non-linear factor methods can be applied, appropriate transformations are made (Appendix B), or empirical tests show similar factor structures are derived without transformations.

(b) Normality and transformations

The mathematical theory of common factor analysis is based on the assumption of multi-normal distributions (Harman 1976), a requirement that is difficult for most investigators to test. Yet if it can be shown that individual variables are normally distributed (in other words, the distribution of observations on a variable corresponds to the familiar bell-shaped curve) there is a strong likelihood that the multi-normality condition is satisfied. It is worth noting that Rummel (1970 p.275) emphasized that normal single or univariant distributions do not necessarily produce bi-variate normality. The importance of this distributional assumption is that any transformation of variables to normality increases the possibility of linearity in a data set (Rummel 1970, p.275), as it has frequently been shown to improve the general clarity of the structure of data sets as revealed by factor analysis (Cattell 1978 p.454). In the case of the hypothesis testing procedures of common factor analysis the tests for significance of loadings and factors obviously assume that the normality requirement is satisfied. Without this requirement (one that spans the whole range of linear parametric methods) it is difficult to accept the utility of the standard significance tests associated with common factor models. In such circumstances it is not surprising to find that the author of one of the most comprehensive texts in the field of factor analysis stressed:

'the requirement that each observed variable be normally distrib-uted. While considerable latitude might be allowed, nevertheless a variable which is distinctly non-normal should not be included in the analysis.' (Harman 1976 p.386)

The phrase 'considerable latitude' obviously gives flexibility to the investigator − one that can be added to Cattell's (1978) point about the robustness of the factor model in accounting for the general structure of relationships, even where linearity was not satisfied. Nevertheless since the importance of the normality requirement must not be ignored, Appendix B discusses the ways in which the normality of variables can be tested, and provides examples of the way in which transformations can be used to adjust the distributions of individual variables. Which of the various strategies described in Appendix B should be adopted if an investigator finds a variable does not correspond to a normal, or approximately normal distribution remains the prerogative of the researcher. Again it is not possible to offer unequivocal support for one particular approach, since the analysis of different data sets can be used to support alternative conclusions, whilst there is considerable debate in the geographical field about the effectiveness of transformation. The issues are complicated by the fact that most work on the problems caused by violations of normality have been carried out in the regression — correlation literature — not in the factor analysis literature — and in econometrics and psychology rather than geography or ecology. However, it is worth beginning by noting Maddala's (1977) point that the problem of normality is really a two-fold one in terms of regression studies, applying first to the derivation of estimates for regression weights, and second to the question of conducting significant tests and establishing confidence limits. Maddala observed that normality is not necessary for the first of these issues, but is a requirement for the second, although Kmentra (1974) has noted that if the stochastic or random disturbance term in regression does not depart drastically from the normal distribution, many of the statistical inference measures such as t or F tests may be safely utilized. Statistical experiments with variables drawn from a large number of different distributions have supported this general point. However, other authorities (Kendal and Stuart 1959) have indicated that significant departures from normality will make themselves felt principally on the test procedures.

As yet few of these issues have been extensively reviewed in the geographical or ecological literature, so it is difficult to draw final conclusions. It seems sensible to err on the side of caution, and not to apply the inferential test procedures of common factor methods unless the data is a sample and has been shown to have been normalized, or evidence is provided on the robustness of these tests — as in the bi-variate cases noted above. In terms of the effects upon the distribution of eigenvalues, and the interpretation of factor loadings, more work has been carried out in ecological and geographical studies. But different conclusions have been reached. For example, Moser and Scott (1962) maintained that a logarithmic transformation of their socio-economic variables on British administrative areas did not really alter the interpretation of the resulting components. Similarly, Roff's (1977) investigations of 54 variables for the 1971 British census showed that even though approximately half the variables were non-normally distributed, individual transformations did not significantly alter the correlations or components. Such results parallel the work of investigators in allied fields such as psychology.

By contrast, Clark (1973) tested the results of raw, fully normalized and logarithmic transformed data in a study of telephone flows in Wales. He showed that broadly similar levels of explanation were produced for the axes, although the fully normalized data showed higher levels of explanation for the largest axes (Table 4.9). Far greater contrasts were found in the scores, as a summary of his evidence shows. The highest score found in the raw data results is for Cardiff (-7.29)

Table 4.9
The effect of normality upon component solutions: South Wales

(a)   Explanations

	Cumulative Percentage Explanation Component						
	1	2	3	4	5	6	7
Raw data	19.73	33.90	47.21	58.76	67.60	75.16	81.15
Fully normalized matrix	23.84	36.23	46.36	55.04	62.44	69.12	74.89
Log. normalized matrix	22.98	35.55	47.04	55.53	62.59	68.82	74.01

(b)   The effect of normality upon factor scores:
       unrotated solution

Raw data		Fully normalized		Log. normalized	
Component 1					
Monmouth	2.31	Monmouth	3.16	Monmouth	3.06
Llandeilo	2.09	Llandyssul	2.87	Cardigan	2.81
Shirenewton	1.98	Pontypool	2.83	Pontypool	2.80
Llandyssul	1.91	Lampeter	2.73	Llandyssul	2.75
Llanarth	1.86	Llanarth	2.71	Lampeter	2.66
Haverfordwest	-2.09	Neath	-3.28	Merthyr	-2.99
Carmarthen	-2.63	Cardiff	-3.40	Neath	-3.24
Swansea	-3.57	Ammanford	-3.51	Milford	-3.57
Merthyr	-5.56	Merthyr	-4.02	Llanelli	-4.53
Cardiff	-7.29	Swansea	-5.91	Swansea	-5.22
Component 2					
Neath	4.17	Cardiff	3.38	Cardigan	2.83
Swansea	3.87	Haverfordwest	3.34	Haverfordwest	2.69
Merthyr	3.25	St. Clears	2.02	Cardiff	2.43
Newport	2.16	Cardigan	1.93	Pontypridd	2.33
Abergavenny	1.82	Merthyr	1.86	Bridgend	1.94
Cardigan	-1.62	Ammanford	-2.30	Swansea	-2.50
Cardiff	-1.67	Shirenewton	-2.77	Carmarthen	-2.54
Llandovery	-2.20	Newport	-2.92	Shirenewton	-3.02
Haverfordwest	-4.14	Abergavenny	-3.39	Abergavenny	-3.14
Carmarthen	-4.93	Swansea	-3.44	Newport	-3.45

Values are in standard scores.   The five leading positive and
negative scores on each component are shown.

Source: Clark (1973)

114

on Component I. But it is only the fourth largest score axis in the fully normalized set and does not even appear in the first ten scores in the logarithmic transformed solution. Such major differences are, of course, more likely to occur in the pattern of scores, since the data matrix is multiplied by the factor loading matrix, ensuring that deviations will be enhanced. It is probable that flow data matrices will be more subject to variation in the results, because of greater extremes in the pattern of relationships. Differences such as these may be domain-specific but cast doubt upon the use of scores.

Confronted with this conflicting, yet sparse evidence, it is difficult to provide any final advice for the would-be investigator since the pattern of relationships in the data set to be investigated is unknown. It does seem important to stress that the type of blanket transformation proposed by Moser and Scott (1962) is not a useful procedure. It is possible variables that are normal may become skewed for a log 10 transformation applied to a normal distribution will create skew. Pringle (1976) has illustrated this problem in an analysis of a geographical data set, and confirmed Rummel's (1970 p.275) opinion. A more sensible procedure is to check the <u>individual</u> variables for normality, and perhaps to only apply logarithmic transformations for the most highly skewed indicators (Murdie 1969), or more correctly to apply a range of transformations to produce the required result as shown by Berry (1972), Warnes (1973) and Jansen (1974). This procedure may be the most appropriate in purely statistical terms, but it is worth returning to Harman's strictures (1976) about the difference between 'statistical' and 'content area' importance. Statistically, the transformation <u>is</u> a sensible procedure, but the obvious difficulty here is that the factor model is then expressed in terms of these transformed variables. So it may become difficult - in any common-sense interpretation - to understand what a factor is measuring. For example, suppose a data set consists of the square root of variable x, the cube of y, the logarithm of 3 and the reciprocal of w. Such indicators would be difficult to comprehend in substantive terms. In addition, Cattell (1978) has reminded investigators that if any transformed variable is used, the investigator must remember to translate it back to the original values after use. At this stage in our knowledge of factor structures in geographical and ecological situations it may be better to simply lower our sights, and transform only if a variable is really badly skewed, assuming it cannot be rescaled or replaced. This recommendation comes partially from the pragmatic principle of acknowledging the difficulty of interpreting fully transformed data, but also from accepting that:

'divergence in factor structures between studies might be the result of transformations having been applied in one study and not in the other, rather than a consequence of real substantive differences in the data.' (Rummel 1970 p.280)

Comparative research may, therefore, be hindered rather than helped by blanket or unthinking transformations at this stage in our knowledge.

Such recommendations will undoubtedly be challenged by exponents of a more mathematically minded approach to factorial studies. Hopefully, such pragmatic attitudes may only represent a temporary stage in our accumulation of knowledge on the dimensions of towns and regions.

There is no doubt that in the long run the use of data with normal distributions or data having been transformed to such characteristics is to be encouraged. Nevertheless it is worth noting that investigators such as Dudzinski et al (1975) have emphasized that if enough structure exists in a data set normality is not essential. Non normal distributions often produce comparable results. These findings are encouraged on practical grounds, but should not be used to avoid what can be called data 'screening or grooming' to improve the fit of variables to the requirements of the statistical models. In any case, the application of non-linear factor analysis (Macdonald 1967) may prove appropriate in the future, by reducing the effect of the linearity and normality assumptions built into the currently popular procedures.

(c) Homoscedascity

Another assumption of the relationship between any two variables in linear statistical models is that of constant variance. in the case of Pearson's Product Moment Correlation coefficient the assumption is made that the variances of the residuals should be equal; in other words, the relationships should be homoscedastic. This means that the variance between any two variables should be uniform over the whole distribution. Frequently, however, when the values for one variable are plotted against the value for another variable, unequal variances can be found such that the variability in A for high values of B is much greater than for low values. So if investigators attempted to predict A from B the estimates are going to be less reliable in the higher ranges, since the distribution of error will not be uniform.

This type of problem is one common to all linear models, and occurs before factor analysis is applied to the similarity matrix; unfortunately little discussion of the issue can be found in the factorial literature. The problem is, of course, very familiar in econometric texts (Maddala 1977), where it has been shown that the problem is particularly important when different sets of data at a micro level are used: for example, households with high and low incomes, large and small businesses, and different sized areas. By implication the problem is certainly likely to be influential in biasing the results of factorial ecologies. One suggestion that has been proposed has been to divide the units by the size of the areas or to use logarithmic values of the data. As Appendix B shows, this has the effect of squeezing the extreme values together and reducing the variability of these values.

(d) Multi-collinearity

It has already been observed that social scientists do not have as much control as natural scientists over the data they use for much of the information is derived from non-experimental sources. The result is that it is difficult to be certain that any two variables are really independently derived. The situation where two or more independent variables are so highly correlated with one another that their separate effects upon a dependent variable cannot be distinguished is described as a condition of multi-collinearity. In essence, it means that one variable is an exact linear function of another variable. Although this problem is primarily described in relation to regression analysis

(Blalock 1963, 1971; Johnston 1979) implications exist for factor analysis since the same linear basis is assumed.

Multi-collinearity is usually shown by high levels of intercorrelation, but it may be disguised by the fact that the variable is a linear combination of several other variables, and as such has really been described in the previous section. The result is that the estimation of the individual regression coefficients in the regression model becomes uncertain because they have very high standard errors. Econometricians, in particular, have faced up to this problem, and suggested ways around the problem. The simplest device is to exclude some of the collinear variables, thereby reducing the error to manageable proportions, or, even better, to use additional indicators as surrogate estimators. An authority in econometrics observed:

'without the luxury of an additional data source, multi-collinearity ... may be best thought of as posing a decision problem for the investigator. On the one hand theoretical considerations suggest the inclusion of relevant variables ... (but) ... if one variable is removed the model becomes incorrectly specified, leading to biased results. On the other hand, removing one of highly intercorrelated variables would generally be expected to result in increased precision of estimation for the partial regression coefficient of the 'left in' variable. The 'appropriate' trade-off between bias and greater precision for the coefficient on the 'left in' variable is the matter of question.' (Aigner 1971 p.95)

The effect of multi-collinearity in a data set used by any factor analyst is to produce a major constraint upon the ability of investigators to derive as many factors as there are variables in a matrix. For example, if a six variable matrix has multi-collinearity occurring between three of the variables, a three-factor model would be the maximum size of solution that could be produced. To appreciate this one must remember that matrices with rows or columns having entries that can be predicted exactly from one another cannot be evaluated, and the determinant will be zero. Until further work is carried out on the problem in the factorial situation the appropriate recommendation must be to exclude the offending variables if they have very high simple intercorrelations, or to redefine them so as to resolve the problem.

(e) Auto-correlation

Auto-correlation is the name given by statisticians to those circumstances in which the value of one variable at one observation (such as 22% of black ethnicity in enumeration district 11) is related to the value at an adjacent observation (such as 25% black ethnicity in enumeration district 12). The problem is that in statistical terms these two values are unlikely to be independent; they have probably been produced by the operation of the same underlying process. In the case of ethnicity the common link could be the fact that run-down inner city housing attracts high proportions of black groups because many of these people are unable to afford property elsewhere, or have been segregated because of racial prejudice. The failure to obtain independence in the samples means that yet another assumption of linear statistical models is breached. Tests of significance in the common factor model are based on this requirement so it seems that many

geographical and ecological data sets cannot conform to the requirements of the rigid statistical models. At first sight one simple way
of avoiding the issue would be to take samples of the data, rather than
using the complete data source. However, if the objective is to
describe a complete population – such as all the areas of a town – the
objective of the study runs counter to the purely statistical requirements, and to the accepted norms in psychology where the various tests
and 'rules of thumb' for practical interpretation are based on samples
from a homogeneous population. In any case the use of random samples
to exclude contiguous units may still not solve the problem. Most
human geographical processes operate by diffusion or spread effects, so
the influence transcends the neighbouring observation. In the last
resort it can be argued that since geographers and ecologists are
primarily interested in those spatially associated indicators, the
whole attempt to use the inferential test procedures of statisticians
(which assume independence of samples) must be challenged. Indeed
Gould (1970) posed the question in a geographical journal that the
inferential approach was a 'wild goose chase'. Yet investigators
continued to use factor analysis in his studies of mental maps and
perception surfaces, so it is obvious that they must have considered
that the procedures can be applied in non-inferential contexts.
Presumably, the use of factor analysis in the descriptive context of
the common factor model is regarded as a rewriting of the data set, in
which the auto-correlation problems are not important. Purists will
disagree with this view, but on practical grounds it does appear to be
a way out of the dilemma. Yet it must be emphasized that methods do
exist for calculating the degree of spatial auto-correlation. Texts in
econometrics (Kmentra 1971 p.295) describe the appropriate tests for
single variables, whilst geographers (Cliff and Ord 1973) have proposed
tests to determine the presence of auto-correlation among regression
residuals in two dimensional situations. Extensions of this work to
the multi-dimensional case appears to be some distance away. Haining
(1980) has provided an extensive review of the difficulties involved,
but ended with a rather pessimistic conclusion:

'in view of the circumstances in which the problem arises it is
unlikely that we shall ever see clear procedural directives for the
research worker who just wants to be told what to do. Understanding the auto-correlation problem is a central issue in quantitative
spatial analysis ... at best we can only hope to take reasonable
precautions.' (Haining 1980 p.40)

The auto-correlation problem and other distributional assumptions of
the linear models of which factor analysis is a part has led some
workers in the field to suggest that non-metric methods such as
multi-dimensional scaling (MDS) may be more appropriate approaches to
these overall geographical problems. These methods have been described
at length by Rushton and Golledge (1972). The difficulty here,
however, is that MDS methods do not provide the dual approach of factor
methods – in the R mode case measuring the collectivities of the
variables by the loadings, and the scaling of areas by the scores.
Instead, they scale the observations, and as such are more properly
regarded as variants of the area taxonomy tradition represented by
cluster analysis procedures. Moreover like all ordering methods a
great deal of the variability of an interval scaled data set will be
lost if it is only treated in terms of order relationships.

4. ALTERNATIVE SIMILARITY MATRICES

Factorial studies in geography and ecology have followed the example of most applied factor work by using the Pearson Product Moment Correlation in well over 90 percent of analyses. Such restriction to a single measure of association inevitably means that only one of a possible range of relationships in data sets have been explored by factor analysis. Theoretically, of course, there is no reason why other correlations or similarity measures cannot be employed. The problem, however, is that there is such a variety of alternative similarity measures that even experienced workers refer to the choice as 'bewildering' (Cattell 1978 p.470) or leading to 'confusion and despair' (Rummel 1970 p.302). As usual, the variety of alternative problems appropriate for factorial analysts means that it is difficult to provide rigid guidelines for any investigator – apart from emphasizing that the coefficient should be the one that is the most appropriate for the particular problem dealt with by the analysis, subject to the obvious data constraints faced by the research worker. Given the constraints of space in this study, the reviews provided by Rummel (1970) and Cattell (1978) and the original classic statement by Carroll (1961), provide basic references to the issues involved, and go beyond the brief summary provided here which concentrates principally upon the problems of using correlations in the first section and co-variance or cross-product matrices in the second section.

(a) Correlations

Despite the criticisms often made about the continued use of the Pearson Product Correlation Coefficient, there seems little doubt that its popularity is based on some solid advantages. The most obvious is that it can be applied to data measured in different units and scales. Moreover, by dealing with interval scale measurements it uses much more of the variation in the data than those based on rank order or dichotomous measures. The very familiarity of investigators with the measure is an additional advantage, whilst the development of a comparative body of literature in factorial studies is helped because similar patterns of relationship are being explored. Finally, the growth of interest in higher order factor analysis (Davies and Lewis 1973; Davies 1980), in which successive factoring is based on the correlation matrix between the first order factors, provides an additional boost to this method since it is advantageous to have the relationships between variables and factors, or high and low order factors expressed in the same measures.

Nevertheless, it cannot be denied that there are some restrictions on its use. The most obvious is that the correlation is only useful insofar that it provides an appropriate description of the relationships between two variables for it is restricted to dealing with the linear relationship. In addition it must be remembered that the measure only deals with the pattern similarity of two distributions; the size differences are excluded. Since geographers and human ecologists often deal with units unequal in size the exclusion of size variations may be a sensible research strategy since it ensures a comparative measure. But the exclusion of size must still be considered to be a restriction. In addition it must be emphasized that the correlation measure assumes homoscedacity. Rummell (1970 p.217) also demonstrated how the coefficient may have a restricted range – in other words it might not run between +1.00 and –1.00. This can occur when two distributions are skewed in opposite directions: the positive range is restricted and correlations under +1.00 are produced (see also, Carroll 1961). Although there is a consensus among applied factor analysts (Rummel 1970; Cattell 1978) – as opposed to mathematicians – that the correlation can be applied to non–normal data without serious distortion in the results, such measures can only be used in a descriptive context. The significance testing approach of the common factor methods demand that the data must conform to bivariate normal distributions if the interential tests possess any validity. It has already been observed that the data can often be transformed to produce normality, but the point about the loss of simple understanding of the meaning of variables has already been discussed. Also, there is always the further problem that back-transformations have to be made once factor scores are derived if the investigator wants to get back to the original units of analysis (Cattell 1978 p.470).

(b) Other correlations

The restriction of the Product Moment Correlation to interval or continuous scale data does not mean that factor analysis cannot be applied to other correlations. If the data is in rank order form, or if the investigator transformed his interval data to rank relationship, then Spearman's rank order or Kendall's tau measures can be used. It is worth remembering that these coefficients often given different values, so a choice has to be made between them. The obvious disadvantage of the use of rank order correlations is that a lot of information is lost compared to interval scale statistics. However, the approach does give undue weight to the middle range of a distribution, since the difference between 50 and 51 is treated the same as that between 1 and 2 or 90 and 91. This has led experienced analysts in psychology (Cattell 1978) to virtually discourage its use – unless the data makes it unavoidable. In this context it might be noted that there has been a long debate in sociology as to the equivalence of rank order and product moment coefficients. A useful recent review is provided by O'Brian (1979) who observed that if the variable is uniformly or normally distributed, and there is a large enough number of cases, the use of rank order causes little distortion in the results. In the factorial literature several researchers from the time of Moser and Scott (1962) onwards have pointed to the similarity in loadings produced by the use of rank order measures in place of product moment correlations. This demonstrates that the strength of many dimensional patterns outweighs the change in metric. Given all the problems associated with distributional assumptions it may be worth advising researchers to check if they obtain the same structures in their factor results when rank order statistics are used. In the descriptive context, at least, it provides a useful degree of confirmation in the stability of the results.

Factor analyses can also be carried out on dichotomous (Christoffer-son 1975) or nominal scale data, where it is scaled 0 or 1. A whole range of alternative measures exists, and the utility of many of these statistics – particularly the phi, phi-over phi-max and tetrachoric (r_t) – have been the subject of great debate among psychologists as can be seen in reviews by Carroll (1961) and Cattell (1978 pp.467-475). This means that extra care should be taken in the use of these measures. It has been shown that the tetrachoric over-estimates the value of the correlation if the data are derived from a normal distribution. Comrey and Levonian (1958) demonstrated the advantages of the tetrachoric in factor studies since their study showed the phi-over phi-max inflated communalities. Many other correlation procedures suitable for particular relationships, such as the point biserial coefficient for measuring combinations of continuous and binary data, could be used by factor analysts if the problem was appropriate. Obviously investigators should take care to justify the choice of measure in relation to the problem of the study, and compare the results with more traditional measures if in doubt.

(c) Scaling

This term is used to describe the adjustment of all the data in the matrix, rather than the individual variables changes produced by transformations of each variable. To some extent the procedure has already been covered in the section on correlations. In the product moment correlation measures each variable is changed to its standard score to remove the effect of size and to ensure that the units are equally scaled, in the sense of being measured in units of zero mean and unit variance. A variety of alternative scaling possibilities have been identified by Rummel (1970), although few have been employed in geographical or ecological work. As direct factor methods (Chapter III) become increasingly popular more attention is likely to be given to scaling in order to change the data into a form suitable for factor analysis. One approach which has a great deal of potential is called 'bounding' since it 'bounds' or reduces the data to a suitable form. For example, trade or journey-to-work flows between nations and cities may be characterized by considerable size variations between the units. Although the standard correlation approach with a Q mode can be applied to produce factors which pick out places with similar sets of origins, an alternative would be to scale the data - perhaps by dividing by the highest trade or journey-to-work interaction such that 1 becomes the highest flow and 0 means no flow at all. Uncorrelated functional entities (trading blocs or commuting zones) would be delineated by the analysis. Probably the most extensive use of this type of approach can be seen in the work of Russett (1967) whose study of international regions in voting, trade and distance terms was subtitled as a 'political ecology.'

Again, a cautionary note is in order. Since these scales are so unfamiliar in geographic and ecological studies, it is worth calcu-lating the relationships by hand in a couple of cases, and exploring the results in known situations to ensure that an appropriate scaling has been carried out for the specific problem, rather than going straight to the complicated multivariate cases.

(d) Co-variance and cross-product matrices

Many variables used in factor analysis are measured in quite different ways, and one of the advantages of the correlation approach is that the initial conversion to standard scores ensures that the influence of the size of the variable is ignored, since the mean of

each variable is zero and the deviation is one. Instead, the correlation measures the similarity of two variables over a set of observations in the R mode procedure. It produces factors which have a mean
of zero, and a total eigenvalue sum that is scaled according to the
number of variables, since each variable adds an equal unitary weight
to the model. Cattell (1966, 1972, 1978), one of the leaders in the
application of factor methods to problems in psychology, has long been
an exponent of the need to produce what he calls real base factoring,
in the sense that the size of variables are taken into account in the
model. Individual factors can, therefore, vary in size instead of
being unit length derivates of unit variance variables. The size
differences involved in real base factoring must not, however, be
confused with the factor sizes associated with ordinary factoring. In
this case the mean variance of the factor is produced from all the
variables, and is usually calculated by squaring the factor loadings,
and summing by column or factor. As yet comparatively few researchers
have utilized this possibility, even in psychology, although Bargman
and Bock (1966) provide a useful discussion on the testing of
hypotheses in co-variance structures. Johnston (1973), however, has
brought Cattell's ideas to the attention of geographers by using a
small data matrix of population migration in western Europe derived
from Magee (1971). The results of correlation, co-variance and
cross-product factoring are shown in Table 4.10. Since the data set
itself is a simple one it is also shown in Table 4.10 in order to
provide a framework for the interpretation of results.

Table 4.10
Use of cross-product matrices: migrant labour in Europe

(a) Data

	Origins: Greece	Turkey	Yugo- slavia	Portugal	Spain	Italy
Destinations:						
Germany	187	133	65	14	183	372
France	5	8	20	103	346	340
Switzerland	7	4	4	1	79	500
Belgium-Lux'g	8	7	0	2	32	95
Austria	2	5	19	0	2	3
Netherlands	2	6	1	1	6	8
Sweden	0	0	5	0	0	5
Totals(000's)	211	163	124	121	648	1,323

(b) Principal axes component analysis:
    correlation based results

| Origins | Loadings | | Communalities |
	I	II	
Yugoslavia	-92 (-91)*	28 (-18)*	98
Greece	-88 (-99)	47 (00)	99
Turkey	-87 (-99)	47 (01)	98
Spain	-72 (-16)	-69 (-95)	99
Italy	-67 (-33)	-34 (-62)	56
Portugal	-41 (-20)	-86 (-97)	91
Eigenvalue	3.51	1.85	
Variance	58.6%	29.8%	

| Destinations | Scores | |
	I	II
W. Germany	197 (-226)*	109 (-33)
France	67 (36)	-207 (-205)
Switzerland	12 (25)	-31 (-17)
Austria	54 (27)	45 (69)
Belgium-Lux'g	54 (37)	22 (50)
Netherlands	71 (50)	28 (65)
Sweden	73 (50)	33 (71)

* Rotated solutions in brackets.
  Decimal points are removed: -92 is -0.92.

(c) Cross-product results: rotated solutions

Origins	Loadings			Destina-tions	Scores		
	I	II	III		I	II	III
Italy	−197.7	−94.3	76.5	Switzer-land	−2.50	−0.02	0.03
Spain	−30.3	−125.8	39.3	France	−0.41	−2.60	0.18
Greece	−1.7	3.3	71.0	Germany	−0.63	−0.48	2.63
Turkey	−1.3	0.4	50.5	Belgium-Lux'g	−0.36	−0.11	0.17
Portugal	−0.1	−39.7	−2.0				
Yugo-Slavia	−0.9	−5.8	23.5	Nether-lands	0.03	−0.11	0.06
				Sweden	−0.02	0.01	0.01
				Austria	0.24	0.00	0.10

Source: Revised from Johnston (1973)

Johnston's Principal Axes results in Table 4.10 show that rotation disaggregates the generality of the unrotated solutions based on the correlation matrix into two axes. Using these rotated axes the loadings on one axis pick out migrants from Yugoslavia, Greece and Turkey as being associated with one dominant destination: West Germany. The other identifies flows from Spain, Portugal, and a lesser extent Italy, which is the dominant destination of France. The low communality for Italy demonstrates that a lot of the variance of its flows is not associated with the two model explanation – implying that this solution may be under-factored. (Chapter VI discusses the under-factoring issue.) Indeed, the strong linkages of Italy with Switzerland and Germany in the raw data matrix are ignored by the simple two model results. A further problem is that the use of corre-lations does identify patterns of similarity, not size, as can be seen by the fact that the relatively small flows of southeastern European countries in Component I are given as much importance as the larger flows from the south and southwestern countries.

The three axis results of the rotated solutions based on the cross-product matrix solve these problems of interpretation, since the size of flows influence the pattern. Thus, Spain and Italy, and marginally Portugal, send workers to France, the dominant score on this axis. In the third axis, the dominant destinction of Germany, has Italy, Greece and Turkey, and to a lesser extent Spain and Yugoslavia, associated with the vector. These results obviously uncover the basic structure associated with size of flow in the data set. It must be noted that

Johnston's results based on the co-variance matrix are very similar to the cross-product values so they are not shown here.

Objections will be raised that the factor analysis results shown in Table 4.10 represent unnecessary complications, since the structure of the migration matrix can be easily identified by flow diagrams. This is not the point. Johnston (1973) deliberately used a small matrix so that the factor results could be set against those derived by other means. Obviously, if similarity of pattern is all that is required the correlation based analysis is satisfactory; if flow magnitudes are important the cross-product matrix seems to be the most appropriate input for the analysis of flows. Berry (1967) provided one of the few other examples of the use of the cross product matrix as a prelude to factoring. His study of shopping centres in Spokane produced a 36 centre x 34 retail establishment data matrix, with each cell coded 1 if a retail type was present and 0 if it was not. A 34x34 cross product similarity matrix compared the distribution of each column, the retail establishments with one another, the factor results producing a scaling of the centres according to the numbers of retail types. Experiments by the author with the use of this type of approach in the study of shopping centres have not been particularly successful in terms of the variance explanations. It is suggested that cluster methods may represent appropriate procedures for determining the similarity of sets of centres. Potter (1981), however, has recently applied standard factor methods with success in a study of Stockport shopping centres. Nevertheless this must not detract from the general utility of the use of cross-product matrices where size variation is needed. It is unfortunate that few package programmes allow investigators the flexibility to use cross-product or co-variance matrices, since they certainly waste less of the information than the correlation measures.

(c) Other measures

The discussion on alternative procedures in Chapter III pointed out that a wide variety of alternative information could be factor analyzed, especially when the Direct Factor approach is used since it is applied directly to the data set which has to be symmetrical. In the geographical field Berry and Barnum (1962) have used the approach in studies of central place systems whilst Gould (1968) provided an early example of the approach in studies of the connectivity of road systems, where 1 was used to indicate a direct connection between places and 0 indicated no connection. Clark (1971), however, demon-strated an example of another type of similarity measure to geographers and ecologists, one that he claimed was particularly useful for analyses of flow matrices. The cos theta measure is a measure of proportional similarity between cases i and n, and variables j to p.

$$\text{Cos}_{in} = \frac{\sum_{j=1}^{p} (x_{ji}\ x_{jn})}{\sum_{j=1}^{p} (x_{ji})^2 \cdot (x_{jn})^2} \qquad \text{Eq. 4.2}$$

The cos theta measure was used originally by geologists (Imbrie 1963; Imbrie and Van Arsdol 1964) in factor analytical studies of sediments and particle sizes. Imbrie (1963) pointed out that the use of percentage data followed by Product Moment Correlations produced 'geologically trivial' results because size differences were repressed and closure was imposed on the range of values. Table 4.11 provides an example of the difference between cos theta and correlation measures in terms of a small data set (Clark 1971).

Table 4.11
Cos theta and correlation measures

Raw Data	To: 1	2	3	4
From: 1	50	25	60	50
2	30	15	40	50
3	20	10	30	50
4	0	0	10	50

P.M. Correlations	1	2	3	4
1	1.00	1.00	1.00	0
2		1.00	1.00	0
3			1.00	0
4				1.00

Cos Theta	1	2	3	4
	1.0	1.0	0.98	0.81
		1.0	0.98	0.81
			1.0	0.88
				1.0

Source: Clark (1971 Figure A-5)

126

It can be seen that cos theta produces a value which reflects the gradient between the fourth observation and the others. In general cos theta produces a value of 1.00 for identical proportions and 0.7071 for a set of random values with no association meaning that investigators must be careful not to exaggerate the size of relationships produced from factor analytical studies using this measure. Clark (1971, 1973) has stressed the advantage of the measure for geographical work, in particular the fact that it is non parametric and does not necessarily assure the data has to be normally distributed, etc. He suggested the index is particularly useful in studies of flow matrices where size is considered to be an important feature of the relationships to be extracted. Few investigators have followed this example, in part because the differences in size between the areal units on which flows are calculated ensure that flow size reflects areal unit size. So measures of association are usually preferred.

5. SUMMARY

This chapter has summarized some of the basic data problems facing would-be users of factor analytical methods. As with all quantitative methods the results finally obtained depend upon the quality of the data, in the sense that a great deal of care needs to be taken in the preparation of the data, with the choice of variables, areas and measures of similarity before factor analysis is applied. Only by paying special attention to this stage will the final set of factorial results be useful in any substantive, as opposed to a purely technical context. Now that the factor analysis calculations can be carried out so quickly by high speed computers it becomes even more important to spend time on this data analysis stage, such as exploring the effect of different decisions in similarity measures, etc. If there is no a priori reason for choosing one measure, a detailed understanding of the range and type of variability in the initial data set, an understanding that can only come from scrutinizing and exploring the data, is an important stage of analysis if misinterpretations in the factor analysis stage are not to occur. Yet it must be admitted that the identification of so many data problems, and the difficulty of ensuring that the assumptions of the statistical or inferential approach to common factor analysis are not breached has created great crises of confidence for the factorial ecologist in recent years. The extreme position is to deny that the methods are of use in dealing with area distributions. But in practice applied factorial analysts in psychology and other fields have shown that the procedures are very robust and can be used in a wide variety of circumstances with successful results. This literature should provide human ecologists with additional justifications for continuing to derive generalizations by factorial methods. After all, there are always assumptions and leaps of intuition in making any generalization; using factorial methods such issues are much more exposed and their appropriateness can be debated rather than being subjectively ignored.

5 ALTERNATIVE FACTOR PROCEDURES

The geometrical introduction to factor analysis described in Chapter II illustrated one of the fundamental characteristics of factorial procedures: the indeterminacy of the solutions. This means that many different factorial solutions, each mathematically accurate, can be obtained from a data set if different assumptions are made. Unfortunately, human ecologists have investigated few of these alternative procedures, unlike their colleagues in psychology. Instead, the vast majority of studies have been based on the Principal Axes technique using a component model – often called the Principal Component solution. Increasingly this has been followed by Varimax rotation. This restriction to one set of procedures is changing very rapidly as alternative factor procedures become more widely available through computer package programmes and the expansion of university computing facilities. Although this variety of different procedures provides an important element of flexibility to individual researchers – enabling them to choose the most relevant procedure for their particular purposes – it also contains the seeds of destruction for the development of a substantive body of literature in factorial ecology. Added to the problems of comparability caused by data inputs that are spatially and temporally variable, are the difficulties associated with the attempts to integrate results obtained from widely divergent technical procedures. Hence it is imperative that factorial studies in geography and human ecology either justify the basis for the choice of a particular factorial procedure, or demonstrate that the results are invariant across many techniques. Without such evidence the specific set of results are unlikely to make as much contribution as they might to the general body of substantive literature in human ecology and geography.

This chapter provides a summary review of the most accessible alternative factorial procedures that are currently available. The discussion does not attempt to cover the computational details of each factoring procedure since these demand a knowledge of matrix algebra and are already carefully explained in the standard texts (Harman 1976; Horst 1965; Rummel 1970; Cattell 1978), whilst some were summarized for geographers by Mather (1976). Instead the discussion reviews the basic principles lying behind the alternative methods. Five different problem areas can be identified: the factor model; the factor technique or procedure used to extract the factors; the rotation of the initial set of axes; higher order solutions; and the derivation of factor scores. Frequently, however, the final decision on which of these procedures should be adopted depends on the interpretation of the results, and on the purpose of the analysis, rather on strictly internal technical considerations. Issues such as number of axes to extract, the interpretation and significance of the loadings and the factors, and the comparison of axes are all important problem areas related to the procedures described in this chapter. Unfortunately, they cannot all be treated simultaneously, so all these problems are dealt with in Chapter VI under the general heading of 'Problems in Interpretation and Choice'.

1. FORMAL DESCRIPTION AND ALTERNATIVE METHODS

Previous chapters have already shown that the factor equations represent a way of <u>linearly</u> reproducing the correlations between a variable and all other variables in the data set. This linear reproduction was achieved by using a limited number of mathematical 'constructs' called factors, each of which are weighted according to the amount of the variability of the variable explained by the factor. The resultant factor equations for variable V_1 from the 4x4 correlation matrix shown in Table 2.5 was:

$$V_1 = 0.7071TT_1 + 0.7071TT_2 \qquad \text{Eq. 5.1}$$

Where: TT_1 and TT_2 are the factor axes and 0.7071 represents
the weights.

Given the need to express these relationships in a more formal and general algebraic way, so that the alternative models and techniques of factor analysis can be more easily appreciated, it is possible to obtain:

$$Z_j = a_{j1}F_1 + a_{j2}F_2 + a_{j3}F_3 + \dots a_{jq}F_q \qquad \text{Eq. 5.2}$$

Where: $a_{j1}F_1$ represents the factor coefficient (or loading)
of variable j on Factor 1

This algebraic approach is, of course, less easy to visualize, but has many computational advantages over the long winded geometrical approach described in Chapter II.

Equation 5.2 is similar in form to the standard multiple regression described in the basic texts on statistics in geography (Johnston 1979) where each variable is also described as a linear combination of the variables plus a residual which accounts for the variance unexplained by the equations. But, unlike the regression model where each variable is also described in terms of the <u>original</u> variables, each variable in the factor equations is described by a <u>new set</u> of vectors called factors. In essence, therefore, the objective of any factor analysis is to determine the number and position of these new axes or vectors in the multi-dimensional space of the original variable set.

So far the term 'factor' has been used in a very general way to cover two very different models of factor analysis, namely component analysis and common factor analysis. Since they make very different assumptions it is important to clarify the differences. In addition, the objectives of the models differ quite drastically: the component model is best suited for data description and exploration, whilst the common factor approach is best suited for hypothesis testing.

(a) Component model

The normal objective of this approach is to extract the maximum amount of variance from any data set. In essence it is a linear re-arrangement of the original data matrix since <u>all</u> the variance in a data set is assumed to be capable of being extracted. Hence the diagonal entries of the matrix are given the maximum values of 1.00 because it is assumed they can be maximally reproduced by the factor solution. The mathematical expression for the component model is given by Equation 5.2. In the best known factor extraction technique, the Principal Axes method, each component successively accounts for a decreasing amount of variance after the dominant or principal source of variance is extracted. This method, originally proposed by Pearson (1901, 1927), and fully developed by Hotelling (1933), is called Prin-cipal Axes Component analysis because the first axis to be abstracted accounts for the largest amount of variance of the linear relation-ships; the second describes the second largest amount of variance from an axis that is placed at right angles to the first, and so on. As such the method is mathematically determinate.

(b) Common factor model

An alternative to the component model is the common factor approach, often summarized simply as factor analysis in its strictest sense. This approach attempts to maximally reproduce the correlations of the data matrix, according to the assumption that the variability of each variable can be split into two parts called common and unique variance. As a result, the equations are rather different from those of the component model, as can be seen in the following equation.

$$Z_1 = a_{j_1}F_1 + a_{j_2}F_2 + a_{j_3}F_3 \ldots + d_jU_j \qquad \text{Eq. 5.3}$$

The terms $a_{j_1}F_1 \ldots F_3$ represent the common factors whilst d_jU_j is the expression for residual error, a term that is similar to that used in the multiple regression equations. This residual error value, however, is usually split into two parts, the specificity (b_jS_j) and the error or unreliability (e_jE_j). The former is that part of the variance associated with a set of specific factors that may be produced by the particular selection of variables, or because of the place-particular features of a town or region. The latter, the error proper, is the 'unexplained' or residual variance. A succinct way of summariz-ing the expressions in the common factor model is as follows:

$$\text{Eq. 5.4}$$

$$Z_1 = a_{j_1}F_1 + a_{j_2}F_2 + a_{j_3}F_3 \ldots + b_jS_j \ldots\ldots\ldots\ldots + e_jE_j$$

$= \langle$ — Common Factors — $\rangle \langle$ — - Uniqueness ----------------$\rangle$

$= \langle$ — Common Factors — $\rangle \langle$ -Specificity $\rangle$ .. $\langle$ .. + Error ..$\rangle$

$= \langle$ — Common Factors + Specific ----------------$\rangle$

$\qquad$ (reliability or reproducibility b_j^2)

Harman (1976 p.20) has shown that an index of completeness (C) of factorization may be expressed as $C_j = 100h_j^2/h_j^2 + b_j^2$ or 100 communality/reliability.

From the previous discussion of the factor equations it should be apparent that the sum of the common factors $\sum_{i=1}^{n} a_{ji}F_i$ for any variable is equivalent to the communality (h_j^2), the amount of explanation of a variable accounted for by the solution. Unfortunately there is no simple way of arriving at the reliability of any equation; the values can only really be obtained by repeated and replicated studies of the same problem. Moreover, it must be noted that the specific part of any solution is, in part, a function of the data set. If additional variables are added to the data they might combine with one of the indicators related primarily to specific characteristics, thereby defining another new factor or a general source of variation.

(c) Differences between the models

The distinction between these two models has not been clearly drawn by empirical studies in factorial ecology. This is rather unfortunate since in theoretical terms they make different assumptions and fulfil different objectives. The component model is basically deterministic since it reorganizes or rewrites the data in a new and unique form. Its biggest advantage is that no assumptions are made about the size of the communality, as 1.0 is entered into all the diagonal cells of the similarity matrix. This means that each variable is given equal weight in the analysis. As such the model is best utilized either for re-writing data in a concise, parsimonious form as a prelude to some other analysis, or for simply exploring or describing the relationships in a data set. In theory the method is not particularly appropriate for hypothesis testing since components are combinations of common and specific variance. But these can be filtered out, so the method may be more appropriate in the earliest stages of investigation, the current position of factorial ecology.

Cattell (1966) pointed out that the specific variance becomes larger as successive components are abstracted. Table 5.1 (from Davies 1975), shows this effect in relation to a set of British towns where the smaller axes (smaller in terms of total explanation) are primarily associated with specific variance, whilst the axes for the smaller towns are also more general and contain greater amounts of specific variance. The biggest disadvantage, however, is the fact that the component model is a closed one. By entering unities or 1.0's in the diagonals it assumes all the variation in a data set is explained.

By contrast, the common factor approach begins by making assumptions about the level of common variance. (Incidentally, Cattell (1978 p.17) prefers the term 'broad' factor as the separate variables are broadly related to different axes, rather than being common to all axes. A specific factor is unique to one variable.) Values that are considered to be estimates of the common variance for each variable are placed in the diagonal of the similarity matrix. Many alternative values can be used, for example: estimates based on the judgement of the investigator; or values derived from the analysis, such as the highest correlation between a variable and all others (Rummel 1970). The most usual value used as the default option in standard packages such as SPSS (Nie et al 1975) is the squared multiple correlation coefficient. This measures the proportion of the variance of a variable accounted for by all the other variables.

Table 5.1
Common and specific variance in the social dimensions
of British cities, 1966

Dimensions	Percentage variance explanation of the axes (specific variable contribution in brackets)			
	Leicester	Southampton	Swansea	Small Towns
General Axes				
Life Cycle	9.0 (0.2)	8.9 (1.3)	8.7 (1.2)	10.2 (4.7)
Socio-Economic		4.8 a(3.1)		
status	14.1 (2.0)	6.1 b(2.5)	14.4 (3.7)	10.6 (2.5)
Mobility	6.2 (1.0)	6.4 (0.5)	6.7 (0.4)	Migrant
Ethnicity	3.6 (1.0)	4.2 (3.8)	5.0 (2.3)	4.5 (8.8)
Urban Fringe	2.4 (0.9)	2.2 (1.2)	1.4 (1.5)	not found
Economic				
participation	2.8 (1.6)	4.3 (0.9)	not found	3.8 (3.8)
Residentialism	3.1 (1.2)	0.7 (2.3)	2.4 (2.4)	2.0 (3.2)
Specific Axes				
Substandardness	0.0 (7.1)	not found	not found	not found
Females	0.0 (0.9)	not found	not found	not found
Number of dimensions				
rotated	9	9	6	5
Total explanation	72.2%	68.9%	63.3%	60.6%
Variance Contribution				
Common variables	41.2%	37.6%	38.6%	31.2%
Specific variables	15.9%	15.6%	11.5%	23.0%
Residual variables	15.1%	15.7%	13.2%	6.4%

a: General Socio-Economic (S.E.) axis
b: High-Medium S.E. axis
Source: Davies (1975 Table 3)

The fact that differential values are placed in the diagonals of a similarity matrix means that each variable is given an unequal weight in the common factor approach. One feature of this is that the model becomes indeterminate in the sense that different solutions can be obtained from the various estimates. A consequence is that common factor scores have to be estimated.

The component model has its supporters, particularly in the purely mathematical context, but it is also worth noting that some physical scientists such as Klovan (1966) seem to prefer the approach. Nevertheless, most statistically minded social scientists argue for the common factor approach. It is usually pointed out that the variables are likely to be a sample of the universe of possible variation, so it is unlikely that one model can account for all the variation in a data set. Errors are bound to occur. This means that the common–unique variance distinction is a vital methodological distinction. Geographical texts on quantitative methods have tended to echo this latter view (King 1969; Mather 1976; Taylor 1977), although much of the empirical work in the field has ignored the point and concentrated upon the component approach. The strength of this commitment to the common factor approach can be judged from the strength of recommendation by Cattell (1978) who is one of the major workers in applied factor analysis:

'if one wants only an immediate statistic given within the private, isolated, self contained world of a single experiment, then the seeming precision with which components, relative to factors, will calculate factor scores from test scores may appeal to the pure statistician. But the scientist who wants factors corresponding to stable concepts ... that will re-appear in other experiments ... will face the difficulties of the somewhat more complex factor model and the problem of getting the truest communalities.' (Cattell 1978 p.67)

Yet it must be admitted that Cattell (1978 p.64) used the component model in a very strict way, since he assumed 'the method must produce as many components as variables.' By contrast, most applied workers in human ecology are prepared to adopt various 'rules of thumb', and to interpret only a limited number of the largest components. The result, in practical terms, is that the empirical difference between the two models may not be great. Indeed it has been shown on numerous occasions that the results from a common factor solution, using the Principal Axes method with squared multiple correlation coefficients as estimates of communality, produce the same substantive results as those from a component analysis using the same factoring procedure. Table 5.2 shows an examples of alternative common factor results. The first is produced from the simple 6x6 data matrix on Calgary used in Chapter II. The table illustrates that there are few substantive differences from the results in Table 1.3 where the component model was used. In 5.2b in a 14 variable study of Calgary it can be seen that the loadings confirm the general pattern of the axes but most loadings are lower, especially the ethnic axis. A straightforward explanation for the phenomena can be given. Ethnic variations are often very unique and cannot be exactly predicted from other variables in the analysis. Since the squared multiple correlations used as the communality estimates are lower the common factor loadings are relatively small. In such circumstances one cannot expect the level of variance for the variable in the final solution to be high. Investigators should be aware of the possibility of this type of result. It is not the factoring model per se that has produced the difference; instead it is the communality estimate. Unless one is careful one might forget that the variance associated with this source of variation may be lost. Mabogunje's (1968) study of Nigerian towns provides a good example of

133

Table 5.2
Differences between solutions: Calgary 1961 and 1971

(a) Common factor patterns: Calgary 1961

| Variables | Loadings after Direct Oblimin Rotation | | | |
| | Gamma = 1.0 | | Gamma = 0.0 | |
	Factor 1	Factor 2	Factor 1	Factor 2
1. Occupation	−0.947	0.066	−0.948	0.052
2. Education	0.925	−0.182	0.924	−0.169
3. Fertility	−0.220	0.875	−0.207	0.873
4. Women at work	−0.036	−0.922	−0.051	−0.924
5. Single family dwellings	0.096	0.973	0.111	0.976
6. Ethnic	−0.937	−0.176	−0.941	−0.190
Factor correlations between Factors 1 and 2		−0.087		−0.117
Initial simplicity criterion		5.609		2.666
Final simplicity criterion		3.246		0.269

(b) Common factor and component loadings: Calgary 1971

| Variables | Principal Axes, Varimax | | | Squared Multiple Correlations |
	Vector	Component	Common Factor	
1. Children	1	−80	−84	94
2. Young adult	4	86	77	72
3. Old Age	1	87	90	94
4. Female participation	4	91	86	79
5. Substandardness	5	84	79	79
6. Large dwellings	2	−86	−82	77
7. Unemployed	5	85	74	77
8. Manufacturing	2	89	81	74
9. Non-migrants	3	94	90	83
10. Immigrants	1	89	78	71
11. Canadian	7	93	69	52
12. British	2	−71	−64	79
13. Ukrainian	6	95	71	49
14. Recent housing	3	−91	90	86

Decimal points are removed: −80 is −0.80.
Only largest loadings are shown. More detailed interpretation
is provided in Table 5.4.

Source: revised from Davies (1978)

134

the problem. The communality for the size variable in the study was well under 0.5 in the common factor analysis. Since so little of the variability of the indicator was incorporated it is not surprising that a size axis did not emerge, as shown by other studies of urban dimensionality (Berry 1972). Without some explicit recognition of this problem inter-nation comparisons of urban systems can produce some quite erroneous conclusions.

At this stage in our knowledge of urban and regional differentiation a strong case can be made for the continued use of the component model, so long as it is not used in Cattell's strict context, and is followed by rotation, etc. In a strictly statistical sense such procedures cannot really be justified, as Johnston (1978) has reminded us. But on practical grounds the familiarity of the approach and the fact it makes fewer assumptions, means that initial investigations are less prone to error. Perhaps a way of resolving the dilemma between the scientific assumption of common variance, and the practical consideration of a less error prone analysis, is to use the component analysis as as prelude to the common factor approach. Such a descriptive summary of the data can be carried out prior to the more exhaustive scientific evaluation of the hypothesis involved, in which the use of the battery of inferential tests can also be applied if the data justifies the procedures. Yet one must not be carried away by the apparent power of the common factor approach. So long as the unique variance approaches zero there will be little substantive difference in the results. Nevertheless, we must recognize that in the last resort the common factor approach is different, in that it tests the adequacy of the initial hypothesis made about the number of axes needed to describe the data or the explanation of each variable. This might seem to be the most productive type of procedure in a confirmatory sense. But the statistical tests used in the factor estimating procedures are based on assumptions that may not be appropriate to many ecological investigations where a complete population is being dealt with. Such problems are discussed in the next chapter.

2. ALTERNATIVE FACTORING TECHNIQUES

(a) Types

The difference between the component and common factor models is only one of the problems facing investigators, since large numbers of very different procedures have been used to extract factors from any data set. Although this study is not designed to deal with the mathematical differences between the techniques, it is important to understand the basic differences between the principal procedures. A useful, if simplistic way of differentiating between those approaches is to review them in four categories.

The unadjusted general solutions are those procedures which use a raw correlation matrix without recourse to any estimation. The earliest attempts at obtaining factor solutions fell into this category, so they have a solid place in the history of development. However, they are not particularly useful today since they attempted to achieve as much generality as possible by describing each variable in terms of one or two general factors.

135

The second category consists of techniques that make some assumptions about the amount of common variance to be analyzed. As such they are called communality estimating procedures. The important decision to be taken is very largely one of deciding which values should be placed in the diagonal of the correlated matrix.

The third category, factor estimating procedures, contains a series of recently derived techniques which make inferences about the number of common factors to be abstracted. Once the decision about the number of factors to be abstracted is made, several different types of procedures may be followed. It must be noted that the initial decision restricts the factor space to a specific number of dimensions. An important aspect of these techniques is that a statistical test of the adequacy of the solution is usually employed. Hence they are particularly useful as hypothesis testing procedures.

The fourth category is reserved for a series of more sophisticated techniques that search for the most general solution, in conditions where estimates of communality and the number of factors have to be made. They are differentiated because of an assumption about the raw correlation matrix, whether, for instance, the matrix was derived from a sample of variables or observations. As they start by adjusting the initial matrix in the off-diagonal elements, as well as the diagonal elements, they are called adjusted matrix procedures.

(i) Unadjusted general solutions. A detailed discussion of these solutions would represent a review of the historical development of the field for they were derived in the first decade of the century by the pioneers of factor analysis. Although these techniques have been superseded — largely because of their simplistic assumptions — they are worth reviewing here simply because their basic objective lay in abstracting factors with as much generality as possible: a laudable objective for all disciplines that search for generalities. Two approaches may be recognized, the so-called unifactor methods that describe each variable in terms of one factor, and the bifactor methods that postulate two factors. Any variance not explainable in terms of one factor in the former case, or two in the latter case, is assumed to be unique.

Two unifactor methods have been proposed. In the first, each variable is only associated with one factor, and these factors are composed of different variable sets. In the second, the two factor theory of Spearman (1904), each variable in the set is assumed to be composed of one general factor; the rest of the variance that is unaccounted for is assumed to be unique. It is not surprising that the universality implied by these methods (particularly the two factor theory) attracted criticism. Indeed Spearman (1927) eventually used the bifactor method to resolve this difficulty. In the characteristic equations for the bifactor solution each variable is composed of one general factor and one specific factor, the rest of the variance being considered to be unique. It may be noted that several specific factors can be derived, each one associated with a certain number of the variables (Harman 1976 p.114).

These techniques are rarely used in the social science literature, having been superseded by more sophisticated analysis, although three geographers, namely King, Jefferies and Cassetti (1972), have used the bifactor approach. This seems to have been adopted because the method focuses upon the generality of a set of variables, a requirement of their study. But one must note that it also assumes that the rest of the variance to be accounted for is associated with one other factor that is composed of several other variables. Such an assumption may, of course, be untenable, although in the example of King et al (1972) satisfactory evidence was presented to justify this conclusion. Nevertheless, one must conclude that the technique depends upon the assumption of group effects in the set of specific variables, and that membership of these groups must be calculated before the technique can be applied. This means the method is likely to have very limited value in future ecological research. Moreover, the assumption of a positive correlation matrix as input for the technique represents another, although less distracting limitation.

A further extension of these techniques is provided by a set of multiple group methods which extract several common factors in one operation, instead of a single axis at a time, as in the currently popular techniques such as the Principal Axis method. Harman (1967 p.234) pointed out that these methods were the subject of an often acrimonious debate among the first major proponents of factor analysis, namely Holzinger, Guttman and Thurstone. But in view of the lack of interest by geographers in these solutions there is little point in summarizing these procedures; the interested reader is referred to existing authoritative sources (Horst 1965; Harman 1976). Yet it is worth noting in passing that in contrast to all other procedures in this section these multiple group procedures need not produce orthogonal factors; oblique factors can be derived since the factors may be correlated. However, a weakness of the approach is that the initial variable set needs to be grouped prior to the application of the factoring procedures.

(ii) Communality estimating procedures. The most popular of the preliminary solutions are the Centroid and Principal Axes techniques, with the latter being much more widely used now that high speed computers are available to make the basic calculations. The objective of these approaches is to extract the greatest amount of variance from the observed correlation matrix. It is usually accepted that the solution is most suitable when the variables are measured in the same units, so it is customary to express these variables in standardized form by means of standard scores which eradicate the size differentials. Appendix A provides a small workbook example which demonstrates one approach to the calculations involved.

Two different models of this Principal Axes technique can be recognized, the Principal Component and the Principal Factor solutions. In the component solution, values of 1.0 are placed in the diagonal of the original correlation matrix, so that the whole of the variance associated with a variable set is analyzed. It has already been pointed out that other values such as the communalities of an initial component solution, can beused in the diagonal of the matrix, the factor solution, are used in the diagonal of the matrix, to produce a factor solution then only deals with the amount of variance that can be predicted from what is a reduced correlation matrix, since the error variance has already been omitted.

Many other decisions can be made about the values to be placed in the diagonal of the matrix. But the Principal Component-Principal Factor analysis distinction is basic to the understanding of the varieties of this technique. The popularity of the component solution is not surprising because it is one of the few factor procedures that produces invariant answers for the same data input. The derivation of one general factor, and a series of bipolar factors in which an equivalent number of positive and negative factor loadings appear, frequently makes interpretation difficult. Rotation techniques are commonly employed to derive a simple structure that is easier to describe.

One variety of the Principal Factor technique is also worth mentioning in view of its acceptance in the social science literature, as well as its availability in one of the standard package sources for computer programmes, namely SPSS (Nie et al 1975). This is a common factor solution with an iteration procedure. In this routine a component solution is derived, and the technique automatically replaces the diagonal elements with a communality estimate (in the SPSS case the squared multiple correlation between any variable and all the others). An estimation is made of the number of factors required in the component solution, and the variances accounted for by these factors are obtained. These replace the initial communality estimates. Then another estimate of the number of factors is made, the variance is abstracted and the iteration continues until there are negligible differences between the two successive communality estimates.

The only other important technique requiring an estimate of the matrix diagonal values (and hence the amount of variance to be analyzed) is the Centroid solution. Child (1970) provided a detailed workbook example of the way the Centroid axes can be derived, and the basis has been described in Chapter II. The main advantage of the procedure was its comparative ease of computation in a pre-computer era, for it was very popular in the pre-war and immediate post-war period. As it only provides an approximation to results that can be quickly derived today from the Principal Axis technique it need not concern us any longer. Its place today lies in the history, rather than the current practice, of factor analysis.

(iii) Factor estimating techniques. The third broad category of preliminary solutions distinguished above are common factor solutions. Unlike the communality estimating techniques, these begin by assuming that a certain number of common axes are present, usually based upon some prior hypothesis of the number of common factors. As such they are described as factor estimating procedures. Once the decision is made to hypothesize a certain number of axes a variety of techniques are available, each deriving a solution in a particular way. Although few of these solutions are familiar outside the psychological literature it has been suggested that the onset of powerful computational facilities will make them more popular, particularly as they are usually accompanied by statistical tests that confirm or reject the adequacy of any hypothesized solution.

One of the most popular procedures is the Minres solution. The name comes from the objective of the solution, which is to minimize the residuals of the factor solution. The technique achieves this by searching for a solution that maximally reproduces all the elements of the correlation matrix, except for the diagonal entries. This is achieved by minimizing the off-diagonal residuals once the factors are extracted from the similarity matrix. Since the preceding section has indicated that the diagonal values are parameters to be determined by the investigator, such a focus upon the off-diagonal entries is likely to absorb more variance in a solution and may, therefore, become a powerful technique in the future. It must be noted that if the total matrix (including the diagonal entries) is minimized it leads to the conventional Principal Axis solution.

Unlike the Minres, or Principal Axis solution, the Maximum Likelihood technique (Lawley 1943; Joereskog 1967) uses the differences between the correlation of each pair of variables to produce factors that are maximally related to the variables. In addition, it adopts an explicitly inferential approach. The assumption is made that the original measurements of the observations, the areas in ecological analysis, were sampled from the universe of such observations. Hence the calculated loadings are considered to be estimated values. A chi-square test is then used to determine whether the estimted factor loadings differ from the assumed universe factor loadings by more than chance. One important feature of the Maximum Likelihood method is its dependence upon an initial set of trial values for the loadings. Principal Axis loadings are often used in this context. Of more importance to this summary review is the fact that the factor loadings are independent of the scale of measurement, since the loadings that are calculated are proportional to the standard deviation of the variables. Although statisticians point out that the method is simple in concept, the solution requires a great deal of computation, and has only recently become available on standard package programmes such as the SPSS. The objective of deriving factors that are maximally related to the variable set is one that factor analysts are constantly trying to achieve, so many statisticians maintain this is the preferred method. The Generalized Least Squares solution (Joereskog and Goldberg 1972) is similar to the Maximum Likelihood method, but uses a different least squares solution (Aitken 1937) as its basis for producing factor loadings.

(iv) Adjusted matrix solutions. The fourth set of procedures are similar in that they are all common factor methods. They differ from the previous methods by making certain assumptions about the data upon which the correlation matrix is based, and then adjust the whole correlation matrix, not just the diagonal values.

1) Variables as universe. Rao's Canonical factoring technique (Rao 1955) is similar to the Maximum Likelihood technique in that the objective is to maximize the correlation between a set of hypothesized factors and the set of data variables, although the communality estimate is an important additional input to the model. The crucial feature of Rao's approach is the assumption made about the generality of the final solution as expressed in the correlation matrix. It is assumed that the variables of the data matrix form a universe, but that a sample of observations is involved. The usual procedure is to insert the squared multiple correlation for each variable in the diagonal of the correlation matrix, then to rescale it by the unique parts of the data, before estimating the factors. The reasoning behind this procedure is that those variables that have a great deal of the

139

variance in the dimensions of the common factor space should play the greater role in any estimation of these dimensions, since the lower the uniqueness the more weight their correlation will have. The approach then seeks the minimum number of factors needed to account for the observed matrix, and provides a statistical test for the closeness of fit between the data and the hypothesized factor. A useful feature of the technique, like the Harris (1971) variant of Image Analysis is that it is scale free. Unlike the Principal Axes factor solution, where the results obtained from factoring the correlation and co-variance matrix may be quite different, the factors are the same in Rao's procedure. The biggest problem associated with the technique, however, is its dependence upon the number of variables used to estimate the unique pattern of the variance. By increasing and decreasing the number of variables the statistical test of significance will be altered, for when more variables are sampled more factors will become significant.

2) Variables as samples. Alpha Analysis (Kaiser and Caffrey 1965) approaches the problem in a rather different way by making the opposite assumption to Rao's method. In this case it is the variables that are considered to represent the sample from the universe of variables, and it is the observations that are fixed. As a result, it is difficult to conceive of the technique leading to the usual conception of statistical inference since the observations are fixed. Instead, common factors are obtained, so that they have maximum correlations with those in the corresponding universe of variables. This enables one to make generalizations about the common factors existing for the universe from which the variable sample is drawn. The measure of this maximum generalizability is known as Kuder-Richardson's reliability coefficiency or Cronbach's alpha. The reliability, it must be re-emphasized, is the non-error variance contribution composed of unique and specific variance. The actual calculations are quite complex and require a series of interactive procedures. They begin by either using trial values for communality estimates or squared multiple correlations. Like Rao's method, the variables are rescaled by dividing by the standard deviations of the communalities, rather than the unique variance estimate. Alpha factors are retained if the eigenvalues are greater than 1.0, since it would be difficult to assign meanings to axes whose generalizability is less than the contribution of one variable. Again a major advantage of the approach is that the Alpha factors are measurement scale free.

3) Image analysis. A third approach may be noted, although it is now recognized that the term does subsume a wide variety of procedures. Image Analysis, as developed by Guttman (1953) is similar to all common factor approaches in that it attempts to partition the variation in variables into common and specific variance. Unlike Principal Axes common factor analysis, or other matrix adjusted procedures, it provides a theoretical rationale for the common parts of the data. The theory is that if the universe of variables was used in any study, then the communality of any variable would be equivalent to the square of the multiple correlations between that variable and all other variables. Hence the calculations for the technique involve placing the squared multiple correlations of the variables in the diagonal, and then adjusting the off-diagonal correlations to maintain the properties of the matrix described as Gramian, that is, ensuring that they produce determinants equal to zero (see Horst 1965; Harman 1976). From the

adjusted matrix an image matrix is derived – which consists of the portions of each variable that can be predicted from all the remaining variables – and an anti-image matrix which is similar to the regression residuals of the matrix since it consists of the parts that cannot be predicted. The dependence of Image Analysis upon the universe of variables really gives it a great deal of generality. Indeed it has been observed by Cattell (1978) that common factor analysis is really a special case of Image Analysis; the two solutions should be identical when all the variables in a situation can be measured.

Although the image procedure ensures generality, the production of common axes is usually based on the n/2 principle, that is half of the variables in the set. This means that a lot of minor factors often appear in the solution. These are usually discarded before any rotation is carried out, but the reasons for this decision are rarely justified. The obvious limitation of the technique is whether the variable set represents a good estimate of the universe of variables. A useful variant of Image Analysis has been introduced by Harris (197), one that makes the results scale-free. The basis of the procedure is to apply proportionality constraints to ensure that the same average factors will be found, irrespective of whether the correlation or co-variance matrices are used, or even whether data is standardized.

(b) Choice among techniques

Given the variety between these very different factor procedures the inevitable problem facing investigators is that of determining the most appropriate technique for their problem. In view of the variety of alternative problems that can be dealt with, there cannot be any single best solution. After all, the techniques depend upon quite different procedures and make very different assumptions that may be appropriate only for particular circumstances. In any case, recent reviews by applied statisticians point to the primary need for 'meaningful interpretations' in the choice among alternative approaches (Harman 1976 pp.4–5; Cattell 1978). Yet even if the decision is made to choose one method over another, the ever present question of the extent of technical dependence of the results must still be faced. In other words to what extent is the set of results a product of the particular technique used? If the results vary by technique, then a really convincing case for the choice of one technique over another must be made. This means that the investigator carries a great deal of responsibility in finding out, and demonstrating whether his results are technique dependent. All too frequently many of the published articles in the field have simply chosen one approach, so the utility of the results cannot be conclusively determined.

The general problem of invariance of factor methods is one that still produces a great deal of debate in those academic fields with much greater experience of the factorial approach (Velicer 1974 and 1977). In geography and human ecology less attention has traditionally been paid to the issue, despite Berry's (1971) early warning of the critical nature of the problem. Exceptions are provided by Davies (1973), Giggs and Mather (1975), Hunter and Latif (1973), Conway and Haynes (1977) and Davies (1978). In the case of the middle three of these papers, few substantive differences in the results obtained from different factoring techniques were identified for the major axes, apart from variable inter-factor correlations. These confirmed the results of Velicer (1977), and contradicted previous reviews by Rees (1972) in urban ecology and Thorndike (1970) in applied statistics. Part of the differences in the results may be due to the inherent structure of the data set as well as the limited number of factor axes identified by these authors, in the sense that only variables linked to the most important sources of intra-urban differentiation were used.

Table 5.3
Variance explanation of 14 variable data set, Calgary 1971

Vari-able	Com-ponent Commun-alities	Squared Multiple Correl-ations	Final Communalities of 7-axis Common Factor Solutions						
			Principal Axes	Likeli-hood	Minres	Alpha	Rao	Image
1	95	94	99* (93.1)†	100	100	100	96	92
2	87	72	65 (46.8)	69	63	60	91	65
3	94	94	97 (91.2)	100	97	96	95	91
4	87	79	77 (60.8)	89	78	81	93	72
5	94	79	91 (71.9)	100	100	100	90	77
6	91	77	85 (65.5)	93	91	80	90	71
7	95	77	82 (63.1)	77	76	67	86	74
8	90	74	78 (57.7)	80	75	82	80	69
9	94	83	88 (73.0)	82	84	78	91	79
10	87	71	72 (51.1)	70	71	71	74	70
11	97	52	63 (32.8)	48	100	83	38	43
12	89	79	97 (76.6)	100	95	86	92	72
13	96	49	59 (28.9)	39	52	68	41	42
14	95	86	95 (81.7)	100	100	100	93	84
Eigen-value	12.91	10.66	11.48 (8.96)	(9.04)	(9.02)	(8.90)	(9.04)	(7.87)
Variance explan-ation	92.2%	76.14	82.0% (63.9)	(64.6)	(64.4)	(63.6)	(64.6)	(56.2)

*Decimal points are suppressed. Hence 99 represents a value of 0.99. These values represent the amount of the common portion of each variable that is explained. Hence 99 is 99% of the common variance of variable 1 which was previously calculated by the squared multiple correlation as 94% of the original variance. The component and squared multiple corelation values are percentages of the original variance.

†Figures in parentheses represent the percentage explanation of the original variance of each variable in the principal axes solution, e.g., 99% of 94 = 93.1%.

Source: Davies (1978 p.235)

Davies (1978) explored the problem of invariance in factorial ecology, using a set of 14 indicator variables summarizing axes obtained from a preceding 60 variable analysis of 83 community areas in Calgary based on 1971 census data. The communalities from these results are shown in Table 5.3. It was noted that the Principal Axes (common factor and component), Maximum Likelihood, Minres, Alpha, Rao and Image results are bound to vary to some extent since the amount of variance accommodated in the various solutions varied. Table 5.3 shows that the component method (which makes no assumption of the difference between common and unique variance) explained 92.2% of the variability of the data set, whereas the sum of the squared multiple correlations used in the common factor methods produces a maximum of 76.1% of the original variance. The various common factor methods, therefore, will only account for a certain proportion of this variance, depending upon the efficiency of their procedures. Not surprisingly, perhaps, the Minres (minimum residuals) method produces the largest explanation of 65%, although this is not much greater than the Principal Axes value of 63.9%. Image Analysis proved to be the least efficient estimator of common variance among the methods that were described.

Of more importance than the total variance explanation is the amount accounted for by individual variables. These values vary considerably, with the greatest variation shown in the ethnic variables: from 97% (Canadian Ethnic) and 96% (Ukrainian) in the component approach, to 52% and 49% when a standard communality estimate such as squared multiple correlations was used as the measure of common variance. From this basis some of the common factor methods only explained half this amount - the values being particularly small in the Image Analysis and in Rao's method, meaning that the level of explanation of these variables is less than a third of their original value. Since ethnicity is often highly localized in cities, such specific patterns are clearly not going to be accounted for by such generalizing procedures as common factor analysis. Indeed these results point to the continued utility of the component approach and the Principal Axes method in descriptive analysis by maximizing the variance of all variables in the study, not just the common portion of the variables.

Table 5.4 shows the distribution of loadings of each of the axes according to the various methods. The variation in the order of extraction confirms Giggs and Mather's point (1975) that this will vary by method. In general, the results from the various methods are very similar, although they vary more than in previous work by Giggs and Mather (1975) or Hunter and Latif (1973), but in both these cases fewer axes of differentiation were dealt with. The largest vector is an Age and Family axis. Despite its overall consistency it does contain variables linked to non-affluent status on the old age side of the vector in the Image solution. Three of the other factors, Socio-Economic Status, Migrant Status, and Young Adult-Participation are also very consistent, whilst the British-Ukrainian distinction is found on all methods, except the Image solution where there is only a marginal linkage to the Ukrainian variable. Rather more complex is the construct associated with the variables measuring Substandard Housing and Unemployment. This factor is found in six of the seven methods, albeit with medium loadings for the unemployment variable in the Maximum Likelihood, Minres and Generalized Least Squares solutions. Again, the Image Analysis only barely displays this separate axis, a

143

Table 5.4
Varimax loadings* for various factoring methods: Calgary 1971

Factor Title and Variables	1a Comonent†	1b Principal Axes	1c Common Factor	2 Alpha	3 Minres	4a Maximum Likelihood	4b Maximum Likelihood	5 Generalized Least Squares	6a Rao's Canonical Factor	6b Rao's Canonical Factor	7a Image	7b Image
1. Age and Family												
1. Children††	(1)† -80**	((1)) ((-72))	(1) -84	(1) -84	(1) 85	(1) 86	(1) 7.81	(1) 86	(1) -80	((1)) ((-57))	(1) -80	((1)) ((-76))
3. Old age††	87	((79))	90	91	-91	-91	-5.25	-91	88	((68))	80	((81))
10. Immigrants	89	((93))	78	78	-79	-79	-4.04	-77	84	((68))	80	((71))
5. Substandardness	(45)	((-))	(49)	(49)	(-48)	(-50)	(-2.27)	(-44)	(45)	-	(78)	-
7. Unemployed	(32)	((-))	(39)	(39)	(-42)	(-47)	(-2.09)	(-41)	(37)	-	(70)	-
11. Canadian citizens	-	((-))	-	-	-	-	-	-	(-31)	((-52))	-	-
2. Socio-Economic Status												
6. Large dwellings††	(2) -86	(3) ((88))	(3) -82	(3) -82	(2) -85	(6) -83	(6) -10.13	(6) -86	(2) -82	(7) ((-87))	(2) -74	(3) ((-78))
8. Manufacturing††	89	((-89))	81	80	76	62	2.10	80	85	72	79	((67))
12. British	-71	((54))	-64	(-62)	(-59)	(-35)	(-2.15)	(-61)	(-63)	((-31))	-77	((-40))
13. Ukrainian	-	-	-	-	-	-	-	-	-	-	(42)	-
3. Migrant Status (Community Stability)												
9. Non-migrants††	(3) (94)	(3) ((93))	(2) 90	(3) 90	(3) -88	(4) -89	(4) -9.24	(1) -89	(3) 93	(4) ((99))	(3) 88	(2) ((-84))
14. Recent housing††	-91	((-89))	90	-91	94	95	24.54	95	-88	((-91))	95	((77))
11. Canadian citizens	-	((-))	(33)	(34)	(-30)	(-41)	-	(-41)	(48)	-	(40)	-
4. Young Adult Participation												
2. Young adult ††	(4) 86	(4) ((-88))	(4) 77	(4) 77	(4) 76	(3) 76	(3) -3.62	(5) 74	(4) 81	(3)(6) ((-93))	(4) 77	(4) ((-81))
4. Female participation ††	91	((-91))	86	85	87	92	6.00	95	90	((-86))	83	((-71))
6. Large dwelling	(-32)	((-))	(-32)	(-32)	(-32)	(-31)	(5.90)	(-31)	(-33)	-	(-32)	-
1. Children	(-41)	((-))	(-43)	(-43)	(-43)	(-41)	-	(-40)	(-40)	((-30))	(-42)	-

Table 5.4
(continued)

Factor Title and Variables	Principal Axes 1a Comonent	1b	1c Common Factor	2 Alpha	3 Minres	4a Maximum Likelihood	4b	5 Generalized Least Squares	6a Rao's Canonical Factor	6b	7a Image	7b
6. Ethnic 2-British-Ukrainian Ethnic												
13. Ukrainian ††	(5) 95	(5) ((99))	(5) 71	(5) 70	(6) 67	(1) 61	(1) 0.78	(2) 55	(6) 64	(2) ((65))	(5) 46	(5) ((65))
12. British ††	(-48)	((-43))	(-60)	-66	-69	-88	-6.33	-72	-64	((-73))	(-)	((-41))
8. Manufacturing	-	((-))	-	-	(31)	(57)	-	(30)	-	-	-	-
5. Non-Affluent												
5. Substandardness ††	(6) 84	(7) ((93))	(6) 79	(6) 81	(7) -87	(5) 84	(5) -3.65	(3) 89	(5) 79	(5) ((92))	(7) (28)	(7) ((82))
7. Unemployed ††	85	((96))	74	73	-66	63	-2.68	69	78	((89))	(30)	((86))
3. Old age	-	((-))	-	-	-	-	-	-	(31)	-	-	-
7. Ethnic 1 – Canadian Ethnic												
11. Canadian citizens ††	(7) 93	(6) ((96))	(7) 69	(7) 67	(5) -93	(7) -45	(7) -2.74	(7) 45	-	-	(6) 47	(6) ((55))
12. British	(32)	((35))	(37)	(41)	(-30)	(-24)	(-2.03)	-	-	-	-	-
2. Young adult	-	((-))	-	-	-	-	-	-	-	-	-	-
10. Immigrants ††	-	-	-	-	-	-	-	-	(40)	-	-	-

*Not found with loadings > 0.3.

* Loadings in parentheses are the second or third most important loadings. Loadings in double parentheses are for the oblique solutions. Column 4b for the ML method used the covariance matrix as input. Columns 6b and 7b show the oblique solutions for the Rao and Image results.

† The order of extraction of the axes.

†† Expected loadings > +0.3 on the axes.

** Factor loadings with decimal point removed: -80 represents -0.80.

Source: Davies (1978 p.288)

not surprising result since these indicators had previously been linked to the Age and Family dimension. Even more different are the results for the expected Canadian Ethnic dimension, which is only indexed by a large loading for the Canadian variable in half the methods. Since it is not identified in the Maximum Likelihood, Generalized Least Squares and Image Analysis results, the axis is one which is much more clearly technically dependent upon the method used to process the results.

One further source of variation between the results of the alternative factorial procedures is shown in Table 5.4, namely the effect of using covariance instead of correlation matrices as input to the common factor analysis. Johnston (1973), following Cattell (1966), suggested that the removal of the size effect, by the exclusive use of correlation matrices among factorial ecologists, may be an important source of bias in our understanding of ecological dimensions. Giggs and Mather (1975) observed that these considerations are not important as far as common factoring procedures are concerned, since the Maximum Likelihood, Alpha, Generalized Least Squares, and Image Analysis methods are scale-free; in other words, the same set of factors ought to be found regardless of the particular measurement units used to scale the variables. The issue was explored in Davies's (1978) study. Column 4b of Table 5.4 reports the results of a Maximum Likelihood factor analysis of the covariance matrix calculated for the 14 variables. Since size is an important part of the covariance matrix, the resultant factor loadings do not add up to 1.00, and they vary considerably in size. By discounting the variation in size, and identifying the first, second, and third ranking variables in Table 5.4, a clear impression of the factorial structure can be derived. The results show that the factors produced from the covariance matrix are substantially the same as those derived from the other methods for this data set, with the same spread of second and third ranking loadings on the two largest axis, Age and Family, and Socio-Economic Status. The conclusion from this study is that the size of the variables does not significantly distort the results derived from the data set used in this analysis. Such a conclusion may not, of course, be produced in other data sets, particularly those dealing with flows in which major size differences occur (see Chapter VIII).

Although it must be accepted that there are variations between the different methods, the overall structure of the results demonstrates that the major axes of differentiation are relatively stable for most methods. It is in the minor axes that differences occur. To a very large extent this may remove the fear of many factor ecologists that their results are technique dependent. It must be emphasized again that the difference between component and common factor methods is a vital conceptual distinction, but it as well not to exaggerate its substantive effect on the description of axes – at least as regards the Principal Axes method. Component loadings are always higher since they contain unique and common variance. However, the exclusion of unique variance, by the habitual use of common factor methods, removes an important part of the differentiation found in cities, particularly the specific variation associated with the idiosyncracies of cities. This point may be ignored through a failure to recognize the very different amounts of variance of each variable placed in the diagonal of the similarity matrix by the use of ssquared multiple correlations. Table 5.3 has shown that the difference between the component and common factor results is over 16% of the original variance. More importantly, however, the unique variance is differentially distributed between the variables, particularly those with a localized distribution. In this study the unique variance is high for the ethnic variables, a finding parallelling Rees's Chicago results (1970). Since ethnic composition often varies quite substantially between most cities, the exclusive use of common factor methods will underestimate the degree of ethnic social differentiation in western cities. The amount of common ethnic variance accommodated by common factor models is often too small to cope adequately with the variation in this source of differentiation.

Table 5.4 shows that the use of 'measurement free' solutions, such as Maximum Likelihood, Image Analysis, and Principal Axes covariance factoring, confirmed the structure of the major urban social dimensions. In other words, the results were not dependent upon the scale of measurement for the data set used in this study. So at least in this example one potential source of invariance reported by Johnston (1973) appears unimportant. Nevertheless, it must be admitted that the findings shown in Table 5.4 cannot be considered to be completely definitive because of the difficulty of applying the confirmatory procedures to a data set which is a complete population and in the absence of rigorous analytical matching methods. Yet as the author concluded:

'to base general conclusions upon the results of a single example is obviously presumptuous; quite possibly different areas or data sets may not be as well structured as the ones used here, so that invariance due to technical differences may be more of a problem. It is recommended that urban ecologists should, as a matter of routine, obtain several solutions to confirm the stability of any particular structure rather than depending on a single procedure – especially if the component model is being used. In addition, the debate among applied statisticians about the appropriateness of particular factor models and rotations, especially for measuring the significance of loadings and identifying approximate numbers of axes, should be more closely monitored. Further, one must take into account the restriction of this study to the linear factor analysis approach.' (Davies 1975 p.295)

Nevertheless, if the results of the case study reported in Table 5.4 have any generality, particularly when linked with the conclusions of Giggs and Mather (1975) in urban ecology, and Cattell (1978) in psychology, then it can be suggested that the Principal Axes, Minres, Maximum Likelihood, Generalized Least Squares, and Alpha methods can be recommended as providing distinctive axes. They have few of the complications associated with other factoring procedures, and have some marked advantages over the other techniques. The first method is well known in either the component or common factor form, and provides as good a solution as other methods; the second produces some of the largest loadings by minimizing the residuals; the third and fourth are 'measurement-scale free' and have associated chi-square tests; the fifth does not require any preliminary factor estimates and is specifically designed to deal with a complete population of cases. In terms of disadvantages, the Minres, Maximum Likelihood, and Generalized Least Squares methods appear very sensitive to the redundancies that are invariably found in the large data sets which are frequently used by geographers and geologists. Hence these procedures may be more appropriately used to confirm axes on the basis of a small number of key variables derived from the larger data set. Moreover, Nosal (1977) has shown that the Minres method fails if the starting point of the iterations is not well chosen.

Both Rao's method and Image Analysis are less attractive on the basis of these results because they provide lower variance levels and seem very dependent upon the choice of rotation. Yet Image Analysis permits quick computation, requires fewer initial decisions from the investigator, and is scale-free. This demonstrates that each of the particular methods has its own advantages. However, if one is primarily interested in determining whether there is a stable structure in a data set, and whether any set of factor results is technique-dependent, then the best strategy for investigators may be to derive Principal Axes component and factor solutions and apply oblique rotation, using a sequential process to extract factors. These results should be checked against a scale-free method such as Maximum Likelihood or Image Analysis solutions. Davies (1978) suggested using the Principal Axes and Image solutions in the search for the degree of stability. Velicer (1977) found the greatest difference occurring between the Principal Axes and Maximum Likelihood methods in his test of alternative methods, although it must be emphasized that these differences were far outweighed by the similarities in the solutions. If the experimental results are compared then we can be very confident the factors are not as invariant as a result of variations in the common variance contributions, the choice of measurement scale, or the rotation type. This led to the conclusion that:

'on the basis of this evidence, therefore, the Principal Axes component approach can still be regarded as occupying a valuable, although perhaps not unique place in the methodology of factorial ecology. Any move towards the use of common factor methods must be justified by a specific need to exclude the unique variance of axes of differentiation.' (Davies 1978 p.294)

Such explicit recommendation for the component approach does, of course, conflict with the conventional wisdom among factor analysts in the other social sciences. Cattell's (1978) plea that has already been quoted about the problems of the component approach must be re-emphasized, namely: that they are restricted to a single example with little prospect of generality. Such authoritative opinions must be respected. Nevertheless, the fact remains that little substantive difference in the common factor and component results have been reported in these studies of intra-urban social dimensions. It can also be argued that most geographers and ecologists are as much interested at this stage in their knowledge in summarizing their data sets for a complete population of enumeration areas, as in dealing with inferential problems related to a sample of units. If the latter is the objective, then the common factor approach is mandatory. As regards techniques, however, the continued utility of the Principal Axes method must be stressed. Certainly most applied statisticians will emphasize the superiority of the Maximum Likelihood method, but the flexibility and robustness of the principal axes approach gives it advantages which we should not ignore. Moreover, Cattell (1978) has admitted that the procedure can still be used with profit; at least in the early stages of an investigation. Also Conway and Haynes (1977 p.145) provided an index which measured the extent of data reduction or measure of parsimony among several methods and concluded that:

'Principal Axes factor analysis is superior to other methods in terms of the parsimonious reduction of data matrices.'

Yet despite these opinions, and the finding that many of the factor procedures give very similar results from a data set, it must be emphasized that these conclusions are based on the requirement of a strong structure in the data set. It is, therefore, important for an investigator to justify to other workers the fact that a stable solution has been obtained. This can usually be achieved only by an explicit demonstration of stability by presenting the result of different technical solutions. Only where the problem of an individual study specifically points to the use of a particular procedure is it safe to ignore this general principle.

3. ROTATIONS

The discussion of the principles of rotation in Chapter II showed that once an initial component or factor solution has been produced the individual axes can be moved or rotated to new positions in the reduced multivariate space defined by the factors without any loss of explanation or variance. Since the factor solution defines the number of variables, this rotation does not, of course, change the relative position or configuration of the individual variables. All that happens is that the axes which describe the variables provide an alternative description of these points. Some solutions, such as the Principal Axes technique, are mathematically determinate as each axis abstracted has to account for the highest, second highest, and successive amounts of variance. The 6x6 Calgary example in Table 2.11 demonstrated that this particular approach does produce general axes, with most variables loading on the largest axes. In practice, investigators usually prefer descriptions in which the axes are associated as uniquely and as simply as possible with individual sets of variables so that the variables have high loadings on some axes, zero, or close to zero loadings on the others. Indeed, Thurstone (1935), one of the leaders in the early development of factor analysis, produced a set of 'simple structure' criteria designed to help researchers fulfil this basic objective and these are usually reproduced in the standard factor analysis texts (Harman 1976 p.98). By closely describing the clusters of interdependent variables through individual axes, the expectation is that the researcher is more likely to pick the causal influences in the data which affect the variables in different ways. How, therefore, can the investigator best identify these clusters and solve this problem? Unfortunately, there is no simple answer since the answer to 'best' depends upon the objectives that the researcher must set. Given the multiplicity of alternative positions for any axis in the factor space, it is up to the investigator to choose the solution which fulfils the purpose of the analysis. In the early days of factor analysis, when two or three axes solutions were widespread, simple graphical solutions were used, and machines were built which could reproduce the configuration of variables allowing 'hands' or 'sweeps' to be manually rotated to various positions. As 'n' factor solutions become more widespread, and the need to produce rigorous quantitative solutions to the problem of rotation became apparent, such approaches were dropped in favour of analytical rotations designed to produce determinate solutions, once certain assumptions were made. A large number of alternative rotation types can be identified in the literature. In the blind or Analytical types two very different types can be recognized depending upon whether the axes are rotated in conjunction (so that each vector is orthogonal, at right angles, to all others), or whether the axes are rotated at an angle, the oblique category. Another category of methods, called Procustus or Target Fitting (Brown 1967; Cattell 1978 Chapter 6) because the expected relationships are defined and the results are measured against the standard, but such approaches have been rarely used by factor ecologists to date so are not described here. Orthogonal (Schoenemann 1966) and oblique (Gower 1975) versions of this procedure exist.

(a) Analytical rotations

In the pure world of the statistician rotations may only be appropriate for common factors as Mather (1971a) has emphasized. In practice rotations clarify component solutions just as well (Cattell 1978 p.115).

(i) Orthogonal types. A number of rather different orthogonal rotation schemes have been produced. One of these, the Quartimax, has already been described in Chapter II. Although the Quartimax solution produced a parsimonious solution from the 6x6 variable Calgary example in Table 2.4, the technique does not seem to have been used by geographers and ecologists since in cases without a very strong and stable structure, the technique fails to discriminate betwen factors. Equation 2.12 showed that the fourth powers of the loadings are summed along the rows of the matrix, and the rows are added together to produce a final measure of parsimony. The production of high and low loadings for each variable frequently results in all the high loadings being associated with one or two factors, producing solutions that often approach the generality of the initial component solution. If generality is the objective of the study, the rotation may be useful in clarifying the initial solution. Most factor analysts are interested in disaggregating the generality of the solution in favour of a series of specific effects, so that each factor is associated as uniquely as possible with a distinct set of variables. Kaiser (1958) showed that one way of fulfilling this objective is by maximizing the variance of individual factors, the so-called Varimax method, in which each variable has as high loadings on as few factors as possible.

The computation details of the construction of the Varimax technique may be found elsewhere (Harman 1976 p.290). Here it is enough to observe that the Varimax technique depends upon the calculation of the variance of individual factors; in other words the mean of the sum of the squared deviations from the means:

$$V_p = \frac{1}{n} \sum_{j=1}^{n} (a_{jp} - \bar{a}_{jp})^2 \qquad \text{Eq. 5.5}$$

As it is the square of the factor loading (a_{jp}) that produces a measure of the amount of overall explanation of the variables, in practice the squared loading is used in Equation 5.5. Multiplying out the equation produces:

$$V_p^2 = \frac{1}{n} \sum_{j=1}^{n} (a_{jp}^2)^2 - \frac{1}{n^2} \left(\sum_{j=1}^{n} a_{jp}^2 \right)^2 \qquad \text{Eq. 5.6}$$

When this is applied to the whole series of factors it means multiplying each term by the sum of the factors p from 1 to m.

Eq. 5.7

$$\sum_{p=1}^{m} V_p^2 = \frac{1}{n} \sum_{p=1}^{m} \sum_{j=1}^{n} a_{jp}^4 - \frac{1}{n^2} \sum_{p=1}^{m} \left(\sum_{j=1}^{n} a_{jp}^2 \right)^2$$

In the initial factor solution the communalities of the variable are likely to be different. In other words each variable contributes a different amount of variance to the total explanation. So, Kaiser (1958) suggested dividing each element in the factor matrix by the communality (h_j) of the variable. Rewriting Equation 5.7 and adding Kaiser's normalization produces Equation 5.8. A constant multiplier has no effect on the process of maximization, so the equation is multiplied by n^2. For simplicity the Varimax procedure (V) is written as

Eq. 5.8

$$V = n \sum_{p=1}^{m} \sum_{j=1}^{n} (a_{jp}/h_j)^4 - \sum_{p=1}^{m} \sum_{j=1}^{n} \left(a_{jp}^2/h_j^2 \right)^2$$

Successive applications of the formula produce different values of V, and the iterative routine stops when the maximum value of V is derived. When this point is reached the matrix consists of a set of loadings in which the variance of individual factors has been maximized.

Table 5.2 has already shown that the difference between the Quartimax and Varimax procedures are quite minimal in the 6x6 Calgary example. This is, of course, a function of the structure of the configuration of points. In general, Varimax is a much more satisfactory procedure. Until the mid 1970s Varimax was the only rotation that was used extensively by geographers and ecologists. In most cases the justification for the choice of Varimax is that it represents an invariant solution. Yet there are some misunderstandings about the type of invariance involved. Kaiser (1958) showed that most, though not all, of the same factors could be recognized when different number of variables (in his case, psychological tests) were added to a test data set and the initial factor solution was rotated. The fact that the factors remained fairly constant in experiments was taken as an indication of the stability of the technique. It implied the same basic results could be obtained by different workers using various sets of tests. Although this finding represented a major breakthrough, in data analysis terms, it should not be allowed to obscure the fact that this is a different type of invariance to 'invariance under factor rotation'

151

in which the objective is to find the most stable factor solution from the successive application of different rotation schemes. Moreover, the fact that Varimax can disaggregate the generality of the initial component or Quartimax solution means that it is an inappropriate procedure in those studies that attempt to find one or more factors that provide the most comprehensive or general description of the data set.

   (ii) <u>Oblique rotations</u>. So far the discussion has been phrased purely in terms of rotating the axes or factors as a rigid frame so that each axis is orthogonal to the others. This certainly eases the computational problems, but the orthogonality assumption does represent a major constraint since on a priori grounds one would expect that the axes of differentiation are more likely to be oblique to one another. Also the parsimony principle would be more effectively satisfied if each axis is allowed to rotate independently. A hypothetical example in Figure 5.1 illustrates the issue. In this diagram two sets of points can be recognized intuitively. The most concise

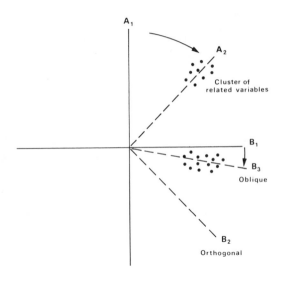

Figure 5.1   Orthogonal and oblique axes

description of the pattern is achieved by allowing the axes to be oblique to one another as in A_2B_3 in order to pass through each set of points. This will ensure that the loadings on each of the axes would be as high as possible, rather than confusing the pattern by adopting an orthogonal framework such as A_2B_2 or A_1B_1. Once this principle of oblique axes is adopted, however, the description of any point in space becomes complicated, and three different types of matrix can be produced. Before dealing with the various types of

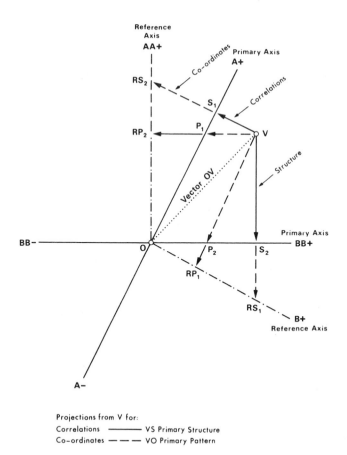

Projections from V for:
Correlations ———— VS Primary Structure
Co-ordinates — — — VO Primary Pattern

Figure 5.2   Reference and Primary axes

oblique solutions, these matrices need to be carefully distinguished by
reference to the case of a single variable V in a two-factor space.

1) Oblique structure and pattern.   In the orthogonal case in Figure
5.2 the description of point V with respect to the two axes was
achieved by projecting coordinates parallel to the axes, namely VA and
VB.   Since the length of OV, the communality, is known, and the triangles are all right angles, the length of the coordinates (OA and OB)
can be used to calculate the angles between S_1OV and S_2OV, that is
OV and the various axes.   These values can be obtained by dropping
perpendiculars from V to each axis to produce the necessary right
angled triangles of OVS_1 and OVS_2.   The result is that two separate
measurements are needed on each axis; for instance, OP_1 and OS_1 in
the case of the first primary axis.   The coordinates OP_1 and OP_2
measure the loading (or so-called saturation) of a variable, since they
measure the position of the variable in space.   In other words, they
provide an indication of the amount of variance of a variable uniquely

associated with each factor. These saturations comprise the primary pattern matrix and are regarded as performing an equivalent role to regression coefficients in a multiple regression equation. Values derived from the perpendiculars, OS_1 and OS_2 represent the correlations between the variable and each axis and constitute the primary structure matrix.

In orthogonal solutions the perpendiculars and coordinates will be identical. As axes become oblique differences occur between the measurements of saturations and correlations; they will be partially influenced by the correlations between the factor axes, so they are not convenient measures of the saturation of a variable with a factor. As correlations, they cannot exceed +1.00, although very oblique axes can produce values of over 1.00 in the primary pattern matrix.

2) Oblique reference or primary axes. The second type of rotation matrix stems from the fact that two different kinds of descriptive oblique axes have been used by factor analysts. The first set of axes are the type which could be drawn in directly on a graphical solution, and are called primary axes. The second set may be regarded as a series of axes, each of which is at right angles to a primary axis. These reference axes (Figure 5.2) were originally devised to ease the computational difficulties of oblique solutions in a pre-computer age. Today they seem to represent mathematical abstractions, but it might be observed that their properties still appear to be more well-known than the primary axes, and at this stage in our knowledge need not necessarily be rejected out of hand.

The possibility of confusion is further enhanced by the fact that the terminology of the reference axes solution is the reverse of the primary axes. Thus, the reference structure provides a measure of the coordinates or saturations of the variables and factors, whereas reference pattern produces the correlations between variable and coordinates. The geometrical reason for this reversal can be seen in Figure 5.2. VRS_2 is parallel to the reference axis AA. It measures the reference structure (the coordinate of the variable and factor) but because it cuts the primary axis A as a perpendicular, VS_1, it measures the correlation on this axis, i.e., the primary structure.

3) Oblique factor correlation matrix. The third matrix lists the correlations between the individual factor axes, and is produced by the relaxation of the orthogonality assumption. It should be noted that if orthogonality characterizes the factor axes, then the correlations between the axes will be zero. Hence all the orthogonal solutions should be regarded as subsets of the oblique rotation procedures.

The results of this survey shows that a complete analysis for either a reference or primary axis solution should provide: a factor correlation matrix, a structure matrix, and a pattern matrix is usually sufficient for most investigations. Even though the primary and reference axes are oblique to one another, the principle of parsimony already established in the orthogonal case can be used as the basic measurement. The position of variable V in the primary axis space of Figure 5.2 will be fixed by the coordinates OP_1 and OP_2 and these measurements will be found in the primary factor pattern matrix. In the case of the reference axis space the position will be fixed by the coordinates ORS$_1$ and ORS$_2$ and the analagous measurements will be found in the reference structure matrix. Working by analogy from the Quartimax solution Carroll (1957) proposed one of the first oblique solutions, the Quartimin method. Using Equation 2.12 as a basis, in which Q + 2M equalled a constant, Carroll used the minimization of m as the basis of his formula. Instead of maximizing the fourth power of the loading, as in the Quartimax orthogonal approach, the sum of the squares of the cross products of the factor loadings was chosen as the

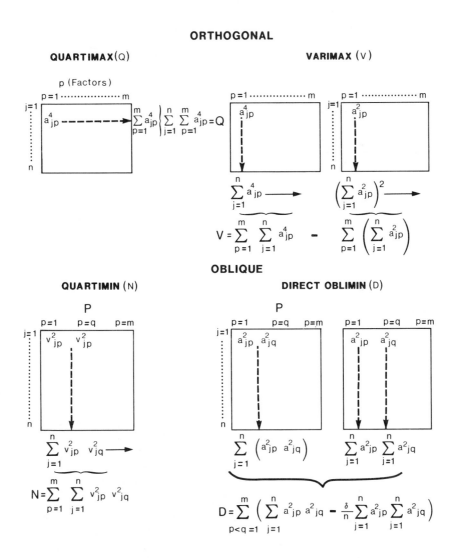

Figure 5.3 Computational bases for the rotation equations

appropriate relationship to deal with (Figure 5.3c). Symbolically, this is translated into the following equation:

$$N = \sum_{j=1}^{n} \sum_{p<q=1}^{m} a^2_{jp} \cdot a^2_{jq} \qquad \text{Eq. 5.9}$$

In computational terms each axis may be envisaged as being rotated separately, so that the values of one factor (or column of the matrix) are altered whilst the others remain fixed. It is apparent that the labour involved in these calculations ensures that such solutions are only feasible with advanced computers. For those unfamiliar with algebraic notation, Table 5.3c shows the matrix calculations required to produce the results compared with the orthogonal cases shown earlier.

It is probable that the analogy with the Varimax solution led to the oblique variant of the technique. Called the Covarimin technique, it involved a minimization procedure, using the difference between the sum of the cross-products of the squared loadings and the sum of the squared loadings. The relevant formula is:

$$\text{Eq. 5.10}$$

$$C = \sum_{p<q=1}^{m} \left(n \sum_{j=1}^{n} a^2_{jp} a^2_{jq} - \sum_{j=1}^{n} a^2_{jp} \sum_{j=1}^{n} a^2_{jq} \right)$$

where: v^2_{jp} is the squared reference loading structure of the jth variable on the pth factor.

Recent work has shown that the two oblique solutions are not independent solutions but are part of a broad <u>family of oblique solutions</u> (Harman 1976 p.311). As all these <u>oblique</u> rotations involve <u>minimization</u> in the iteration procedure, they are called 'oblimin' solutions. The relevant formula uses a variable parameter y and derives normalized loadings by dividing the factor loading by the communality of the variable (h_j):

$$O = \sum_{p<q=1}^{m} \left[n \sum_{j=1}^{n} (a^2_{jp}/h_j^2)(a^2_{jq}/h_j^2) - \right.$$

$$\text{Eq. 5.11}$$

$$\left. y \sum_{j=1}^{n} (a^2_{jp}/h_j^2) \sum_{j=1}^{n} (a^2_{jq}/h_j^2) \right]$$

when: y = 0.0 a solution equivalent to the quartimin is derived;
y = 1.0 the solution is equivalent to the covarimin.

Several authorities have shown that quartimin methods constantly produced factor axes that are highly correlated, whereas the covarimin solutions produce a bias towards orthogonality. A compromise solution, using y = 0.5, has been the most widely used oblimin example. Not surprisingly this was called the Bi-Quartimin technique (Carroll 1957).

This generalization of the original oblimin functions to a whole range of solutions using the arbitrary parameters, provides additional flexibility in the rotation analysis. In recent years, factor analysts have become impatient with the rather cumbersome reference axes approach. They have produced a series of additional techniques that are applied directly to the original loadings, rather than transforming these to the reference axes. The most popular of these schemes are the Direct Oblimin and Promax techniques, both of which have flexible parameters that enable the research worker to produce a series of rotation solutions.

The Direct Oblimin technique developed by Jennrich and Sampson (1966) seems to have achieved the greatest popularity initially because of its availability in the University of California (Los Angeles) Biomedial package programs (Dixon 1968) and more recently in the SPSS (statistical package for the social sciences, Nie et al 1975). The formula for the factor pattern (FP) without normalization by the communalities is shown in Figure 5.3 and is written symbolically as:

$$D = \sum_{p<q=1}^{m} \left(\sum_{j=1}^{n} a^2_{jp} \cdot a^2_{jq} - \frac{\delta}{n} \sum_{j=1}^{n} a^2_{jp} \cdot \sum_{j=1}^{n} a^2_{jq} \right)$$

Eq. 5.12

where: a^2_{jp} is the squared factor loadings of the jth variable on the pth factor;

$a^2_{jp} \cdot a^2_{jq}$ is the cross-product of the squared factor loadings;

δ is the variable parameter.

As the name implies, iterations continue until D, the factor pattern of the oblique primary axes, is minimized. The appearance of the parameter in the formula ensures that a series of different solutions are possible with different values of δ. In practice it must be a negative quantity, with values between 0.0 and -1.0 being favoured. Otherwise the iterations do not produce convergence, although experiments have revealed that it was possible to produce satisfactory rotations with a parameter up to +0.4. Figure 5.3 shows how the formulae can be calculated from the factor loading matrix.

Apart from these basic rotations, a number of other blind analytical methods have been invented and reviewed. Factor analysts such as Cureton and Mulaik (1975) have compared results from weighted Varimax and Promax solutions whilst Gorsuch (1970) compared four oblique and orthogonal measures. Hakistian and Abell (1974) have also compared many of the methods and concluded that

'no single computing procedure - general paradigm or specialization ... can be expected to yield uniformly optimal oblique solutions for all kinds of data.' (Hakistian and Abell 1974 p.444)

This means that factor ecologists should, as a matter of routine compare the results from several different rotations. It is useful to check the difference between Varimax and Direct Oblimin, to explore the effect of using 0.0, -0.5 and -1.0 in the latter in relation to the degree of obliqueness and size of loadings, as well as evaluating the effect of higher order solutions.

(b) Choice among solutions

Understanding the differences between the major types of rotation schemes does not, of course, solve the basic problem facing any investigator, namely: which type of rotation should be chosen. This is a familiar problem. In the same way that the multiplicity of objectives among factor analysts could lead to the choice of different factor techniques, so the requirements of individual investigations can lead to the choice of different rotations. In other words, there is again no single 'best' solution. The choice depends upon the objectives of the analysis. However, there are some basic considerations which are useful in guiding an investigator to a choice between the various types of rotation.

Table 5.5
French regions: direct variance associated with factors

Solution	Direct Variance on Each Factor					Total
	1	2	3	4	5	
Principal Axes: Unrotated Component	54.0	14.0	8.9	5.5	4.8	87.8%
Rotated:						
(i) Orthogonal - Quartimax	53.5	11.9	11.4	6.2	4.9	87.8%
Varimax	50.2	11.4	14.3	6.9	5.1	87.8%
(ii) Oblique - Direct Oblimin (0.0)	49.3	11.2	12.2	6.7	5.1	84.4%

The first issue is whether rotation should be used at all. In some circumstances the structure of the variables in the multi-factor space of the axes may be such that rotation does not add a great deal to the interpretation of the results. Table 5.5 provides an example of this issue by means of a case study of a 33 variable x 22 area data set of socio-economic indicators for French planning regions obtained from Hansen (1968) that are described in detail in Chapter VII. It can be sen that the first factor (which will be shown to measure the importance of the Paris region in Chapter VII) is so dominant, and is such a characteristic of the data set, that rotation had a marginal effect upon its importance. Indeed, the differences in the variance explanation of the oblique solution are primarily a consequence of the fact that the table only shows the direct variance between each variable and factor. In oblique solutions the rest of the variance is accounted for by the joint variance (Harman 1976 p.269) which is linked to the degree of correlation between any pair of factor axes (p and q). This is usually obtained from:

$$J_{pq} = 2c_{pq} \, T_p T_q \sum_{j=1}^{n} a_{jp} \cdot a_{jq} \qquad \text{Eq. 5.13}$$

where: a_{jp} is the jth loading of the primary pattern of the reference structure and c_{pq} is the correlation between factor p and q.

However, the results in Table 5.5 cannot be assumed to be a general one; it is a product of the particular structure of the data set. Normally the effect of rotation is to disaggregate the largest unrotated component and to redistribute the variance elsewhere in accordance with the simple structure principles. A good example is provided by the 6x6 Calgary example (Table 5.2). The effect of rotation was to replace a general first axis by two specific vectors linked almost exclusively to two sets of variables; associated with socio-economic and family considerations respectively. In substantive terms rotation certainly eased the description of the two axes.

It is often claimed that factor analysis should always rotate the results from their preliminary solution. In general, this is probably a sensible procedure, since it does clarify the relationships between the variable clusters. Nevertheless, there are circumstances in which unrotated solutions may be the most appropriate structures to interpret, as for example when investigators may wish to have as much variance as possible associated with the first axis. An area in which this is a useful principle is in the scaling of central places, where Berry (1968) and many others have used factor analysis to reduce a matrix of commercial functions over n places to a more concise description. Unfortunately, the application of factor methods to this type of problem has not been without errors. For example, in Abiodun's (1968) study of Ijebu province in Nigeria (Berry and Horton 1971) the first factor was shown to account for a total of 52.7% of the original variance in the data set. Yet the author specifically pointed out that:

'the lack of any strong loading for any of the primary variables on the first Principal Component makes it difficult to label.' (Abiodun 1970 p.184)

Such a statement is not compatible with the variance figures so a replication of the study from the correlation matrix stage was carried out using different types of rotation (Table 5.6). The results of the unrotated solution showed identical variance amounts, but demonstrated that over 19 variables had loadings over ± 0.5. Scrutiny of Abiodun's diagrams and tables revealed that the errors probably came from interpreting a normalized matrix of factor loadings, not the actual

159

Table 5.6
Nigerian hierarchies:
direct variance associated with the factors

Principal Axes: Component Solutions		Vectors					Number of Axes $>$ +0.25 Factor Correlations	Variance Total (5 Factors)
		1	2	3	4	5		
a) Abiodun		52.7	15.3	8.7	5.4	4.3		86.5%
b) Replication		52.7	15.3	8.6	5.5	4.4		86.5%
Orthogonal								
Quartimax		36.9	19.6	11.8	5.1	3.9		77.2%
Varimax		30.6	23.9	13.1	5.4	3.9		77.2%
Oblique								
1. Reference Structure (y) Oblimin:								
Quartimin	(0.0)	34.8	33.4	14.8	5.5	4.0	3	92.5%*
Biquartimin	(0.5)	29.6	19.8	11.5	5.2	3.9	2	69.9%
Covarimin	(1.0)	35.5	35.1	15.65	5.9	4.1	0	96.2%*
2. Primary Pattern (δ) Direct Oblimin:	(0.0)	28.7	20.4	11.6	5.3	4.0	2	69.9%
	(−0.5)	28.0	20.2	11.9	5.4	3.9	2	69.4%
	(−1.0)	27.5	20.1	12.1	5.5	3.9	2	69.1%
	(−2.0)	26.3	18.7	12.3	6.7	4.0	2	68.0%
	(−3.0)	25.1	17.2	12.1	8.5	4.0	4	66.9%

* Loadings $>$ 1.0 appear in the structure matrix so meaningful eigenvalues cannot be derived. Calculations by author.

Table 5.7
Factorial interpretation of British overspill schemes

(a) Correlations*

		Aldridge			Macclesfield	
		I	II	III	I	II
Aldridge	I	100	79	72	75	63
	II	79	100	57	60	53
	III	71	57	100	50	49
Macclesfield	I	75	60	50	100	75
	II	63	53	49	75	100

(b) Two factor solutions: Principal Axes technique

		Components		Communality	Varimax Rotation	
		I	II		I	II
Aldridge	I	93*	−15	88	78	52
	II	83	−22	74	76	40
	III	77	−45	79	87	20
Macclesfield	I	86	36	87	38	85
	II	81	46	86	27	88
Explanation		70.8%	12.5%			

* Correlations) have decimal points removed
  so 100 is 1.00; 93 is 0.93.

(c) Common factor solutions

Vari-ables	Commun-alities	Unrotated			Varimax			Promax		
		I	II	III	I	II	III	I	II	III
1	0.83	90	−04	12	52	−43	−61	27	−20	−50
2	0.86	82	−04	−44	33	−26	−83	−11	10	−107
3	1.00	79	−56	24	26	−92	−29	−05	−108	08
4	0.96	87	42	21	92	−19	−30	111	09	09
5	0.59	73	21	13	66	−25	−29	72	06	−0

(d) Correlations between promax factors

Axes	Correlations		
	I	II	III
I	1.00		
II	−0.64	1.00	
III	−0.78	0.73	1.00

(e) Higher order solutions

Axes	Eigenvalues	% Variance	Primary Axes		
			I	II	III
I	2.43	81.1%	90	37	−25
II	0.36	12.1%	87	47	−13
III	0.21	6.8%	93	09	36

Source: Davies (1971) for (a)(b);
        Mather (1971) for (c)(d) and (e)

161

values. Table 5.6 also shows the way in which the application of rotation disaggregates the first factor and redistributes the variance over the other vectors. If, as is usually the case, the objective of factor analysis in this type of study is to produce the most general scaling of the centres by the factor scores, the unrotated solution would appear to be the most appropriate solution.

Another example of the same problem can be seen in Table 5.7. The data set consisted of three English overspill communities in Aldridge and two in Macclesfield (Davies FMT 1971) in which 59 indicators of community characteristics were used as the basis of similarity between the different aged estates described in order of age as: I or newer and II older. The correlations (Davies 1971) show that in general the greatest similarity is between schemes in the same town, but the newest estates were also very similar. A Principal Axes component analysis of the data confirmed there is an overall measure of similarity, since 70.8 percent of the variation could be accommodated on one axis. However, if the data is subjected to a Varimax rotation, two components are extracted, each of which are associated with individual towns. Davies (1971) attempted to bring the problem of automatic and un-critical use of Varimax to the fore. He observed that the superficial result was to destroy the overall similarity of the schemes, at the expense of the differentiation based on individual towns.

An interesting interchange between Mather (1971a, 1971b) and Davies (1971b, 1971c) on the interpretation to be placed on this set of results demonstrates the problems facing investigators in the field. In retrospect, the initial issue of whether component and common factor methods were confused proved to be of little substantive importance since Davies (1971b) admitted the term 'factor' was used in its general sense. He showed the common factor results led to the same possibility of confusing an unrotated solution displaying generality, with rotation producing specific common factors (or components) (Table 5.7c). More important was Mather's demonstrtion that the two or three first order factors were obliquely related to one other, leading to a single second order axis. For Mather (1971b) the interpretation problem was really one of scale. This really extended the purpose of the initial paper by confirming Davies's (1971a) initial point that there was a problem of potential misinterpretation when either an unrotated component solution or a Varimax solution were uncritically used, the standard procedures of researchers at that time. The important issue of the interchange was the demonstration of the need to routinely scrutinize several solutions at different orders to avoid mistakes in interpretation.

In many ways the most fundamental decision faced by many investiga-tors is the choice between orthogonal and oblique solutions. In some of the examples used so far (especially the 6x6 Calgary example in Table 5.2) the differences in the substantive interpretation of the loadings are relatively minor. In other words, the structure of the data is such that orthogonal relationships are found. Such evidence cannot be used to justify the use of orthogonal solutions in general. On a priori grounds, it might be expected that most axes of differ-entiation are, in fact, more likely to be obliquely related to one another. An advantage of the use of oblique rotations, therefore, is that the approach does not impose a particular structure on a data set. If orthogonality is present it will emerge from the oblique analysis.

The reverse is not true. If obliqueness among axes is present, it will not be identified by orthogonal rotation. In addition, of course, the use of oblique rotations produces a factor correlation matrix, which can itself be factor analyzed to produce second or higher order factors (Thurstone 1947), an approach providing even more general conclusions, since the axes will be themselves related.

This preference for oblique rotations should, however, be tempered with caution. Geographers and ecologists have had much more experience with Varimax rotation. So it is always worth comparing the oblique results with those obtained from the Varimax procedure. In part, this recommendation is based on the fact that Varimax is a simple procedure. But, in addition, it must be noted that most of the generalizations in the literature were obtained from analyses that used Varimax, so the results are made more comparable for a wider body of literature. In this way, the extent of deviation of the oblique results from the imposed orthogonality of the Varimax method can be measured. Given the fact that the differences between the Varimax and Oblique solutions are dependent upon the structure of any data set, it is difficult to produce definitive conclusions about the bias in our results, consequent upon the allegiance to Varimax rotation. What is comforting for the creation of a body of a substantive literature is that those geographical and ecological studies that have compared Varimax with Oblique solutions (Hughes and Carey 1972; Davies and Lewis 1974; Davies 1978) concluded that there are comparatively few substantive differences in the pattern of loadings – at least for the major axes of differentiation. Usually the Varimax solutions are more general, in that minor loadings are found on the axes. However, Davies (1978) showed that there were variations betwen the oblique and orthogonal versions of the Rao and Image Analysis results in a comparison of seven different factoring methods, leading to the conclusion that: 'for certain factor methods the choice of rotation is important.' (Davies 1978 p.290)

So far, this discussion has downplayed the problems caused by the possibility of variation among the oblique rotation schemes. At this stage it must be admitted that the experience of factor ecologists with the range of alternative oblique rotations is very limited. What is apparent is that some of the reference structure solutions, namely Quartimin and Covarimin, can no longer be recommended, since both have displayed consistent biases – the former towards extreme obliqueness, the latter towards orthogonality. The Biquartimin, however, produces results that seem comparable (if occasionally slightly orthogonal according to Hakstian and Abell (1974)) with the Direct Oblimin using the parameter of $\delta = 0.0$. The latter is normally considered to be the standard value. However, Table 5.6, the case study of the Nigerian central places, showed that the Direct Oblimin axes become more oblique once the negative values for this parameter are increased, whilst tests showed it was difficult to derive a solution for this data set when the parameter was increased above +0.4. Bailey and Guertin (1970) have documented the advantages of the Direct Oblimin approach in the applied statistics literature, and the method has been adopted by one of the major standard package programme series, SPSS (Nie, Brent et al 1975). It is a mark of the lack of precision in parts of the series that the method is only described as an 'oblique' rotation. Yet this is likely to lead to the continued dominance of the Direct Oblimin approach among geographers and human ecologists, although it is worth noting that the Promax procedure also has its adherents (Mather 1976; Perle 1977). The next is that is is a moot point whether the Direct Oblimin (with δ = 0.0) or Promax (with the use of the fourth power) will be recognized as the standard oblique solution, or whether this role will be taken over by newly developed oblique rotations. What is important for any investigator is the need to demonstrate the extent to which the results obtained from these analytical or blind rotation technique are relatively invariant, and represent a good fit to the pattern of points in a multivariate space. Alternatively the Procrustes transformation (Cattell 1978) could be applied but this demands some hypothesis of what structure is expected. In this context the fact that more or less oblique results can be obtained by varying the parameter values may be noted. Indeed investigators have been encouraged in the basic factor analytical texts (Harman 1976; Cattell 1978) to routinely try several methods or different parameter values to uncover the best interpretation of the data:

'in the last analysis the use of the methods should view them only as aids towards his scientific objectives: if any procedure fails to produce a meaningful result, discard it and try something else; if a particular method yields a meaningful solution, retain it.' (Harman 1976 p.335)

Such advice should not be viewed as a carte blanche for an uncritical use of procedures. The final decision to use a particular method should always be fully justified and tested against other results, a relatively routine matter now that high speed computers make the analysis part of an investigation relatively easy. In this context, however, it is also imperative for an investigator to go beyond the first order results and look at the second or higher order relationships between the factors.

4. HIGHER ORDER SOLUTIONS

One of the major disadvantages of the use of orthogonal rotations is the fact that orthogonality is imposed upon the axes. In geometrical terms it means the factors are located at right angles to one another in the multi-factor space. Intuitively, however, it seems much more logical to assume that the axes of differentiation are correlated with one another, unless we have evidence to the contrary. This implies that the orthogonal model is restrictive. As Cattell (1968) expressed the point, the assumption of oblique factors:

'rests on the general scientific proposition that in an interacting, unsegregated universe most influences will tend to show some correlation ... If we insist upon entities which are statistically uncorrelated they may well be conceptually contaminated.' (Cattell 1968 p.223)

Fortunately, the rapid development of oblique rotation methods in the last decade have allowed investigators to easily test these general impressions about inter-factor correlation, although studies in geography have been dominated by the Varimax model. Investigations of the extent of correlation of one axes with another, however, is only

164

part of the utility of oblique methods. By factor analyzing the factor correlation matrix, it is possible to derive even more succinct summaries of the relationships in a data set. The first order factors are themselves generalized into second order factors, whilst if these axes are still oblique, third or even higher order axes can be produced. The process usually stops when the factors prove to be orthogonal to one another. Most of the traditional 'rules of thumb' can be applied to the 'number of axes' problem at each scale of generalization, although the author has found that a useful - if subjective - minimum stopping criterion is to extract factors up to the stage at which 50% of the variance of each lower order axis is accounted for by the higher order model.

Higher order analyses are well-known in psychology and go back to Thurstone's (1944) pioneering work, but they were rarely pursued in the geographical-ecological area until the last decade. Now the potential has been demonstrated in intra-urban studies (Davies and Lewis 1973; Davies 1975, 1977; Perle 1977), inter-urban or regional classifications (Davies and Welling 1977; Palmer et al 1977) and particularly in studies of functional regions (Davies and Musson 1978; Davies 1980). In this latter case a valuable feature of the method lies in the identification of overlapping and successive levels of generalization. Despite these and other examples, this comparative lack of attention to the higher order possibilities is unfortunate, since Rummel (1970) and Cattell (1978) both point out that it is likely that many of the major theories and relationships identified in the social sciences relate to second or higher order rather than to first order links. For example, Marx's concept of the primacy of the economic system may be more clearly seen at a second order, since this would allow first order political and social relationships. In psychology, Pawlik and Cattell (1964) have illustrated how the Freudian ideas of Id, Ego and Superego could be identified at the third order from a variety of first and second order axes produced from objective personality tests. In the field of urban social geography Davies and Lewis's (1973) identification of the three second order axes shown in Table 5.8 (namely Social Status, Family Status and Ethnicity-Migrant) suggest close parallels with the Shevky-Bell (1955) conception of three axes of differentiation. This means that the Shevky-Bell model may be more relevant to the second rather than the first-order of relationships, although the overlap between the axes, and the presence of additional higher order axes in other countries (Davies 1975) produce modifications of the simple ideas that are explained in Chapter I.

Figure 5.4 shows one of the possible sets of relationships between first and second orders that could be identified. This type is best described as a stratum or hierarchical model with slightly overlapping links between the first and second orders in place of what could be called a monarchical hierarchical system with unique linkages between each low and higher order factor. Cattell (1978) as well as Rummel (1970) have attempted to summarize a suite of possible relationships - the former using eight models, the latter four categories. Cattell's range seems the more comprehensive and is shown in Figure 5.5. Within this set of types a broad distinction can be drawn between the stratum or hierarchical type, with a series of layers each linked to lower order axes by a variety of alternative linkages, and the recticular type where all factors are linked together. At this stage in the

(a) Successive levels of generalization

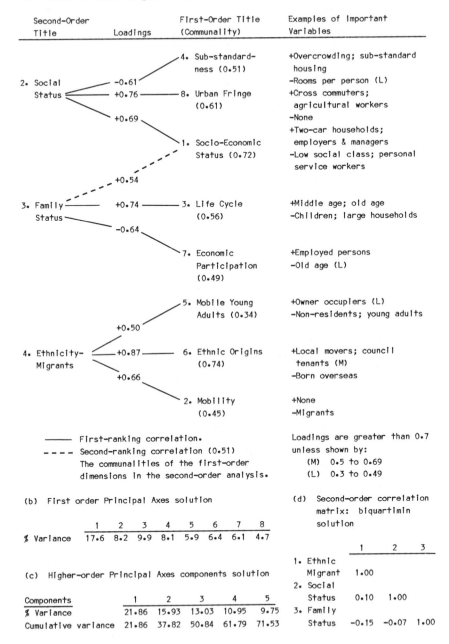

Second-Order Title	Loadings	First-Order Title (Communality)	Examples of Important Variables
		4. Sub-standard-ness (0.51)	+Overcrowding; sub-standard housing
2. Social Status	−0.61		−Rooms per person (L)
	+0.76	8. Urban Fringe (0.61)	+Cross commuters; agricultural workers
	+0.69		−None
		1. Socio-Economic Status (0.72)	+Two-car households; employers & managers −Low social class; personal service workers
3. Family Status	+0.54 +0.74	3. Life Cycle (0.56)	+Middle age; old age −Children; large households
	−0.64	7. Economic Participation (0.49)	+Employed persons −Old age (L)
		5. Mobile Young Adults (0.34)	+Owner occupiers (L) −Non-residents; young adults
4. Ethnicity-Migrants	+0.50 +0.87 +0.66	6. Ethnic Origins (0.74)	+Local movers; council tenants (M) −Born overseas
		2. Mobility (0.45)	+None −Migrants

——— First-ranking correlation.
- - - - Second-ranking correlation (0.51)
The communalities of the first-order
dimensions in the second-order analysis.

Loadings are greater than 0.7
unless shown by:
    (M) 0.5 to 0.69
    (L) 0.3 to 0.49

(b) First order Principal Axes solution

	1	2	3	4	5	6	7	8
% Variance	17.6	8.2	9.9	8.1	5.9	6.4	6.1	4.7

(c) Higher-order Principal Axes components solution

Components	1	2	3	4	5
% Variance	21.86	15.93	13.03	10.95	9.75
Cumulative variance	21.86	37.82	50.84	61.79	71.53

(d) Second-order correlation matrix: biquartimin solution

	1	2	3
1. Ethnic Migrant	1.00		
2. Social Status	0.10	1.00	
3. Family Status	−0.15	−0.07	1.00

Figure 5.4   Association between the social dimensions of Leicester
Source: Davies and Lewis (1973)

166

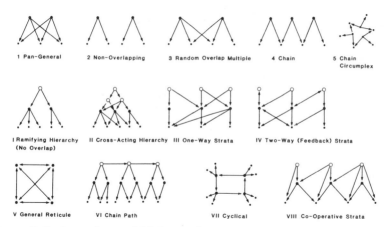

Figure 5.5 A typology of higher order types.
Types 1 to 5 show linkages of variables with primary axes;
Types I to VIII show the variables with all higher order
factors. Source: Cattell (1978 p.200)

development of higher order relationship in factorial ecology studies
it is difficult to know which is the most appropriate type, although
the evidence so far suggests that the hierarchical model with overlapping linkages is the most common type of intra-urban relationship.

The most comprehensive summary of the potentialities of higher order
analysis to data has been provided by Cattell (1978), building upon
previous work in the field. He pointed out that higher order factors
are best considered to be determiners, in the sense that they represent
higher order influences spread over lower order axes, although he also
envisages a kind of spiral feedback process (Cattell 1978 p.224) involving the mutual interaction of high and low order axes. In addition, Cattell has proved to be the pioneer in trying to separate these
various influences, in the sense that higher order relationships might
be partialled out to leave what are called stub-factors at each level.
In the study of intra-urban dimensions in Canadian cities (Davies
1978a, 1978b) it is possible to postulate the presence of a second
order Non-affluence axis linked to separate Immigrant and Substandness
first order axes. By partialling out the effect of Non-affluence in
the first order axes, one would presumably separate out the individual
influences attributed to the ethnic immigrants and the quality of
housing. In this search for causal linkages, path coefficients (Wright
1960) may prove to be particularly rewarding.

An important aspect of the search for greater precision in the
understanding of second and higher order relationships is the need to
directly calculate the values of the higher order factors from the
variables. Cattell (1978) has described the technical procedures
involved in such calculations of the so-called Cattell-Wright formula,
and refers to the alternative methods proposed by Schmid and Leiman
(1957). As yet such methods have not been applied to geographical and
ecological problems, although it is worth noting that Royce's (1976)
study of mouse emotionality maintained that much clearer higher order

patterns are produced by such methods than from the factoring of factor correlation matrices.

In general, therefore, the restriction of most factorial ecological work to the orthogonal rotation model has meant that much of the understanding of the dimensionality or urban and regional systems via higher order analyses has been lost. This can be best appreciated by pointing out that the first order axes frequently reflect the initial mix of variables. Hence a variable set with a lot of ethnic indicators will probably produce several orthogonal ethnic dimensions, but will fail to tell the investigator whether these axes are correlated; a vital consideration. The standard use of higher order analysis in the future will help resolve this dilemma, and may help the evaluation of theories of social area variation. Obviously problems with the application of higher order methods will be found, not the least of which may stem from Giggs and Mather's discovery (1975) of the instability of the higher order solutions under different factoring methods. Nevertheless, higher order solutions should be calculated as a matter of course if the factorial approach is to be used to its full potential in urban and regional ecology.

5. SUMMARY

One of the most basic misinterpretations made in the early applications of factor analysis in human geography and ecology was that it was a single procedure, not a family of related techniques in which a number of different solutions could be obtained. Although this difficulty has now been resolved there seems little doubt that most of the results produced by factorial ecologists were derived from a single interpretation of the data, rather than results produced by a deliberate choice from among many different solutions whether of factoring procedure or rotation in which the reason for the choice is given. For those of an ideographic persuasion in ecology the unique interpretation may be appropriate. Beauty, or whatever, may be in the eye of the beholder! For those interested in scientific method the reasons for the choice of techniques, and interpretation must be given if the results are to be acceptable, alongside the opportunity to replicate these results (Davies 1968). Moreover, when package factor routines are being employed it does seem vital to check the results from one package against those from another to ensure that consistent results are obtained. By justifying the choice of solutions or demonstrating stability the results of a single study will have value in a substantive context, in building up a body of scientifically acceptable literature; factor ecologists still have a long way to go when judged by this standard.

6 THE INTERPRETATION OF FACTOR PROCEDURES

The previous chapter has shown that the technical decisions implicit in the choice of one type of factor procedure over another pose a number of difficult problems for any investigator. Complementing these difficulties are the issues that relate to the interpretation of factor solutions. Five major sets of problems are involved. The first is the question of how many factors should be used. A variety of very different approaches have been adopted to deal with this question. The second is the interpretation of the loadings, the individual weights of each variable on the various factors or axes. The third relates to the methods adopted in naming and interpreting the factors themselves. The fourth issue is concerned with the comparison of the individual factors or the whole factor matrix. The final section deals with the calculation and use of factor scores. The five sections of this chapter deal with each of these problem areas in turn. Unfortunately, it must be admitted that compared to psychology the factorial ecology field is still relatively unsophisticated and in many of the areas definitive tests or guidelines are not yet available despite reviews by Gould (1968) and Clark, Davies and Johnston (1973). There is a great deal of flexibility in the choice of procedures and subjective decisions have to be made at various stages in the interpretation of factor methods. Confronted with this situation, this introductory discussion summarizes the advantages and disadvantages of the various approaches as specifically applied to the human ecology field. In general it is rare to find that any single method is sufficient to resolve all the problems involved in these various areas so a recommended policy is to employ a variety of alternative approaches. This usually minimizes the effect of the potential technical bias of particular methods, or the subjectivity often implicit in preferring one method over another. On purely pragmatic grounds - as opposed to the rigidly inferential procedures favoured in common factor methods - it is suggested that when the results from different procedures coalesce, relatively safe conclusions about the interpreted solution can be drawn. In the last resort, however, it must be admitted there may be a strong element of subjectivity in the final interpretation of factorial ecologies. This is no different to the situation found in allied disciplines for the importance of meaningfulness in interpretation has been acknowledged:

> 'it appears foolish to allow analytic decision criterion to override a research sense of the data.' (Rummel 1970 p.357)

Similarly, it has been observed that:

'Factor analysis ... yields solutions which are convertible from one to another, and a preference of form must depend upon appropriate content criteria.' (Harman 1976 p.109)

1. THE NUMBER OF AXES PROBLEM

The question of how many factors to interpret is the most basic issue in factor analysis and is still the subject of debate (Crawfod 1975). The heart of the problem is a function of the oft repeated principle that factorial procedures are invariant. As many mathematically valid and alternative factor solutions can be derived from a data set, the problem is one of choosing between solutions. Given such circumstances, it is the responsibility of the investigator to provide careful arguments to justify the particular choice of the number of factors in the final solution. All too frequently results are published which fail to provide satisfactory reasons for the 'cut off' decision about the number of factors. This leads to doubt about the suitability of the solution in the mind of the readers, thereby reducing the value of any set of results in contributing to the wider body of literature. Unfortunately for those who like a series of basic guidelines to solve this problem, there is no single acceptable solution to this problem. Instead a large number of alternative procedures have been proposed as being useful in solving this particular problem. These procedures can be summarized as falling into three distinct categories based on quite different assumptions: inferential test procedures or statistical tests; indices of completeness of factorization or statistical indices; and a series of guidelines normally called 'rules of thumb'. Normally the first two sets of methods are applied to common factor solutions - although there is one exception - whereas the empirical 'rule of thumb' approaches are used primarily in component analyses. In addition to the use of these various procedures most investigators also make use of the 'meaning' of the factors - as identified in the interpretation of the various axes. Such issues are dealt with in the third section of this chapter. Given such a plethora of alternative methods, and the absence of a consensus about the utility of the various approaches, it will be argued that an investigator would be most unwise to depend upon a single approach to define the number of axes to interpret. Instead, a combination of the various approaches may be the most appropriate procedure.

(a) Statistical tests

These tests stem from the basic assumption made in common factor analyses (not component analyses) that the individual measurements, and hence correlations of the variables, are subject to sampling errors, since they were derived from a sample of cases out of the whole universe of cases. This means that the correlations, and subsequently the factors derived from them, represent estimates of the true values of the relationships that could be obtained from all possible measurements of the variables. Hence the objective of the statistical test associated with common factor analysis is to confirm whether the number of common factors abstracted from any matrix provide a satisfactory reproduction of the observed correlation matrix - subject to the sampling variation contained in this matrix. Most of the common factor tests, therefore, are based on the principle that once the common factors are abstracted, the subsequent factors left in the residual correlation matrix are too small to be anything but chance variations. This decision is based on a defined probability level, for example 0.01 (0.001) which means that the observed value could only have occurred by chance 1 in 100 (1000) times. Since the detailed representation of the test statistics really demand a knowledge of matrix algebra there is little point in reproducing them here; they are carefully documented in the standard statistical texts (Harman 1976; Gorsuch 1974). Yet notwithstanding this point it is worth describing the basis of many of the methods. Rippe's (1955) test for the Minres solution, and Wilks's test (Harman 1976 p.205) for the Maximum Likelihood procedure, both depend upon the calculation of a statistic which uses the number of cases in the sample to weight the logarithm of the ratio of the common factor matrix (as measured by its determinant or characteristic value (Harman 1976; Horst 1965)) to the observed correlation matrix. This statistic is assumed to be distributed as χ^2 or chi-square. A further calculation to define the number of degrees of freedom leads the investigator to reject the hypothesis of common factors if the statistic exceeds the chi-square value for the chosen significance level. The same calculation can be repeated with a different number of common factors. If the statistic is now lower than the chi-square value for the probability level chosen, the solution is deemed to be significant at this level. An example of the approach is provided in the results shown in Table 6.1.

Despite the acceptance of the superiority of the common factor model in many statistical texts in geography, significance tests have only rarely been applied (Davies 1978; Giggs and Mather 1975) in empirical human geography or human ecological studies. In purely inferential terms this is unfortunate. But there is little doubt that on pragmatic grounds the utility of these statistical tests should be queried, since there are many problems associated with their applications in geography and human ecology. The most obvious difficulty is that most human ecologists have dealt with a complete population, not a sample. Since the question of sampling errors does not arise, the tests are really redundant – except, perhaps, as a more rigorous 'rule of thumb' (Davies 1978). It is, of course, possible to argue that the observations are not defined as individuals, but come from areas. Since the taxonomic units could be drawn up in different ways, any one set of areas is a sample of possibilities. This argument seems, at best, a very tenuous one. In any case the restriction of the sampling question to the cases (or individual areal observations) ignores the question of whether the variables form an adequate sample from the universe of variables, an important consideration in Q mode analysis. Methods such as Alpha Analysis (described in Chapter V) are designed to deal with this specific question.

Another difficulty is that the utility of many standard factorial statistical tests were questioned in the early 1970s; particularly Horn's test (Crawford and Koopman 1973), Bartlett's Test (Gorsuch 1973) and Parallel Analysis (Humphreys and Montamelli 1975). One consequence is that revised or new test procedures applied to older analyses may lead to different results. Harman's text (1976 p.213) provides a classic example of this problem with respect to his 8x8 matrix of physical characteristics. The two factor model was considered to be statistically significant for this data set for over 20 years. The application of newer inferential tests showed that a three axis model represented a more satisfactory solution – even though Harman maintained that the third axis could not be interpreted in any satisfactory manner. This illustrates that technical progress is often likely to redefine the interpretation of any study. Hence the solution chosen by an investigator may not be as definitive as might be considered at first. Moreover, it points to another dilemma facing even common factor analysts, namely the fact that the meaning of the individual axes is given a large measure of importance in the final decision of how many factors to interpret. The result is that a difference has to be drawn between what may be called the 'statistical' significance of the solution and the 'content area' significance of the axes. Harman (1976 Chapter 1) provides a useful discussion of this problem, demonstrating that even in the apparently rigorous area of common factor analysis subjective issues intrude. It is also worth noting that many authorities such as Gorsuch (1974) and Cattell (1978 p.476) point out that the commonly accepted tests of the significance based on residual correlation matrices, such as Bartlett's method, often reveal the presence of significant factors, even though these are often rejected by investigators as having little meaning in content area terms. This seems to be a problem that is particularly acute if the size of the sample is large. A consequence is the production of an excessive number of significant factors, many of which are not interpreted. So despite the apparent precision of the tests, common factor analysts frequently make additional and subjective decisions about how many factors to interpret. At this stage in the development of factorial methods the conclusion must be that the limitations of these inferential tests (in relation to the problems formulated by most factorial ecologists) must lead one to question the utility of these procedures. They have an illusory air of objectivity. Nevertheless, it is not appropriate to abandon these statistical measures. Rather, we should work towards the day when our problems can be organized in such a form that the full rigour of the approach can be applied.

(b) Statistical indices

Less rigorous than the 'full-blown' statistical approaches are those methods that try to relate the question of how many factors to extract to a single measure that summarizes the 'degree of fit' of the common factor model to the variations in the similarity matrix.

(i) The Reliability Coefficient proposed by Tucker and Lewis (1973) is normally used in conjunction with the Maximum Likelihood solution. It calculates the ratio of the amount of variance associated with the factor model (the common variance) to the total variance of the data set which includes unique variance. As such it summarizes the quality of the representation of the factor model at each step when additional numbers of factors are extracted. The problem here is that like all similar indices the investigator has to decide on what value of the coefficient is acceptable in the analysis. Unfortunately, there is no standard to indicate the superiority of one solution over another. Most investigators, however, accept that an appropriate value is in the region of 0.9, the value associated with the 7 and 8 axes solution in Table 6.1.

(ii) An alternative, and much less rigorous method, is the Index of Parsimony of Thorndike and Weiss (1970) which can be applied to the results of many different factoring methods. This index (P) is calculated from the following formula:

$$P = (1 - q) \sum_{i=1}^{q} \frac{\lambda_i}{m} \qquad \text{Eq. 6.1}$$

P = Parsimony Index; q is the number of factors; m is the number of variables; λ_i are the eigenvalues associated with the q factors.

The index runs on a scale from 1.0 to 0.0. It tends to the latter value when the number of factors approaches the number of variables, and to the former when all the variance is accounted for by a single factor. Although the index represents an objective evaluation of the effectiveness of the 'degree of parsimony' or data reduction of the factor model, no standard value has been proposed which investigators can unequivocally use to justify abstracting a certain number of factors. Moreover, Haynes and Conway (1977) used the method in human ecology as a summary statistic to compare the efficiency of data reduction of several different factoring methods applied to a single data set. Like the Coefficient of Reliability, therefore, this index would seem to have its primary value in providing investigators with additional evaluations of the utility of the factor model – not with any final, unequivocal decision rule.

(iii) Velicer's (1976) Stopping Criterion provides another type of solution to the problem of how many factors to extract. In this case an exact decision rule is proposed, and the method is applied to the component not to the common factor model. The basis of the approach is to calculate an index at each stage in the extraction of components. The index is based on the average of the squared partial correlations after the first few components have been partialled out of the correlation matrix. Velicer maintained that an appropriate stopping criterion is identified when the summary index is minimized, and demonstrated how a minimum value is followed by increasing values. Obviously the method is quite different to the inferential methods of the common factor methods, where the error variance associated with samples is explicitly assumed. Yet Velicer provided a note of caution, pointing out that the method should only be applied to component methods when the objective of the analysis is to express the variance shared among the variables – rather than upon a simple parsimonious reduction of the data set. In his application of the criterion to several textbook factorial examples he obtained fewer numbers of axes than those produced by other test procedures. So it is possible that the approach identifies the axes that are usually interpreted by the factor analyses – rather than the so-called 'excessive' number produced by over-dependence upon inferential significance tests. At this time it must be admitted that the method has not really been fully evaluated, let alone carefully used in human ecology. Nevertheless, there seems a great deal of potentiality in this approach in dealing with 'the number of axes' problem in component analysis, either when the component approach is the final solution, or when it is used as a first stage in some subsequent common factor analysis.

(iv) A fourth approach to the problem is based on a quite different principle, one associated with the identification of factors from a sample of variables out of a possible universe of variables. It has already been shown that Kaiser and Caffrey (1965) proposed the Alpha factor model to solve this problem. Factors were extracted from the data set so long as they had 'Positive Generalizability' in the sense that the eigenvalues must be equal to or greater than 1.0. (This is the variance added by one variable.) Unlike the significance tests based on probabilities from a sample of cases, the addition of more variables does not necessarily increase the chance that additional factors will be shown to be significant. The eigenvalues of the common factors will only increase in size if the vector explains some of the variance of the additional variables.

(c) Content area 'rules of thumb'

A very different approach to the number of factors problem is found among those who favour an empirical approach to make this decision. A series of so-called 'rules of thumb' have been proposed to help guide the investigator. Given the large number of these procedures described in the standard texts on applied factor analysis (Rummel 1970) the only 'rules' described here are those considered to be the most useful, or extensively used.

The fact that the Eigenvalue 1.0 rule proposed by Kaiser (1961) has been the most popular stopping criterion, would imply this is the single best solution to the problem. The value seems to have been derived from Guttman's (1955) lower bound criteria, one of the rules designed to resolve the question of factor indeterminacy. It is most easily understood as implying that there is little point in abstracting axes if they account for less variance than is contributed by one variable. This adds 1.0 (or unit variance) to the analysis. Authorities such as Cattell (1978 p.86) have pointed out the inadequacy of the approach in practical terms, so the continued automatic use of the rule must be questioned. In any case, one must recognize that the rule has a fluctuating basis, one that is dependent upon the number of variables in the analysis so it is not very useful for comparative work. For example, the cut-off would be at the 10% level in a 10 variable analysis, and 1% in a 100 variable analysis. Moreover the application of the rule to a rotated solution can produce different results to those obtained from an unrotated solution. So, despite the logic of the 'one variable' idea, the possibility of deriving very different results in circumstances where different number of variables are being used, would seem to give the advantage — on comparative grounds at least — to the use of some value accounting for a minimum amount of the variance. A five percent 'cut-off' value is often used, since it is difficult to see how axes accounting for much less than this value have any real generality.

Much more useful as a 'rule of thumb' is Cattell's Scree Test (1966). It is not strictly a 'test' since it is applied to the changes of slope found in the plot of eigenvalues against factor numbers. The logic of the procedure is that the factor variance should level off when the factors simply measure random errors. In Figure 6.1 the bottom of the major slope in the distribution of eigenvalues for Edmonton community areas (Davies 1978) is at components four and five. This is followed by a scree to the seventh axis. After this point the slope is very constant. In general, there is very strong support for this approach in the psychology literature (Cattell and Vogelman 1977; Cattell 1978) although it must be admitted that there are grounds for misinterpretation, because frequent changes of slope occur in certain solutions. Similar to this approach is the use of a <u>major discontinuity</u> in the plot of eigenvalues, although experience has shown this usually produces too few factors. For example, in Figure 6.1a application of the procedure would lead to the choice of the fourth axis as marking a major decline in the size of the vectors.

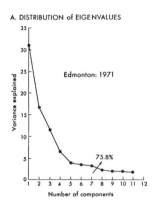

A. DISTRIBUTION of EIGENVALUES

Edmonton: 1971

75.8%

B. COMMUNALITY TIPPING POINT

Number of communalities	Number of components in solution				
	5	6	7	8	9
Greater than 0.7	34	38	42	45	47
Under 0.5	11	7	3	3	2

C. DISTRIBUTION of FIRST RANKING

Number of components in solution	Components								
	I	II	III	IV	V	VI	VII	VIII	IX
5 components	23	16	7	11	3				
6 components	21	15	8	7	5	4			
7 components	18	15	8	7	4	6	3		
8 components	16	15	8	10	2	2	2	5	
9 components	16	13	10	6	2	3	2	5	3

Source: Davies (1978a)

Figure 6.1 Rules of thumb to determine the number of factors:
Edmonton 1971

The fact that the 'Scree Test' does not always produce a unequivocal solution leads to the conclusion that one of the values of the approach is to provide investigators with a general decision zone, within which the investigator can explore the effect of extracting more or less axes. In Figure 6.1 the decision zone covers the whole 'scree' from five to seven axes. What, therefore, is the effect of abstracting axes in this range of solutions? Davies (in Davies and Barrow 1973) proposed two 'rules of thumb' for dealing with this problem. The Communality Tipping Point attempted to pinpoint the number of factors at which the extraction of more and more axes does not really add any more variables into the analysis, in the sense that more than half their variances accounted for by the model. In Figure 6.1 two communality levels (0.7 and 0.5) are used to define this progression. It can be seen that there is a rapid decrease in the number of variables with less than half of their variance accounted for by the model, as defined by the number of components abstracted. By the seventh factor the number of additional variables with most of their variance added to the analysis has declined to a very small, and relatively stable level. In Figure 6.1b the seven axis solution seems to represent an important change in the level of explanation – perhaps not of all the axes, but certainly of the individual variables.

The Factor Complexity approach has been used in ecology (Davies and Barrow 1973) and identifies the number of variables that have their first or second highest loading on a vector. It is helpful for picking out the point at which additional factors are also associated with small numbers of variables. In this context it has proved useful to adopt Burt's '1 in 20' rule (1941) as a stopping criterion. In this an axis has to have at least 1 in 20 first ranking loadings on a vector before it is considered to be worth interpreting. The logic employed here is that if only one variable is being indexed, the axis is not one of 'general' or 'broad' importance in terms of the data set and, therefore, can be discarded. Nevertheless, the discarding of minor axes should be carried out with care. The single indicators may be pointing to additional sources of variation that could be identified as important sources of differentiation if a re-analysis with another domain of variables is carried out. In Figure 6.1 the application of this 'rule' points again to the choice of a seven axis solution. Beyond this level axes are produced which account for only two (i.e., less than 1 in 20) of the highest loadings out of 60 variables.

Experience with the use of these various methods indicates that it seems sensible to compare the effect of the use of several different procedures in the search for the 'final' solution to interpret. This is likely to define the general area of the 'cut off' in the number of factors. Once this has been identified it is important to look at the interpretation of the factors and, if oblique solutions have been used, to investigate the changes in the factor correlation matrix. The addition of new axes that are highly correlated with previous axes, would imply that the factoring has gone too far.

Confronted with all these alternatives it is easy to despair of ever obtaining a solution that would be accepted by the majority of workers. In practice, the problem is not as bad as it seems at first sight. An increasing number of studies have shown that the application of these methods often produces very similar answers when there is a definable structure to a particular set. Cattell (1978), the doyen of factor analysis in the field of psychology, has maintained that the application of the 'Scree Test' to a Principal Axes solution will normally give a satisfactory answer to the 'number of axes' problem. If this is followed by the Maximum Likelihood method, together with its chi-square test based on Wilks' statistic, the adequacy of the solution is confirmed. This is subject, of course, to the fact that the observations fulfil the requirements of the statistical model. For those who are attracted to one or other of the 'Rules of Thumb', rather than to the statistical procedures, one must note that the Maximum Likelihood (M.L.) technique was not widely available until two or three years ago and the solution is still expensive to calculate. Many studies (Cattell 1978; Velicer 1977; Davies 1978) have shown that the results obtained are usually very similar to those derived from the Principal Axes approach, in which a Scree Test has been used to determine the number of axes. It would seem, therefore, that the practical considerations of cheapness and speed should be set alongside the greater theoretical utility of the inferential M.L. approach. So even where the statistical approach can be applied, the greater sophistication of the technique does not necessarily produce substantially more accurate results. They may, however, be better, in the sense that they provide more conclusive proof of the adequacy of the solution. In other words, they are a feature of the phase of justification, not discovery, in scientific methodology. Yet such a conclusion can only be reached if care is taken over the decision of which solution to interpret. It is vital to justify why one solution was chosen over another.

(d) Over and under factoring

The conclusion of the last paragraph leads to a related question within this general 'number of factors' problem. What is the effect of overfactoring (i.e., producing more rather than less axes) compared to underfactoring? In many of the earliest factorial studies a deliberate policy of overfactoring was employed, since the addition of one more factor to those that are to be interpreted clarified the pattern of loadings, particularly in the last axis to be interpreted. Davies and Lewis (1973) provided an example of their ecological study of Leicester (England). Most authorities agree that overfactoring is the lesser of the two 'evils', based on the argument that it is better to have small trivial factors than to lose some potentially important axes by under-factoring. In Table 6.1 the consequence of extracting four to eight factors from a 14 variable x 83 data set of Calgary in 1971 is illustrated (Davies 1978). Although the results shown relate to the Principal Axes component solution, followed by Direct Oblimin oblique rotation, it is worth noting that these results were very similar to those produced by the Maximum Likelihood and other factoring methods.

Figure 6.1 shows that the biggest discontinuity occurs after the fourth axis of the unrotated component solution, and this is followed by a scree running from five to nine components. This suggests a 'cut-off' in the five to nine axes range. The 'eigenvalue 1.0' rule points to the utility of the four axis solution which explains 77.6% of the variance. Unfortunately, two of the 14 variables – Canadian ethnicity and Ukrainian ethnicity – have communalities under 0.5 for this solution. This indicates that the four axis results fail to account for all the sources of variation in the data, although

Table 6.1
Identifying the number of factors: Calgary 1971

(a) Percentage variance for component solutions: Calgary 1971 (for 1-10 axes)

	1	2	3	4	5	6	7	8	9	10
60-variable set	31.9	16.2	12.2	7.6††	4.8	3.5	2.6	2.5	2.0	1.8†
Reduced set	27.9	19.0	16.7	13.9†	6.4††	4.7	3.6	2.6	1.9§	1.2

* Significant at the 5% level.
† Eigenvalue 1.0 cut-off.
†† 5% cut-off.
§ML, Maximum Likelihood method; PA, Principal Axes method; GLS, Generalized Least Squares method; R, Rao's method.

(b) Test statistics for three common factor methods: Calgary 1971§

Number of Axes	Degrees of Freedom	Chi Square			Tucker's Reliability Coefficient		PA Component Explanation (%)
		ML	GLS	R	ML	GLS	
5	31	96.18	71.8	166.5	0.77	0.86	83.9
6	22	53.3	44.1	123.2	0.84	0.89	88.6
7	14	28.5*	25.8*	80.3	0.89	0.91	92.2
8	7	11.3	10.5	42.3*	0.93	0.95	94.8

(c) Loadings on the higher order solutions: reduced data set, Calgary 1971

Axes	Higher Order Axes				Communality
	2	3	1	4	
2. Socio-Economic Status	80				71
6. Ethnicity (Ukrainian British)	83				72
3. Migrant Status		-81			70
5. Canadian Ethnicity		-76			68
1. Age and Family			90		85
7. Non-Affluence			82		81
4. Young Adult Participation				96	93

(d) Oblique component loadings on different number of axes: Calgary 1971*

Table 6.1
(continued)

Expected Axes	Variables	Number of Components Abstracted (Principal Axes) 4 (Image)	5	6	7	8
Socio-Economic Status	6. Large dwellings	70 (-66)	79	82 [Ia]	88 [Ia]	83 [Ia]
	8. Manufacturing	-85 (79) [I]	-92 [I]	-92	-89	-92
Ethnicity (Ukrainian-British)	12. British stock	87 (-79)	83	62 [Ib]	54 [Ib]	53 [Ib]
	13. Ukrainian stock	-67 (52)	-25 [I]	-45 / 99	-45 / 99	-43 / 99
Migrant	9. Non-migrants	92 (-89)	97 [IIa]	94 [IIa]	93 [IIa]	97 [IIa]
	14. Recent housing	-84 (84) [II]	-88	-89	-89	-91
Canadian Ethnicity	11. Canadian citizens	69 (-43)	65 [IIb]	90 [IIb]	96 [IIb (12)]	97 [IIb (12)]
	10. Immigrants	-	-	-	35	34
Age and Family	1. Children	-81 (-82)	-78 [III]	-67 [III]	-72 [IIIa]	-71 [IIIa]
	3. Old age	93 (94) [III]	88	79	80	76
	10. Immigrants	86 (81)	86	74	93	96
Non-Affluence	5. Substandardness	81 (74)	86	96	93 [IIIb]	91 [IIIb]
	7. Unemployed	74 (66)	77	83	96	96
Young Adult Participation	2. Young Adult	-86 (-76)	-86 [IV]	-86 [IV]	-88 [IV]	94 [IVa]
	4. Female participation	-90 (-84) [IV]	-92	-92	-91	-87
	1. Children	- (41)				41 [IVb 6)]
New axes produced by the solution		IIa/IIb	Ia/Ib	IIIa/IIb	IIIa/IIIb	IVa/IVb
Correlations between these axes		-23	-37	+54	+54	-44
Variance explanation (%)		77.6	83.9	88.6	88.6	94.8

*Figures in parentheses are the Image analysis results of a four-axis solution.
Decimal points for all factor loadings are omitted: +70 is +0.70.
Source: Davies (1978)

arguably it does account for what could be described as the variance the variables have in common. Obviously it is not strictly 'common' variance, since a component model was used. The problem of accommodating the specific, localized ethnic variables is again illustrated. Although the test statistics of the Maximum Likelihood and Generalized Least Square procedures cannot be treated as full confirmatory procedures – given the use of the whole population of Calgary communities – they can be used as additional guides. In both cases the chi-square results point to the utility of the seven axis solution, since Table 6.1c shows that the probability of obtaining a value larger than 28.5 by chance is one in 100, whereas for the eight axis solution it is 13 times in 100. Setting the reliability coefficient at the accepted value of 0.9 produces results of eight or seven axes. By contrast, if the significance level for Rao's Canonical solution is set at the 5% level probability level, there is support for the eight axis results. Alpha and Image analysis results were also obtained for the data set, but the test statistics were much less definitive. In the case of the Alpha results the 1.0 'cut-off' from the Generalizability principle suggested a four axis solution although the fifth axis proved to be very close to this level, since its value was 0.98. In the Image solution the same criterion produced a ten factor solution.

In general, these results confirm one of the basic characteristics of common factor test procedures. More axes are usually produced than the investigator cares to interpret in terms of the substantive meaning of the axes. This is very much the case with Image solutions. But it leads to the question of what interpretation can be placed on the various axes. Table 6.1c shows the loadings of the oblique axes for a set of Principal Axes solutions. It demonstrates that the extraction of successively more factors produces more detailed axes, and substantially increases the level of variance explained: up to 94.8% by the eighth factor solution. At this stage it appears that the extraction of axes may have gone too far in the direction of identifying specific or one variable axes, instead of the sought after general vectors. The young adult variable separates from the 'female participation' indicator to form two highly correlated oblique vectors. Generality seems to have been lost. Hence it was concluded that the seven axis solution marks the end of effective factoring. Table 6.1c shows the difference between the seven and four axis solution. The seven axis result separates: an Ethnicity axis from the Socio-Economic vector; a Canadian Ethnicity component from the Migrant vector; a Non-affluence dimension from Age and Family; and perpetuates a separate Young Adult-Participation axis.

The plethora of alternative results in this study demonstrates the difficulties involved in most real world examples – as opposed to the artificial or simple cases used to illustrate the principles of factor analysis in most elementary texts. Support can be found for four, seven, eight and ten axis solutions, although it does seem that a majority of the guidelines point to a seven axis solution. This, of course, contradicts the established 'Eigenvalue 1.0' rule and the 'Scree Test', but such difficulties often occur when small numbers of variables are being processed. What is important here is the low level of variance explanation for the four axis solution, whilst its failure to account for most of the variance of two of the axes represents an additional limitation. Yet the decision to choose seven rather than four axes as the most suitable final result in this example also depended on the provision of another piece of information. Table 6.2d shows the result of applying the same Principal Axes method to the 7x7 correlation matrix derived from the oblique rotation of the axes. The results show that the four axis higher order solution is basically the same as those produced from the four first order vectors. This discussion might lead to the conclusion that the number of axes problem is not such a difficult problem as was first feared. In part we might accept this viewpoint so long as higher order analysis is routinely employed. However, on balance, it does seem more appropriate to choose the seven axis first order solution on grounds other than the test statistics. One advantage is that the seven axis solution identified the minor axes as separate entities – yet they are related by their obliqueness to the major axes. Another is that the relationship between various scales of generalization in the data (between first and second orders) can be more easily demonstrated. Finally, it cannot be assumed that underfactoring, in this case the choice of four axes, would have lost a critical source of the common variance. Nevertheless, an important part of the variability in the data set – the ethnic variables – would have been lost. This was considered to be unsatisfactory in any description of the social dimensionality of Calgary.

To investigators concerned with problems of common factors and common variance, this interest in such specific, localized axes may not be appropriate. It indicates, however, the fact that the descriptive and place-particular interests of the urban ecologists often outweigh the exclusive search for the generalities of a population. In this context the human geographer and human ecologist may have more in common with the geologist in their use of factorial procedures since component methods are favoured in both cases (Klovan 1966, 1976). The conclusion must be that the type of approach adopted here demonstrates the need to constantly re-evaluate the results obtained from a variety of different procedures. No one solution is right or wrong; they are only more or less satisfactory for the problem in hand. Certainly it indicates that investigators in the field have come a long way from the days of running a single analysis, and interpreting one result by recourse to a single 'Rule of Thumb'. In this context it is worth quoting Rummel:

> 'no one study result by itself suffices to establish the best number (of factors). Re-analysis of the same data, involving altered design decisions and replications across cases and variables, will in the long run eliminate factors due to specific design decisions or random influences in the data. Rather than seek a definitive answer to the question of the best number of factors in a single analysis we should expect the question to be answered like any empirical proposition – by the ultimate convergence of many diverse empirical findings.' (Rummel 1970 p.367)

As yet such procedures have not been very common in factorial ecology.

2. INTERPRETATION AND SIGNIFICANCE OF LOADINGS

The same type of debate between exponents of a pragmatic, as opposed to a statistical or inferential approach, to factor analysis is repeated in the way in which loadings are interpreted. 'Rules of Thumb', or

181

subjectively derived 'cut-offs', can be compared with fully fledged significance tests.

(a)   'Rules of Thumb'

In a strictly theoretical sense any non-zero loading can be considered to be of some value, for some variance is being measured by the factor solution.   In practical terms, however, most investigators choose some 'cut-off' value above which loadings are considered important enough to discuss, whereas those below this level are ignored. Unfortunately, no single measure of size of loading has been universally agreed upon.   However, +0.3 seems to have emerged as a generally acceptable level for most contemporary workers, contrasting with the +0.4 or +0.5 used in research in the 1960s.   The +0.3 cut-off level seems to have been adopted because it is the closest whole number to the situation in which 10% of the variance in a variable is accounted for by the factor; strictly speaking, of course, the square of +0.317 is the closest to this 10% cut-off.   Although this level represents the lower limit of loadings usually considered to be worth interpreting, it must be stressed that more emphasis should be placed on loadings above +0.7 (again, strictly, this should be +0.71) since this means that 50% of the variance of the variables is accounted for by the factor.   It is worth remembering that the difference between +0.4 and +0.5 is not the same in variance terms (16% and 25%) as the variation between +0.8 and +0.9 (64% and 81%).

To focus upon the differences in the size of loadings in their study of Leicester, Davies and Lewis (1973) divided the loadings into three types called low (+0.3 to 0.49), medium (+0.5 to 0.69) and high values (+0.7).   At this stage it is worth pointing out that factor ecologists seem much more prepared to define axes where only medium or marginally high loadings are involved.   Most psychologists would be less inclined to accept such solutions as being satisfactory.   This difference can be attributed, in part, to the greater variance involved in many studies of areal distributions, but it is probably also related to the relatively fewer number of factorial studies in urban ecology.   Hence the indicators for the expected axes of differentiation have yet to be conclusively defined and measured.   Questions about the differences in the signs of the loadings are best left until the next section which deals with the naming and interpretation of the factors.   It is particularly important to take care in using the appropriate loadings in oblique rotations, where the factor pattern (structure) loadings of the primary (reference) axes – often called saturations – are the most useful for interpretation.   The factor structure (pattern) loadings of the primary (reference) axes are more akin to correlations.

The size of the loadings alone cannot be the only consideration relevant to the interpretation of factors.   An axis containing only a single large loading cannot be considered to have much general value to those researchers interested primarily in common factors.   If, however, the variable is the only indicator of its type in the analysis, this result could be indicating the presence of an additional source of variation.   This may be strengthened by a reanalysis of the data with the inclusion of additional variables.   This type of interactive procedure prevents an ecological investigation falling into the trap of simply presenting the results of the first set of variables collected

for the problem being investigated. Chapter I has already described
the need for alternative studies for this pattern.

The relative importance of the loading on a factor, compared to its
value in other factors, is also worthy of attention. The easiest way
to screen the results is to convert the loading to rank orders and look
at the distribution of first, second, third rank loadings on each
factor. If a strong structure is present in the data it is rare to
find more than three loadings greater than +0.3 on different axes. If
an axis has less than 1 in 20 first rank loadings (Burt 1952) it is
debatable whether the axis is worth interpreting in the context of the
study, simply because they are so unique. In any case experience has
shown that these factors with few first rank loadings, or those with
small loadings, are not very stable when different factoring methods
are used. Hence the weakness of the axis is confirmed, and a question
mark must be placed over its generality or utility in the study in
question.

(b) Statistical significance

The essentially pragmatic approaches to the interpretation of
loadings described in the previous section can be contrasted with the
concept of the statistical significance of loadings as developed by
applied mathematicians. Obviously 'significance' stems from the idea
that the loading revealed in the analysis can differ from its true
value because of sampling or measurement errors. The objective of the
statistical approach is to calculate the number of times that the value
of the loading could have occurred by chance. If the value of the
loading could have occurred by chance only one in 100 times (1% level),
or five times in 100 (5% level), the investigator is likely to accept
the loading as being statistically significant. The first investigators in the field treated loadings in the same way as correlation
coefficients, and used standard significance tests for these values.
Such approaches are still mentioned in some of the basic factorial
texts (Child 1970; Gorsuch 1974). Since there is little doubt that
loadings are not correlations this is an unsatisfactory approach. Burt
and Banks (1947), however, proposed a specific test for loadings, which
took into account the number of factors already extracted (k), as well
as the number of variables (n) and the sample size (N).

$$S.E._{(r)} = (1 - r^2) \; \frac{n}{(n - k - 1)N} \qquad \text{Eq. 6.2}$$

An example may clarify this procedure. If there were 24 variables
and 24 tests, a loading of 0.86 on the ninth factor would produce a
standard error (S.E.) of 0.07. At the 5% level (2 x S.E.) the standard
error would be +0.14. This means that loadings below + 0.72 could not
be considered to be due to chance. If a loading exceeded a value that
was 2.25 (3) times the standard error, the loading may be considered to
be significant at the 1% level (0.1%), and it could have occurred by
chance only one in 100 (1000) times. Since this test depends upon the
number (N) of cases in the sample, the number of variables (n), and the
number of factors already extracted, the formulation is much more
complex than the simple standard error approach to correlation
coefficients. Yet its restriction to the orthogonal case in the
Principal Axes component model is an important limitation. Child

(1970 p.99) provided a table of the standard error values and these provide a quick check for any investigation. The table clearly shows that the size of loadings needed to reach any level of significance increases with the number of factors extracted - given any number of variables - and it decreases at any level as the sample size increases. Large sample factor analyses, therefore, are likely to produce more significant loadings than those obtained from small sample analyses.

In recent years applied mathematicians have made a great deal of progress in constructing statistical tests for loadings derived from oblique solutions, such as those by Jennrich (1974) and his co-workers. But again the question of whether factorial ecologists can really apply these tests must be raised, for most studies in this field are usually dealing with complete populations. Indeed it must be argued that applied factor analysts should constantly be 'on guard' against the application of these inferential rules to different content areas from those for which the statistical tests were originally derived. This is especially true with the use of variables in Q or P modes, when the inferential statistics were originally based on observations, or cases, in the R mode. The relevance of these points to situations in which cases or observations are towns, or regions, must be noted. They place a further question mark on uncritical reliance upon inferential tests. An additional point is that most tests are restricted to rotations carried out by analytical methods, rather than to the topographic or rotation fitting approaches favoured by Cattell (1978) but, again factor ecologists have rarely dealt with these methods. Hence Cattell concluded:

'decisions on the art of separating significant and insignificant loadings, when both sampling and rotations are involved ... will probably long continue to rest on experience and Monte Carlo studies with selected plasmodes.' (Cattell 1978 p.483)

One Monte Carlo approach would be to take a large number of rotations of known and factored data sets and calculate the distributions of loadings for statistical tests so as to build up the confidence limits (Pennell 1972) for the degrees of significance.

In the last resort, however, the significance of individual loadings may not be the real problem; the whole factor pattern is much more important. This takes one back to the whole question of factor significance discussed in the previous section (see also Cattell 1978 p.482). Although it is important to be aware of these developments in inferential testing, and perhaps work towards the day when these, or revised measures can be profitably applied, it seems more appropriate to echo Cattell's opinion that experience is the most appropriate guide at this time: particularly given the breaching of the statistical assumptions. It is worth noting in this context that the statistical tests proposed by Pennell (1972) and Jennrich (1974) have shown that practically all loadings of +0.3 are validated in their tests. Hence the 'cut-off' criterion that emerged from the pragmatic considerations described earlier appear to represent useful, if conservative, first approximations to the general problem. If anything, only small minor axes may be lost if the size of the loading, rather than the results of a significance test is used as the final decision criterion. Yet although this may be appropriate in what amounts to the early or 'discovery' phases of the application of factor methods to geographical and ecological problems, it obviously produces drawbacks in terms of the 'justification' phase when the acceptability of the results is being evaluated. Pragmatically minded investigators – other than those adopting a rigidly statistical approach – would agree with Cattell (1978 p.485) that:

'a calculated significance value for a single study has less real importance than an average of loadings across replicated studies.' (Cattell 1978 p.485)

Table 6.2
Average loadings for the Migration axes

1 British Studies	Average Loadings	Average Loadings	2 Prairie (Canada) Studies
Five year movers	– 95	+ 91	Low inter-municipal movers
Female movers	– 93	+ 89	Non-migrants
Single person movers	– 82	+ 83	Middle-aged
Local movers	– 67	+ 66	Dwellings built 1951–60
One year movers	– 65	+ 35*	Single adults
Mature adults	– 39		
(Axis is missing in the small town studies. It was found in Leicester, Swansea, Southhampton)		– 61	Mature adults
		– 68	Recent immigrants
		– 71	High inter-municipal movers
		– 85	Dwellings built 1966–70
		*	(Missing in Winnipeg. Present all other centres, i.e., Calgary, Edmonton, Saskatoon, Regina)
Decimal points are removed: – 95 is – 0.95			
Source: Davies (1975, 1978)			

Table 6.2 shows two examples (Davies 1975, 1981) of this principle from two comparative studies, one of six British towns (Leicester, Southampton, Swansea, Colchester, Pontypridd and Llanelli) the other of five Canadian Prairie metropolitan centres (Winnipeg, Edmonton, Calgary, Saskatoon and Regina). In all cases a Migration axis was identified, despite the absence of this construct in Shevky and Bell's (1955) theoretical interpretation of intra-urban dimensions. Yet it must be noted that different variables indexed the dimension in the two cases. In the Canadian Prairie case, all except one of the average loadings for the nine variables indexing the axis were above 0.6; the exception, single adults, was absent in the Winnipeg example, and because its value was 0.35 it was only marginally associated with the axis. A feature of note is the fact that the indicators form a scale, with non-migrant characteristics on the positive side of the axis, and

high migrant features on the negative or opposite side. By contrast, the six indicators in the British study were all on the same side of the axis: five of the variables having average loadings above 0.6 for the three large towns in the 180-391,000 population size in 1966. The axis could not be identified in the small town analyses. This suggests that the presence of the axis is size dependent, at least at the scale of the enumeration level analysis used here. Despite this modification the presence of a migrant related dimension seems clear, in the two countries (admittedly indexed by different variables). Although subject to some redundancy in the set of variables used here a set of tests which reduced the number of indicators did not lose the axis. The axis, therefore, appears to be a very real dimension of urban social differentiation. So the persistence of the axis across a series of studies gives very strong empirical support to the concept of a first order Migrant dimension, although the constrict was used (without real evidence it must be noted) in the well known set of social area models proposed by Timms (1971).

3. NAMING AND INTERPRETATION OF FACTORS

The naming and interpretation of factors is a vital step in any study because it ensures that the substantive content of any set of results is communicated to the general scientific community. As with all naming or classificatory procedures the labelling of a factor gives a summary identity (or meaning) to an axis, although in the last resort the vector is, of course, a collection of weights on a hypothetical mathematical constrict. Arbitrary symbols such as I, II, III can be given to each axis for identification, but most researchers prefer to adopt a different approach to the problems. The most frequently used method involves summarizing the content of the factor by reference to the highest loadings, preferably by a one or two word title since five or more words provide cumbersome titles. The alternative approach is to use some theoretically suggestive title, one that might be linked to the cause of the pattern of loadings or one that might stimulate future work. Unfortunately, the titles of axes of differentiation provided by individual authors frequently leave much to be desired.

Table 6.3 provides an example of the labelling problems. The table is a revision of a factor analysis of the USA metropolitan system (Coughlin 1973). The loadings and scores are rearranged from the original source to help comprehension. Not all the axes of differ-entiation were given precise titles by the author. Moreover, it is suggested that some of those applied lacked precision. For example, it is difficult to see how the title 'middle America' was obtained for Factor II; a more useful summary based on content would appear to be 'Social Conditions'. Factor III was not given a title, but by refer-ence to the pattern of scores as well as loadings the axis can be summarized as 'Degree of Retirement/Tourism' since the axis separates the characteristics of retirement and tourist centres in the south and west, from the northeastern industrial centres. Similarly, the first axis identifies variables linked to urban size, a feature confirmed by the difference in size between the large negative loadings for the major cities, and the large positive values for the small, medium sized metropolitan places. The plain, yet causal title 'Size' would appear to be an appropriate title. It demonstrates that a dimension can be

Table 6.3
Labelling of factor axes: rewriting of Coughlin's (1973) results

Variables and Original Factor Names	Major Loadings	Major Scores	
Factor I - (Traditional Structure - Population Size)			
% Owner occupied	+746	3.93	Flint
		3.08	Orlando
Economic diversification	-533	-----------------	
Number of physicians	-526		
Persons not killed by car	-602	-5.32	Chicago
Bank deposits per capita	-671	-5.38	Newark
% Workers using public transit	-889	-5.36	Boston
		-10.24	New York

Notes on characteristics leading to revised factor label:

I. Size
The scores distinguish the larger centres from smaller places.
They demonstrate that the negative loadings are correlated
with large size, whereas owner-occupation is a smaller town
phenomenon.

Variables and Original Factor Names	Major Loadings	Major Scores	
Factor II - (Middle America)			
Persons not murdered	848	3.89	Patterson
Infant survival rate	708	3.34	Syracuse
% Voting	793	3.25	Buffalo
Two parent families	683	3.23	Hartford
Married women not working	682	-----------------	
Income < $3,000	694		
Negro/white income rates	663		
Dwellings not crowded	641		
Residential integration	593		
Five years schooling	588		
Population/admin. unit	-508	-6.06	Memphis
Segregation index	-557	-6.24	Norfolk
Business repair workers	-582	-7.42	Columbia
Personal service workers	-675	-9.14	Shreveport

Notes on characteristics leading to revised factor label:

II. Social Conditions
A northeast-southern city split is linked to higher survival
rates and greater family stability in the north and greater
segregation and a more service oriented economy in the south.

Variables and Original Factor Names	Major Loadings	Major Scores	
Factor III (No Title)			
Heart disease deaths	585	6.21	Ft. Lauderdale
Unemployment white/non-white	544	4.99	Miami
Hotel workers	415	4.87	Tucson
% in high income industries	−557	4.61	El Paso
Rental occupancy	−518	-------------------	
% Employed in labour force	−591	−2.58	Trenton

Notes on characteristics leading to revised factor label:

III. Degree of Retirement-Tourism
   Higher death rates, more under-employment and low income
   workers are found in Florida and the cities of the southwest.

Variables and Original Factor Names	Major Loadings	Major Scores	
Factor IV − (Clean Air)			
Clean air (mean)	888	4.24	Honolulu
Clean air (maximum)	860	3.78	Albany
Air free of organic matter	854	3.69	Lansing
Air free of organic matter (max.)	772	3.05	Miami
		−3.84	Chicago
		−4.73	Chattanooga
		−4.80	Mobile
		−5.46	Phoenix
		−5.05	Los Angeles

Notes on characteristics leading to revised factor label:

IV. Environmental Quality
   Air quality is better in the major resorts and in centres
   isolated from the cities of the west, midwest and south which
   depend on automobiles for transportation.

Variables and Original Factor Names	Major Loadings	Major Scores	
Factor V - (Urban Stress/Health)			
Survival rate	-767	-4.98	Ft. Lauderdale
Not killed in accidents	-600	-------------------	
Health	-521	4.51	Baltimore
		4.54	Worcester
		4.62	Bridgeport
		6.60	New Haven

Notes on characteristics leading to revised factor label:

V. Health
    Lower survival rates and poorer health are found in older
    industrial centres of the northeast.

Variables and Original Factor Names	Major Loadings	Major Scores	
Factor VI - (Cosmopolitan and Affluent)			
Median years schooling	873	6.28	Washington, DC
% with college education	827	6.05	San Jose
Nos. in arts	815	5.39	San Francisco
High school education	813	5.14	Los Angeles
Professional, tech. workers	751	5.07	Denver
Dining faculties	694	3.93	New York
Sound dwellings	642	-------------------	
Business service workers	640	-5.34	Mobile
Median income families	589	-6.04	Huntingdon
College profs/1000 pop.	366	-6.58	Wilkes Barre
		-7.53	Johnston

Notes on characteristics leading to revised factor label:

VI. Education and Social Facilities
     Higher levels of both education and professional-technical-
     artistic workers pick out the more modern western cities from
     older southern or industrial northeast towns.

identified by the effect of some major influence upon certain indicators, rather than by indicators directly measuring the phenomenon, such
as population or retail size.

It will be obvious from Table 6.3 that the choice of a title is an
intuitive step in the last resort. It is difficult to formulate rules
for the choice since the production of succinct generalization requires
some degree of abstract thought. Normally, the label chosen depends
upon the pattern of the largest loadings, although it is worth noting

in this context the effect of variations in variance contributions; a loading of 0.9 (81%) constitutes much more variance to an axis than three variables of 0.5 (25% each). If necessary there is nothing to prevent an investigator consulting the pattern of scores to help in the assessment of the axis. Indeed, in the Q mode factor analyses of flow matrices (Davies 1979, 1980) it has been shown that one high scoring destination is often associated with a set of places having high loadings on the axis. These are the origins from which the flows emanate. Examples are provided in Chapter VII. If the destination is used as the title, it provides an example of the use of a causal link for naming axes.

Since the choice of a label is so crucial – particularly in compara-tive research – it is as well to exercise a great deal of care over the meaning given to the axis. Table 6.3 shows it frequently helps to summarize the loadings on an axis – so as to focus on the general meaning of the factor – before making the final choice. It is also worth checking back to the raw data to identify the high scoring places so as to make sure the interpretation being placed on the axis is the appropriate one.

Many dangers can be identified in the naming process. These are worth enumerating, given the crucial nature of this decision. One obvious one is when the title chosen is a word with a variety of meanings – especially in the popular vocabulary. Jargon is often unpopular, but if a factor is to be indexed with precision there is often no way out of the dilemma. The following is a clear demonstration of commitment to clarity of idea in psychology:

'what causes confusion is ... when some factorists use popular terms like sociability and extraversion – with their trailing misconceptions and journalistic value judgements – in referring to precise concepts like surgency, affectia, exvia, etc.' (Cattell 1978 p.232)

Factor ecologists are a long way from identifying such a systematic and precise set of concepts as those used in psychology, but the signposts have been provided.

A related problem is the premature labelling of particular dimen-sions, in the sense that the first investigators to identify an axis may not have chosen an appropriate words. Abrahamson (1974), for example, used the labels Isolation and Eccentricity Tolerance in his study of US cities when the highest loadings for the former were linked to an indicator that measured syphillis deaths, whilst the number of nightclubs and number of rapes highlighted the latter. These are rather bizarre, if not tragic titles! So there is no need to follow the original designation of factors if better titles can be suggested. This is particularly true in those circumstances where attempts are being made to build up a cumulative body of literature. The descrip-tive titles suitable for one study may conceal some more general conclusions. Table 6.3 can be used once again to illustrate this point, where the labels 'Size', 'Social Condition', 'Degree of Retirement/Tourism', 'Environmental Quality', 'Health', 'Education and Social Facilities', appear to fit more closely to the established ideas on inter-urban structures (Chapter VIII) than the original descriptive phrases used by Coughlin (1973): Traditional Structures - Population Size, Middle America, No Title, Clean Air, Urban Stress - Health, Cosmopolitan and Affluent. To be fair, however, the author was more interested in identifying goal dimensions, rather than 'structures', and he described the factor axes in general terms in his article, rather than giving them precise, communicable titles. In any case the apparent use of orthogonal axes in the study means that the research is of relatively limited utility as the relationships between the vectors remain unexplored.

Table 6.4
Loadings for the second axis in Colombo (Sri Lanka):
Land Use or Familism?

Loading	Variable Name
+88	% 6-8 member households
+88	Sex ratio
+82	% 6-17 years of age
+79	% 9-11 member households
+68	% 3-5 member households
+58	% under 6 years of age
+44	% Burg
+38	% Houses
+37	% Residential units
-79	% Non-residential buildings
-83	% 18-54 years of age

Decimal points are removed: 37 is 0.37

Source: Herbert and de Silva (1974 p.147)

Similar problems of unrepresentative titles can be identified throughout the factorial ecology literature. Table 6.4 shows an extract from Herbert and de Silva's (1974) otherwise valuable study of Colombo (Sri Lanka) where one of the major axes is named Land Use, when in fact seven of the eight largest loadings (above ± 0.5) index household, family, or age criteria, whilst another variable measures non-residential buildings. A land use association is certainly present in the axis but the size and balance of the loadings demonstrate that the axis is much more a dimension of Familism. This may imply that the total Colombo social structure is much more closely linked to the structures of the western world than to the undeveloped world - despite the lack of exclusive commercialization of the central area.

These problems in factor labelling are not, however, restricted to the factorial ecology field. The creation and application of generalizations is flawed in most areas of research. Perhaps it is an advantage of the factor approach that the label, the generalization, can be linked back to individual weights, rather than being a simple intuitive act without evidence to back up the point. In any case similar labelling problems occur in all areas where factor analysis has been employed and it has been suggested that:

'the history of personality factorings and meanings is strewn with misidentification.' (Cattel 1978 p.232)

The lack of clarity in the naming of many factors reduces the use of the limited evidence we have for urban and regional dimensionality. In such circumstances there seems a particularly pressing need to squeeze as much as possible out of individual studies by devising an index or codification of factor dimensions. In this, as in other areas, psychologists have illuminated the pattern that can be profitably followed. In an appendix to his latest book Cattell (1978) demonstrated how axes can be indexed according to several criteria: content area (T for objective test data; Q for questionnaire data and L for life record); symbol and title for the dimension; first or second order axes (arabic or roman numerals). For example, UIT AP.21 is the first order axis label from test data for the exuberance (Eb) factor, whereas UI(Q) AP IX is the label for the second order, roman numeral indexed axis for Humanism which has been derived from questionnaire data. Obviously, repetitive testing upon standard data sets is needed before an axis can be said to have been positively identified and added to the catalogue of psychological traits. This type of codification, therefore, parallels the work of chemists in identifying the atomic weight structures, and in the indexing of vitamins.

Table 6.5
Loadings for crop dimensions in Ohio

Variable/ Component	King's Labels:						
	General Crop I	Corn, Barley II	Mixed Grains, Barley III	Barley Oats IV	Wheat, Mixed Grains V	Wheat VI	Hay VII
1 (Corn)	45	75	-10	-18	-11	-39	11
2 (Small grains)	-23	16	83	-37	32	-02	-01
3 (Wheat)	73	38	-13	11	44	30	14
4 (Oats)	41	-70	19	34	28	-32	03
5 (Barley)	-26	55	47	61	-18	04	01
6 (Soybeans)	65	-37	38	-15	-47	16	16
7 (Hay)	-93	-09	-14	-04	05	01	30

Decimal points are removed: 45 is +0.45

Source: King (1969 p.179)

In comparative work care should also be taken over the _signs_ of a factor. All too frequently in early factorial studies negative signs were ignored. Yeates and Nicholson (1969) identified a so called French component in their study of Winnipeg, even though the highest loadings were French ethnic (+0.68), Jewish (-0.50), Ukrainian (-0.48) and Polish (-0.46), demonstrating the axis was a French-East European vector. King's labels (1969) for crop dimensions (Table 6.5) consistently downplay the importance of negative loadings, especially in the so-called Corn and Barley axis where the highest loadings were corn (+0.76), oats (-0.70), barley (0.55), wheat (0.38) and soybeans (-0.37). The addition of 'oats' to the title is surely called for in a crop specific label. To use titles such as Wheat and Oats for such small axes as Vectors VI and VII is also misleading.

Table 6.6
Loadings for the Family Status axis in Winnipeg and Calgary

Winnipeg Loadings > ± 0.3	FAMILY STATUS Variable Description	Calgary Loadings > ± 0.3
90	Small households	-59
80	Apartments	-77
78	Female heads of households	-70
75	Female-male ratio	-40
74	Proportion of old aged	
71	Small dwellings	-63
71	Young adult economic participation	-67
58	Female economic participation	-61
56	Adults - divorced	-53
52	High intermunicipal movers	-61
42	Single adults	-64
40	Immigrants	
31	Dwellings built prior to 1946	
	Proportion of young adults	-85
	Provincial migrants	-40
	Native ethnic	-32
-31	Construction, transport workers	
-31	Canadian ethnic stock	
-36	Household size	+37
-59	Mature adults	+40
-70	Single detached dwellings	+71
-71	Owner occupied dwellings	+73
-76	Large households	+41
-88	Children in population	+59
-92	Fertility ratio	+76
-94	Persons per household	+58
-97	Family size	+62
-98	Number children per family	+61

Decimal points are removed: 90 is +0.90
Source: Davies (1975, 1978)

Yet in the last resort these signs should not be regarded as sacrosanct, as Table 6.6 shows. The comparison of the family status axes for Calgary and Winnipeg (Davies 1975, 1978), based on a standard variable set of 67 indicators from the 1971 census, shows a reversal of signs between the studies. This is purely a consequence of the arbitrary allocation of signs for the first loading in the first axis to be extracted. Obviously the scores will be reversed as well. The differences in the size of the loadings in the two axes shows that the axes are not identical in the two cities, although a congruence coefficient (a measurement of their similarity) demonstrates there is a high level of comparability between the axes. These differences tell us something about the two metropolitan areas. In Calgary in 1971 the low family status side contained high proportions of young adults, as well as provincial migrants and native ethnic people, but is without the high proportions of old aged, immigrants or older dwellings found in Winnipeg. By 1971 the downtown area of Calgary had experienced an explosive growth of new highrise apartments, compared to the more slowly growing, less prosperous city of Winnipeg. The differences, therefore, demonstrate the place-particular associations of each city within the context of the general axis of differentiation.

The naming of axes with negative and positive loadings can also be confused by two very additional features. For those axes in which the loadings on one side represent the opposite characteristics to the other (for example, the variables picking out single and old age, as opposed to indicators of young families) the term factor scale is a useful description. Bi-polar axes, by contrast, can be used to describe axes indexing quite different characteristics, in which there is no causal link between the features. For example, a vector that might have been produced by the chance associations of loadings with very different features, or one linked to the fact that unrelated phenomena may co-exist in area. In all cases where combinations of negative and positive loadings occur, it seems appropriate to help the reader's interpretation of the relationships by separating the variables on each sign and arranging each in order, as shown in Table 6.3, rather than arranging the information in the order of the variables.

Another set of interpretation problems come from the calculation of the variance or explanation associated with each axis. For example, in a study of central places in Nigeria it was maintained that:

'the lack of any strong loading for any of the primary variables on the first principal component makes it difficult to label.' (Abiodun 1970 p.184)

Yet her fourth table showed that the axis explained 52.7% of the variance of the solution. This is incompatible with the statement about the 'lack of strong loadings'. It is probably explained by the package computer programme used to produce the results in London in the late 1960s. The loadings for each factor in the ICL programme used were standardized to a zero mean. This means the values were not true loadings, in the sense that they can be squared to produce a variance or explanation value. Parallel problems are often found in investigations using the SPSS (Nie, Brent et al 1975) package programme. The printout of eigenvalues and percentage explanations always relate to the unrotated, not the rotated solutions. The percentage variances for the rotated solutions are not provided by the programme; investigators have to calculate them. This confusion is found in many publications (Carter and Wheatley 1979). A related difficulty can be found in studies using oblique solutions, where the sum of the squares of the loadings on each axis will not produce all the variance because oblique solutions contain direct variance associated with the correlations of the axes (Chapter V). In this context it is also worth remembering that oblique solutions produce two factor loading matrices: the saturations of loadings suitable for description are obtained from the primary pattern matrix of the primary axes (and reference structure of the reference areas); the correlations, from which variance amounts can be calculated, are shown by the structure matrix of the primary axes (or the primary matrix of the reference axes).

Interpretation problems also come from the confusion between component and common factor models. In the former all the variance is analyzed by placing 1.0's in the diagonal of the similarity matrix; in the latter, communality estimates are often used, values calculated from previous component analyses or work. Since these communality estimates vary in size from variable to variable a differential weighting is introduced, as was seen in Table 5.3. However, the eigenvalues for each factor produced by most common factor programme adopt the convention of accounting for 100% of the variance. This is not 100% of the original variance as in the component solution, but 100% of the variance used in the common factor model. This point often leads to confusion. For example, Hodge (1968) seems to have adopted a common factor model in his study of service centres in Saskatchewan where nine axes accounted for 71.1% of the variance. However in Hodge's description (in Berry and Horton 1970 p.220) it was subsequently maintained that only: 'six factors were relevant and the varimax rotation reduced this to three.' The three axes which were named only account for 57% of the common factor variance - which is itself only 71.1% of the total. The scores on these three axes were used in the regression equations to explain changes in the pattern of service centres, and it is apparent that only two-fifths of the original variance was incorporated in the model. It is not surprising to find that low levels of explanation were produced.

A final issue worth considering is the need for the explicit consideration of the domain of the variables used in the analysis (Chapter IV). All scientific investigations only deal with part of the possible range of relationships in the real or laboratory world. Hence there is little point in factor analysts being accused of only dealing with part of the universe of variables; it is a common problem in qualitative as well as quantitative work. When naming factors it is a good idea to consider variables that may be missing from the data, but which might form part of the factor scale. In the intra-urban classification field for example, exponents of 'quality-of-life' indicators (Smith 1972) or more social dimensions (Abrahamson 1974) give the impression that their results are additional to the axes produced from standard socio-economic variables derived from the census (Berry 1972). Intuitively, one might suspect that many so-called 'quality' variables are intimately linked to the socio-economic status or economic base of settlements, so they may be measuring very similar constructs by a different variable battery. Chapter VIII provides an example of the integration of the two domains of variables, thereby providing evidence on this point. Another example is provided by Hunter and Latif's (1973) adoption of 15 variables to index one or other of the three Shevky-Bell dimensions of social over analysis. Although they provide evidence of the utility of the Social, Family and Ethnic dimensions in the data set, they provide no information about whether other axes of differentiation are present in the city being analyzed. Such a procedure, therefore, is not a test of the utility of the Shevky-Bell axes in accounting for intra-urban differentiation. Rather it is a test of the utility of these 15 variables in indexing the three particular dimensions of variation.

All these problems testify to the difficulties that face investigators who attempt to label axes. They demonstrate that the factorial approach, even in its common factor mode, is not a simple technical operation producing unequivocal conclusions. Yet many of the difficulties noted above are the typical 'teething' problems associated with the application of new techniques. The hope is that these problems can be resolved once they have been exposed. Since factor ecologists can draw upon the experience of more advanced work (in psychology in particular) there seems little doubt that those difficulties will represent a passing phase in the history of explanation. Nevertheless, it must be admitted that the labelling process is, in the last resort, an intuitive step. This is not a purely mathematical or statistical problem; symbols to identify the hypothetical vectors will do just as well. Rather, the intuitive step is taken to endow the common factor or component with some meaning relevant to the substance of the study. Since similar naming or generalizing procedures occur in all investigations, there is little profit in singling out factorial studies for being 'subjective' in this context. One advantage of the factorial approach is that the factors contain precise measurements of the respective contributions of the variables to the individual axes.

Finally, the question of the relationship between factors and scientific constructs must be addressed. Constructs are the component structures, or 'building blocks' of theories, and as such are not directly observable. As in all theories these 'building blocks' have to be translated by the correspondence rules of the theory (Harvey 1969) into some specific operational definitions. in the factorial ecology literature it is implicitly assumed that the individual factors stand for, or measure, the underlying constructs, since these factors are themselves not observable but are hypothetical or mathematical vectors indexed by part of the variation of a set of empirical indicators or variables. There is no need for any particular empirical investigation to contribute to a wider body of theory. But the development of a systematic body of literature in the field can only occur if the results from individual studies are linked together in some way. In this context it is certainly profitable to be able to use an underlying theory to produce explanations for the derivation of individual constructs. The identification of these constructs in the real world is one of the tasks of factor analysis, although again it must be noted that there is another inferential leap involved in equating a factor with a construct. Hopefully, however, the theory is expressed in such a form that there is an unequivocal deduction of the axes of differentiation from the assumptions and formulation rules of the theory. These can be directly related to the empirical evidence as summarized by the factors. It seems easier to index the constructs with individual variables. In practice the constructs are not directly observable, since they manifest their effects in a set of individual indicators. For example, Socio-Economic Status is more than just a measure of employment, possessions, or attainment in education. This is where factor analysis - particularly in the common factor model - comes into its own. It provides a mechanism for identifying the underlying structures and these can be linked to the emerging literature on structuralism in human geography (Harvey 1973; Dear and Scott 1982).

4. FACTOR COMPARISONS

If progress is to be made in the development of a systematically related body of literature of the study of the dimensionality of axes and regions, the factors identified in individual studies need to be checked against those derived from independently derived analyses of different axes. The problems involved in the comparison of alternative factor procedures applied to the same data set has been described in Chapter V. Unfortunately the field of factorial ecology frequently resembles an eclectic set of unrelated studies, each with their own variable set. This parallels what Cattell (1978 p.247) called the 'sterile isolationship of unconnectable studies' in psychology. However, if sufficient numbers of common or marker variables are found between studies, it becomes possible to directly compare the results of one analysis with another. In this way a cumulative, and systematically related, body of information can be derived, such that each study contributes to the emerging body of generalizations. Despite early attempts at systematigation by Hodge (1968) and Berry (1972) in the field of inter-urban differentiation, and in the study of intra-urban social areas, this cumulative or incremental approach to the creation of a systematic body of literature has been curiously absent. Inevitably, much of the utility of the factorial approach in urban ecology has been lost.

Before any factor can be considered to be a meaningful dimension in comparative research (in other words, an axis that appears in several cities in the urban context) the factor should have been identified on at least one occasion in circumstances that allow comparison. Four broad categories of factor matching procedures are identified in this section as attempting to fulfill this task. They are called: subjective comparison; quantitative matching; variance allocation; and the joint analysis procedure.

(a) Subjective comparison

The simplest method for comparing factors involves the subjective comparison of the results of two or more studies. No techniques are used; the two sets of factors are verbally described and related to one another. Where completely different variable sets are being used there are few alternatives to this subjective approach. But if identical, or very similar, variable sets are studied, there is no excuse for not using some of the quantitative matching techniques described in the next section. Unfortunately, examples of this approach were still common in the literature in the last decade (Evans 1973). In view of the sophistication of the factorial analysis it is rather strange to find that the final result of the comparative study depends upon such a subjective procedure. The approach cannot be recommended for the comparison of factors, but may point the way to the use of more rigorous procedures.

(b) Quantitative matching

The second approach to factor comparison involves the use of some quantitative measure of comparison. Rummel (1970), Harman (1976), and Cattell (1978) have all reviewed these measures. Two rather different types can be identified: one measuring individual factor vectors, a pair at a time; the other comparing the whole factor matrix of one study with the factor matrix of another.

(i) <u>Vector comparisons</u>. Among the vector comparison techniques the coefficient of congruence (r_{cc}) is the most popular measure since it incorporates the magnitude as well as the pattern of any set of factor loadings:

$$ r_{cc} = \frac{\sum\limits_{j=1}^{n} a_{jp} \cdot a_{jq}}{\sqrt{(\sum\limits_{j=1}^{n} a^2{}_{jp}) \cdot (\sum\limits_{j=1}^{n} a^2{}_{jq})}} \qquad \text{Eq. 6.3} $$

Where: a_{jp} is the factor loading of variable j on factor p.

It may help comprehension to verbally describe the formula as:

<u>The sum of the product of loadings for each pair of variables;</u> divided by the square root of the product of:

    (i) the sum of squared loadings on axis p, and
    (ii) the sum of squared loadings on axis q.

One other advantage of the coefficient is that unlike correlation measures it does not work upon the deviations from the mean, thereby, perhaps changing a positive to a negative value. Instead, absolute values are compared, and r_{cc} only increases when the loadings are the same sign. However, Pinneau and Newhaus (1964) have pointed to a major disadvantage of the method. If two of the factors load highly on a set of variables, a large r_{cc} will be produced, even if the pattern of loadings is different. In addition, the coefficient is more properly applied to orthogonal rather than oblique matrices. It may also be noted that the comparison is usually made over the whole set of loadings, even though only those below a certain 'cut-off' level are rarely interpreted. In minor axes with few high loadings this could produce an important element of spuriousness, though the comparison could be calculated only on the high (>0.3) loadings. Until recently, the index had another disadvantage, namely that no acceptable significance test was associated with the value. Hence different investigators used various cut-offs to measure 'significance'. Korth and Tucker (1975) has provided what seems to be an acceptable table of values which are listed in Cattell (1978 p.253). For example two studies with ten variables in common, and four factors in each study, would require a r_{cc} value of 0.93 at the 5% level, whereas 30 variables would require a 0.46 coefficient in a four factor study and 0.52 in an eight factor study. These values are smaller than those quoted by Kaiser (1974) in his Index of Factorial Simplicity.

An alternative, and nonparametric approach to the problem is called the Salient Variable Similarity Index (Cattell, Balcar, Horn and Nesselroade 1969). This index explicitly recognizes the need to exclude loadings that are considered insigificant (called non-salients). A very conservative value of ±0.10 is often used (Jennrich 1974), although this value will vary with sample size. The number of variables above this critical limit that are common to each study can be extended in cell F_{11} in Table 6.7.

Table 6.7
Calculation of the salient variable similarity index

Factor 1	Factor 2			
	PS	N	NS	
Positive Salient Variable PS	F_{11}	F_{12}	F_{13}	n_1
Hyperplane Variable H	F_{21}	F_{22}	F_{23}	n_2
Negative Salient Variable NS	F_{31}	F_{32}	F_{33}	n_3
	n_1	n_2	n_3	n

These values are placed in the following formula and a value of 1.0 (−1.0) will indicate perfect agreement (the complete reflection of the other) whereas 0.0 is no agreement.

Eq. 6.4

$$S = \frac{(F_{11} + F_{33}) - F_{13} - F_{31}}{(F_{11} + F_{33}) + F_{13} + F_{31} + 1/2(F_{12} + F_{21} + F_{23} + F_{32})}$$

The index has been considered to be a special development of the chi square test, and significance levels for the values of S have been calculated (Cattell 1978 p.258). Inevitably the index has weaknesses, notably the arbitrary decision of what constitutes a salient variable as well as the need to reflect (or change) the sign of any loadings to conform to those produced by another axis if the factors happen to be reversed.

These alternative factor matching techniques apply primarily in circumstances where identical variable sets are being used. Other techniques, which are more suitable for situations in which different variable sets are employed, are summarized by Harman (1976 p.346) and Gorsuch (1974 pp.284-286). Investigators should treat all those factor matching techniques with some caution however; even the ones described here have their limitations. Indeed, again on purely pragmatic grounds, it may be increasingly inappropriate to depend exclusively upon a single measure of similarity. Probably at least two factor matching techniques should be employed in the search for invariant axes, or rather axes in different study areas that are close enough to be considered similar. In all circumstances where different study areas, or different populations are being dealt with, it is likely that place-particular influences will intrude to modify the pattern of similarity. Such influences go beyond the errors of measurement or sample limitations. In common factor methods such local influences will be classified as specific variance.

(ii) <u>Matrix comparisons</u>. In the final resort, however, techniques for comparing one whole factor matrix with another matrix of the same variables, will provide the most rigorous and sophisticated comparisons. The work of Kaiser, Hunka and Bianchini (1971) has proved to be particularly important in helping to solve this problem. Cattell (1978) and Harman (1976) have summarized the calculations involved, and have described alternative procedures. As yet factorial ecologists do not seem to have used these methods. In any case they are more suited to circumstances in which sets of well defined factors are being compared. This review has pointed out that this is still beyond the stage of many research areas in the factorial ecology field. Moreover, since this book is restricted to the factorial study of individual matrices, so the related procedures known as Canonical Correlation methods are not considered here (Cooley and Lohnes 1971; Clark 1972; Johnston 1979). It is worth noting that these are matrix comparison proceures, in which two matrices based on different types of variables are compared in one study. Normally the variable sets relate to quite different domains of information. For example, Gauthier (1968) related indices describing level of economic development with indices of accessibility in Brazil, whilst Clark (1973) used the method to investigate the similarity between telephone flows in Wales and a set of socio-economic indices. Briggs and Leonard (1977) have provided a valuable summary of the problems associated with the use of these methods.

(c) Variance Allocation

Despite the range and sophistication of the comparative methods that can be applied to factors, an eternal problem of comparative factorial ecology still remains, namely, the separation of the general, broad or common features of a data set from the specific or place-particular characteristics. All too frequently it is implicitly assumed that the components abstracted in any analysis represent the 'common variance, and the rest is 'unique' variance as in the classical common factor model. This is not the case. Cattell (1966) has observed that components are hybrid mixtures of common factors and various amounts of unique variance. In view of this fact it would seem that the common factor model is the only way of solving this problem. But common factor models depend upon some prior assumptions of the number of common factors, or of the size of communalities. In exploratory studies there may not be any satisfactory basis for making either decision. At least one pragmatic way around this problem can be suggested. As it involves an empirical approach to allocating the

variance in any comparative study it is called the Variance Allocation
method (Davies 1975).

Table 6.8
The Variance Allocation method applied to the Family Status axis

(a) Variance Allocation

Cities	Explanation (%)	Variable Variance Allocation (%)		
		Common	Specific	Residual
Calgary	17.36%	13.32%	3.2%	0.84%
Edmonton	20.04	17.15	1.7	1.23
Regina	27.16	16.23	10.1	0.88
Saskatoon	26.67	21.08	4.33	1.25
Winnipeg	21.80	19.80	0.77	0.89

(b) Common variables

Average Loading	Variable	(Missing in City Named)
+84*	34) Persons per household	
+84	9) Average number of children	
+83	12) Fertility	
+77	1) Children	
+77	15) Owner occupied	
+72	16) Single detached dwellings	
+66	8) Family size	
+58	23) Large households	
+53	3) Mature adults	
+40	35) Household income	(Negative value in Saskatoon)
-46	38) High intermunicipal movers	
-47	7) Divorced	
-56	6) Single adults	
-69	25) Young adult participation	
-71	11) Females	
-72	14) Female participation rate	(Regina, Calgary)
-73	5) Old aged	(Calgary)
-75	19) Small dwelling	
-81	17) Apartments	
-83	22) Small households	
-83	13) Female household heads	

* All decimals are excluded in this table: 84 is 0.84

Source: Davies (1977)

201

The first stage consists of finding out which sets of components are most similar to one another among the comparative studies, such as by the use of congruence coefficients. Each set of similar components is compared, and all those variables with high loadings (>+0.3) on all the similar constructs are considered to be common variables for that axis. Variables specific to the axis are called specific variables, and those with loadings under the cut-off level of +0.3, can be called residual variables. The contribution of the common variables, specific variables and residual variables to the variance explained by the axis can then be separated. Average loadings for the common variables can be calculated. Table 6.8 illustrates the ideas in the case of the family status axis derived from a comparative study of the social dimensions of the five metropolitan areas of the Canadian Prairies. The variance contributions to each axis in the five centres that are linked to the common variables (called common variable variance), the specific variables (called specific variable variance), and the residual variables (called residual variable variance) are shown in the Table 6.8 (Davies 1977). Obviously the method has a simplistic basis, in that it ignores errors of measurement, and simply averages the values of the variables common to each axis when more sophisticated deviation processes could be used. By comparing loadings in this way the method provides at least one way of demonstrating the similarity of axes, and, more to the point, provides an estimate of the breakdown of common, specific and residual variance on each axis. Another example of this procedure has already been provided in the previous chapter Table 5.1. In this example of three large British centres ranging from 150 to 300,000 population, and three small towns in the 40,000 range, it can be seen that the amount of common variance for any axis is much larger for the bigger places. The small centres also have larger amounts of specific and residual variance on the axes, demonstrating the way in which place-particular influences distort the pattern of associations. These results demonstrate the presence of city size influences in the number, size and variance explanation of component axes.

This use of the Variance Allocation method has explored the relationships existing in a set of factorial ecologies using similar variable sets. However, a more profitable use of the idea would be in the context of data screening, to identify the common variables in a data set. This would provide useful estimates of communalities as a prelude to common factor methods. Although the idea has to be used in conjunction with quantitative factor matching techniques, one advantage would be that the various contributions to the size of an axis are explicitly identified.

(d) Adjusted inputs

The fourth approach to the comparison of factor results consists of altering the inputs to the analysis to facilitate comparison. Unlike the other procedures, which have been derived directly from the work of statisticians working in allied fields, these have been independently discovered by factorial ecologists and several types can be recognized. For example, Wiltshire, Murdie and Greer-Wootten (1973) used partial correlations between pairs of variables to filter out third variable effects. An alternative approach, which relates to the observations (the areas in R mode studies) consists of the joint analysis of individual city sets. This, of course, focuses on the direct comparison of individual city areas, rather than upon the comparison of factors. Instead of treating each city independently, by producing the analyses separately, and then subjectively comparing the results or sampling areas from different cities, all the sub-areas of the cities are combined into one analysis. This method has been adopted by Jansen (1971) in Sweden, and Davies and Barrow (1973) in Canada. The obvious advantage is that the factor scores of an R mode analysis are <u>directly</u> related to each other, making it possible to show how the tract of one city exactly differs from that of another, since they are part of the same study. If followed by analysis of variance techniques, the distribution of sub-area scores on each axis in a city can be object-ively compared to the distributions in another. But a danger does exist. The cities may be so different that the resulting dimensions are blurred, and will lead to averaged aggregates that are difficult to interpret. If there is any doubt on the basis of the comparison between the cities, it is imperative that the Joint Analysis method should be preceded by a study, or at least a comment on, the suitabil-ity of jointly comparing the set of towns. This is an important omission in Jansen's (1971) study of Swedish towns, since his analysis incorporates cities that range from over one million to 30,000 popula-tion and disguises the 'city scale' effects. Davies' (1977, 1983) studies of Canadian Prairie metropolitan areas explicitly showed how the centres formed a separate group in Canadian terms, and provided a justification for their combination into one analysis.

(e) Temporal comparisons

All the comparative methods described to date relate to comparisons of results drawn from different spaces or areas. Another set of difficulties occur when temporal comparisons are involved. In essence, the problem is that in addition to the variations in cases (areas) and variables, the introduction of time produces a third mode (see Figure 3.4). Although the development of three mode factor techniques (Tucker 1966; Bentler and Lee 1979) may help solve this problem in the future, factorial ecologists have had little experience with these techniques. Instead, they have used some of the methods described in the preceding section to deal with the temporal comparison, such as: the subjective comparison of axes drawn from different times (Murdie and Ray 1972); congruence coefficients calculated between the same variable sets through time (Davies and Healy 1977). Johnston (1979) has shown how canonical correlation methods can be used to compare matrices based on the same observation sets through time. Other workers have used change coefficients (Murdie 1969; Brown and Horton 1971) which measure the change in a variable in an area between two years. The matrix is then factor analyzed. Unfortunately a slight absolute change from an insignificant base can produce very large coefficients. Murdie (1969) partially resolved the problem by expressing the index for a variable as a ratio for change in that variable in the city as a whole, but this adds a city-specific bias. An alternative approach is to use shift and share coefficients (Berry and Horton 1970) in which change is parti-tioned into three parts. In the context of employment in industries the decomposition is related to national growth, industry mix and regional change indexing local deviations from the average. The matrix of these indices are then factor analyzed to produce a direct measure-ment of change. Particular care needs to be taken over the choice of similarity coefficients before factoring. The normal correlation coefficient will repress differences in the sizes of the variables in favour of comparing deviations, whereas any measurement of change presumably needs to take size differences into account. This would appear to be a situation in which covariance matrices might be usefully employed as input to the factor analysis. Yet interest in the specific measurement problems should not be allowed to obscure some fundamental questions about the meaning of the indicators through time. It has already been pointed out in Chapter III that all indicators are likely to be time dependent, so their interpretation can vary as societal values change through time. The problems of measuring changes through time are dealt with at some length in the basic texts on applied factor analysis (Rummel 1970; Cattell 1978) whilst an important collection of papers edited by C.W. Harris (1963) is still worthy of study by factorial ecologists. Finally, the issue of short term ecological changes must be addressed. Parkes and Thrift (1977) have demonstrated how factor analysis can be used to interpret these daily ecological changes within a city, whilst King and Jeffrey (1972) have shown now factor methods applied to bi-monthly levels of unemployment in US cities can classify cities according to their cyclical responses to economic change.

5. FACTOR SCORES

Most factor analysts are primarily interested in defining the sources of variation or collectivities of variables in an observation-variable data matrix. So it is not surprising to find that developmental work in the field has focussed primarily upon the derivation of factors and loadings, not of scores, as any statistical textbook in the field will demonstrate (Harman 1976; Horst 1965). However, the calculation and use of factor scores involve a series of problems that are important to investigators. Before dealing with these issues it is worth pointing out that it can be argued that factor ecologists are primarily interested in the spatial patterns of variation. As such, the factor score problems can be avoided by adopting the Q mode approach in the case of the observations-variables data matrix. In this way the factors will identify sets of observations that have similar patterns and the scores will show how important each variable is to the factors or observation cluster. As yet, it is only in the study of flows (such as in intra-urban studies of connectivity described in Chapter VIII) that the Q mode approach has become a standard procedure in any particular problem area in factorial ecology.

(a) Problems of composite values

A basic problem involved in the interpretation of scores is that they are composite measurements, just like the loadings, and like all such values can be produced in several ways. In the introduction to factorial procedures in Chapter II, factor scores were shown to provide measurements of the importance of each <u>observation</u> on the various factors in the R mode approach. This means that a j variable x i observation matrix has been summarized minimally by:

first, a j x p factor loading matrix;
second, an i x p factor score matrix.

Each row of the factor score matrix, therefore, locates each observation in the new mathematical space of the factors p. This means the factor score matrix is a summary of part of the variability in the original data matrix, one that measures the observations in terms of factors. So in the simple case of Booth's (1893) 27 areas x 6 variable data matrix described in Chapter II, Davies (1978) showed the factor score matrix consisted of scores on 27 areas on one factor.

Since factor scores are composite measurements and, like all composite values they can be derived in several ways. An example may help illustrate the point. In Table 6.8 the actual values for each of the six variables in the 1961 data matrix for one of the census tracts for Calgary described in Chapter II. It has previously been shown that a two-axis solution is the most appropriate result for the data set. The factor loadings (actually components in this solution) are usually converted into factor coefficients by dividing them by the eigenvalues in the same way as shown in Table 2.12 so as to produce values weighted by their acual importance to the solution. Loadings could be used directly, for Parkes (1973) found that scores calculated in this way gave more interpretable results in his factorial ecology. These are shown in rows 3 and 5 of Table 6.8b. Clearly the different loadings shown in Table 5.2 for various factorial solutions will produce different coefficients so there is always the potential for instability in the size of the scores. Only in the case of the determinate Principal Axes component solution where the components have a fixed position is there one set of scores. Obviously changes in the value of loadings because of rotation can lead to different scores, so it is important that the choice of factor loading solution is carefully made. The fact that these values are composite measurements must be emphasized, for like all composites they can be made up in different ways. An example from the data matrix for 1961 census tracts in Calgary illustrates the point.

Six measurements for Area 1 on the various variables are shown in Table 6. (row 4) and have been converted into z scores (deviation from mean divided by standard deviation). These are multiplied by the factor coefficients first for Factor I and secondly by Factor II. So the z score of +0.352 for components are derived from a Principal Axes Component solution followed by Bi-quartimin rotation. The occupation variable multipled by the coefficient for Component I produces a value of −0.649, the contribution of the occupation variable to the factor (actually a component) score for this first axis. The addition of the values in row 3 produce the score of 1.18.

In this example, it can be seen that <u>all</u> the variables contribute something to each factor score. However, since the variables are structured into two different dimensions (Component I and II), most of the contribution is likely to be associated with these sets of variables. For Component I it is the ethnic, occupation and education variables as expressed in the factor coefficients that should make the major contribution. For Component II it is the fertility, women-at-work and single family dwelling units that are of major importance. However, for any axis the potential of these relatively high factor coefficients to be translated into a high factor score is only realized if the particular measurements for these variables on the observation in question are also high. Table 6.8 shows that the occupation and

Table 6.9
Calculation of component scores

(a) Descriptive statistics

	6 Ethnic- ity	1 Occu- pation	2 Educa- tion	3 Fertil- ity	4 Women- at- Work	5 Dwell- ings S.F.
Mean	38.66	25.54	56.11	58.19	33.89	66.43
Standard Deviation	8.23	8.65	10.23	20.54	6.72	22.24
Values: Census Tract 1	41.70	41.50	44.60	69.19	31.29	66.60
Deviations from mean	+3.04	+15.96	−11.51	+11.00	−2.60	+0.17
Z Scores	+0.37	+1.84	−1.13	+0.54	−0.39	−0.008

(b) Composition of component scores: Calgary

Row	Component I	6 Ethnic- ity	1 Occu- pation	2 Educa- tion	3 Fertil- ity	4 Women- at- Work	5 Dwell- ings S.F.	Score
1	Score on I							+1.183 +1.197 −0.014
2	Contribution	−0.130	−0.649	−0.384	−0.034	+0.013	+0.001	
3	Coefficients	−0.353	−0.352	+0.341	−0.063	−0.034	+0.057	
4	Z Scores	+0.369	+1.843	−1.125	+0.536	−0.387	+0.008	
5	Coefficients	−0.060	+0.004	−0.049	+0.392	−0.353	+0.373	
6	Contribution	−0.022	+0.007	+0.055	+0.210	+0.137	+0.003	+0.412 −0.022
7	Score on Component II							+0.390

education variables had relatively high values, as shown by the Z
scores over +1.0. (The direction of the signs is a reflection of the
fact that the variables were inversely related, i.e., high levels of

low status occupations were related to low levels of formal education).
So these two measurements contributed most to the composite score. In
this particular area (census tract 1) the contribution of the ethnic
variable was relatively low because of low levels of ethnicity. For
most areas in Calgary, however, ethnicity must be closely associated
with the occupation and education variables, otherwise the single
dimensional structure linked to the three variables would not be
produced.

The same arguments apply to the score on Component II. The fertility, women-at-work and single family dwelling unit variables are of
major importance as weights in the calculation of the scores. This
time, however, the single family dwelling units are quite low, or
rather are near the average for the city. The result is that in this
particular area this particular variable makes an insignificant
contribution to the factor score.

This example demonstrates that since the factor scores are composite
measurements, they are dependent upon the size of the loadings for a
factor on a variable, as well as the original size of the measurement
of the variable in the observation unit. Like all single composite
values, therefore, these scores can attain similar values from several
sources. For example, when one assumes three variables are important
in the contribution of any particular factor score, a value of 1.5
could come from a high contribution of 1.5 from one measurement on a
variable and low values on the other two or three equal contributions
of 0.5. Horn (1973) using King's (1969) study of Ohio showed that the
implication that scores are produced by similar sets of standardized
values is not true; they can conceal wide deviations. In empirical
investigations, therefore, it is always worth checking the factor score
matrix against the original data set to determine why high or low
scores were derived, before particular interpretations are given to the
values, at least in the case of the larger scores. If a variable has a
skewed distribution the extreme values can really influence the size of
the scores. This explains why the transformation procedures (Clark
1977) in Table 4.9 demonstrate that transformation affects scores
rather than the loadings, even in structured data sets. Moreover it is
obvious that if a new variable (A) highly correlated with another (B)
in a vector is added to the data set, the factor structure will not
normally be affected much if it is a strong vector, but the
collinearity will influence the score since the score will be weighted
by an extra amount because the effect of A will probably be doubled by
the addition of B.

The fact that the score for any area on a component is based on all
contributions from all the variables was emphasized by Joshi (1972) in
his study of social areas in Katmandu. He pointed out the incongruity
of using all the loadings, or rather the factor coefficients, on these
variables when the factors themselves were only interpreted in terms of
the highest loading. His suggestion was to calculate scores for a
factor only using the values (factor coefficients) that were
interpreted as part of the factor loading matrix. In most cases, as in
the example in Table 6.8, the minor loadings on an axis are likely to
be too small to have much effect in the calculation of the scores –
except when there may be a large number of variables. In such cases
these individual insignificant values may add up to a considerable

total. Joshi's (1972) point relates to an issue with long history in applied factor analysis. Indeed Cattell (1957) suggested that individuals in psychological studies might be subject to less misinterpretation if a group of variables were selected to represent a factor score instead of calculated scores. The values or scores for such an area would be simply summed from the values of the observation on these variables. Standard scores, of course, have to be used, otherwise variables with large means and deviations will have proportionally greater weight in the final value. In the case of area 1 in Table 6.8, this would have produced an average value of 1.15 for Axis I if the variable signs are co-ordinated by scaling the values in the same way, and 0.56 for Axis II if the same procedure is applied. This is hardly different to the calculated scores and is a function of the fact that the two sets of three factor coefficients on each axis were all so similar. An alternative is to simply choose a key variable and its value in an observation unit as representing the factor score. Although this avoids the problem of differential weighting, like all single indicators it has the disadvantage of only measuring one part of the dimensional structure represented by the factor, and will be subject to its own individual idiosyncracies. Obviously both approaches provide simple approaches to the study of the variations in the areas, but the fact remains that it is useful to weight each measurement on a variable in an area by the contribution that variable makes to the size of the factor dimension. Other workers (Spence 1968; Smith 1972) have advocated weighting each score vector by its associated eigenvalue to take into account the relative importance of each axis although it can be pointed out that the size of the factors may be a reflection of the composition of the particular variable set and not of any real conceptual importance.

(b) Approximation procedures for common factor scores

So far the discussion has been phrased entirely in terms of the component model. When a common factor method is employed the calculation of scores becomes much more difficult and complex. The reason is simply that in the common factor approach the variance is portioned into two parts: common and specific. This can obviously apply to the variation in each row of a data matrix. For each observation or area in the standard R mode approach, there are a set of measurements on the individual variables. Each value or measurement part is attributable in part to the common variance, whilst an additional part is linked to the specific variance and error. In the case of some variables most of the variance may be specific since it cannot be predicted from others. Since the scores for common factor solutions are only based on the common variance it is clear that the scores have to be estimated, the difference from the component scores being the fact that they do not take the specific variance into account. So they cannot be exact; they only estimate part of the original observation values.

In the factor analysis literature Guttman's (1955) review of the determinacy of factor scores is still a basic paper in the field, as is the work of Glass and Maguire (1966), Harris C.W. (1967), Horn and Miller (1966), Moseley and Klett (1964), Schweiber (1967), Schoemann and Wong (1972) and Green (1976). These papers demonstrate that even if a unique rotation is found there are still many potential common factor scores which will account for the variable scores of the cases, although the scores are usually mutually correlated. A number of different methods exist for estimating common factor scores and the results have been compared by Baggaley and Cattell (1956) and McDonald and Barr (1967). Authorities such as Harman (1976) prefer the regression approach in which the data matrix is regressed on the loading matrix, using least square methods. An important assumption of this approach is that the average proportions for the variance of a variable are assumed to be the same for each observation. This ensures that the common factor scores are bound to be estimates of the true values of each observation, with the complication that these observations have proportions of specific and error variance. In the factor analysis literature, Velicier (1975) has described the relationships between common factor, component and image scores whilst Horn and Miller (1966) have described the problems in calculating common factor scores; issues that should be taken into account by factorial ecologists. Before leaving this general problem area it must be emphasized once again that most of the developmental work in factor analysis assumes a homogeneous sample or population is being dealt with. Cattell (1978 p.313) has observed that applied workers may be dealing with several groups with different means and standard deviations. This is certainly the case in most factorial ecologies, where sets of different areas or regions are being dealt with. This compounds the problem of estimating common factor scores and may lead to differential fits of the common factor scores, in the sense that the fit may be chosen for one set, but not very good for others. Until factorial ecologists have a better understanding of the sources of variation in the social character of cities and regions, it can be argued that it may be better to use scores from a component model if the purpose is simply to describe a data set in terms of fewer dimensions. Common and specific variance will be mixed up, the loss will be the error variance. Such a conclusion is obviously anathema to the inferentially minded investigator, but it is a compromise favoured by many empirical investigators. Once enough studies and replications have been carried out in particular areas to more clearly delineate the expected sources of variation, so common factor methods, including scores, will be more and more favoured.

(c) Description of spatial patterns

The most widespread use of R mode factor scores in factorial ecology, whether of the component or common factor variety, is as a scaling or measurement of the observations on the factors. These scores are usually mapped to portray the spatial variation in the dimension isolated by the analysis. As all the data is usually originally collected on the basis of a defined observation unit (frequently an area such as a census tract in intra-urban analysis), it is important to relate the scores to these units. This means that the most accurate method of mapping the data involves some variety of the chloropleth method, in which a series of shadings are used for a range of factor score values. Many different sets of shading schemes and sets of cut-off values can be found in the literature. But since the factor scores are usually expressed in standardized units one of the simplest is to use a set of 0.5 units such as > +1.5, 1.49 to 1.00, 0.99 to 0.5, +0.49 to 0.0, 0.0 to -0.49, -0.5 to -0.99, -1.00 to -1.50, > +1.50, plotting the actual values for the extreme values to identify the places with highest scores on each axis.

Now that census information is available for smaller and smaller areas and computers can handle large data sets, the size of data matrices handled by factorial ecologists has increased. This frequently creates problems for those interested in mapping. Chloropleth maps for so many small areas are difficult if not impossible to reproduce at the normal page size. This is one reason why many factorial ecologists have started to apply isopleth mapping techniques. As Figure 1.1 showed in the case of Calgary, an isopleth is a hypothetical line joining places of equal value. If the factor scores on a factor are first plotted in their areas, the continuity of intra-urban social areas often makes it possible to draw in isopleths for particular values that are estimated between the individual points. Theoretically, of course, one should _not_ use the procedure, since the scores are usually based on average values for areal units. Nevertheless, they provide a good _visual_ interpretation of the highs and lows in areas, refining what amounts to the _social topography of regions or cities_. In studies where a large number of enumeration areas have been analyzed, as in the 541 areas of Cardiff studied by Davies (1983), isopleth maps seem to be the only realistic way of portraying the general social variations of the city. As a generalization technique, therefore, the approach has much to recommend it. More sophisticated generalizing techniques such as trend surface analysis have also been applied to scores (Goheen 1970). This type of approach defines and measures the various trends or surfaces that can be fitted to the distribution of points. As such, it represents a further generalization of the spatial pattern of variation, since the minor or local patterns will be eradicated by the lower order surfaces.

Most investigators go beyond simply using the individual factor score maps as separate portrayals of the social complexity of cities or regions. Taxonomic techniques are often used to integrate the various factor scores into a single set of categories or taxonomies. A variety of different procedures have been used, most popular being the variety of techniques known as cluster analysis (Sokal and Sneath 1973; Wishart 1975). Hierarchical methods used to be the most popular, in which each observation unit is successively grouped with another on the basis of a summary measure of the similarity of the observation units such as a euclidean distance measure across all the scores. A disadvantage is that the units, once formed, remain inviolate. As a result, non-hierarchical methods (Wishart 1975; Bartlett 1977) have proved to be increasingly popular since re-allocation procedures allow individual units to be detached from their initial groupings allowing assignment to the most similar higher order group. Tests based on other multivariate procedures, such as Discriminant Analysis have also been applied in recent years in the search for more definitive types of clusters. However, in the last resort it must be recognized that there is no single or best cluster solution suitable for all purposes.

Apart from their use in mapping, factor scores have frequently been used to determine the precise spatial patterns of particular dimensions. Anderson and Egeland (1961) were the first to demonstrate the methodology, although in the context of social area indices, not factor scores. Rees (1970) provided a comprehensive example of the approach using factor scores. Basically the scores for the areal units are arranged in terms of a set of distance zones and directional sectors from the city centre. Analysis of variance techniques can then be used

to determine whether the scores vary be direction or distance. Rees's
(1970 p.371) Chicago study showed that the major dimensions of Socio-
Economic Status and Family Status vary significantly by both sectors
and zones. Rees suggested that Burgess (1925) and Hoyt (1939) were
both correct when they observed that Social Status varied by zone and
sector. In this case, both hypotheses are relevant for the dimension,

Table 6.10
F ratios and dimensional structures

Area	Sources	1960/1 Size '000s	1. Socio-Economic Status	2. Family Status
1. Chicago-Northwest Indiana	(Rees 1970)	5,959	1.15	2.60
2. Metropolitan Toronto	(Murdie 1968)	1,824	30.00	10.70
3. Indianapolis	(Anderson and Egeland – for 3-6)	639	1.44)	
4. Dayton		502	5.89)	
5. Akron		458	6.03)	29.90
6. Syracuse		333	19.89)	

Socio-Economic Status calculated between sectors and zones.
Source: revised from Rees (1970 p.374)

although many other studies have shown that sectoral variations in the
dimension are more important. When the outermost zones were excluded
in Chicago sectoral variation proved of greater significance. In
contrast, zonal variation (or concentricity) proved to be of more
importance in the case of Family Status variations. In the comparison
of several cities shown in Table 6.10, Rees (1970) concluded that:

'there appears to be a regular ordering, by size of city, of the
relative contributions of zones and sectors to the variation in
socio-economic status and family status. For the larger the city
the greater the importance of zonal variation of socio-economic
status as compared with sectoral variation, although sectoral
variation remains the more important in all the cities. Similarly,
the larger the city the greater the importance of sectoral variation of family status as compared with zonal variation, although
zonal variation remains, by far, the more important in all cities.
The only exception in this size-governed continuum would seem to be
Toronto ...' (Rees 1970 p.373)

These results indicate that there is not a simple equivalence between
Socio-Economic Status and sectors, or Family Status and zones or
concentricities. Subsidiary zonal patterns of Socio-Economic Status
can be found, and these are usually linked to the presence of low

status groups in the inner city. Frequently these groups are culturally different, so clusters of ethnicity are found to be added to the
general patterns. Other spatial patterns are described in Chapter IX.

Table 6.11
Analysis of variance results: Prairie metropolitan centres 1971

Component Titles	F Ratios
1. Family status	3.08 *
2. Socio-economic status	1.31 '**
3. Migrant status	1.58 **
4. Participation-non affluence	7.55
5. Ethnic I (British-East European)	70.43
6. Ethnic III (Canadian-immigrant)	31.14
7. Housing Age	9.87
8. Ethnic II (French-German)	98.66
9. Young adults	5.87
Significance Levels (a) 5%	2.4 **
(b) 1%	3.3 *

**Within city variations greater than between cities at level shown

Source: Davies (1978)

Rees's (1970) comparison between sets of cities using factor scores
as the basic measurement of the areal variation demonstrates how
factorial ecologists can derive generalizations across sets of cities.
Davies (1978) provided another example in his study of community areas
in five Canadian metropolitan centres. A joint analysis study produced
directly comparable factor scores for the 365 areas in the five cities
on nine dimensions. Analysis of variance techniques (Table 6.11)
revealed that only the three largest dimensions, Socio-Economic Status,
Family Status and Migration Status had more important variations within
cities than 'between cities'. In other words, there were higher
variations in the importance of these axes within a particular city
than between the cities. By contrast, the other axes, including the
ethnic dimensions, displayed greater differences between the cities,
indicating the presence of more city-specific variability in these
relatively minor axes. It is worth noting that these values were
component scores, so specific and common variance were combined in the
values. Since the approach needed to identify the city-specific
sources of variation (such as ethnic variation) it does appear that a
common factor approach would not have been so successful in isolating
the differences in these values.

(d) Use in explanatory studies

The primary concern of this study is with the structures that can be isolated from data sets. However, many investigators have been more interested in using factor analysis procedures to summarize or rewrite the data sets in a more parsimonious form so that these new values can be used to predict the variations in some other characteristics. For example, a study of the migration of people from a set of areas could profitably use a series of variables to define the character of these units. These variables can be summarized by factorial methods into a reduced set of vectors and they can be used as independent variables in a multiple regression estimation of migration flows. A frequent advantage of the approach is that problems of collinearity in the data are reduced, or alternatively some of the highly correlated variables on separate axes can be removed. In this latter case the factor method can be used as a screening device to remove unwanted indicators. More typically, however, scores are used to identify the areas of high or low character on some dimension of variation. Then these areas can be used as sample areas in which to carry out interviews of some behavioural characteristic. Robson (1969) and Herbert (1976) provide examples of this sort of approach, in studies of parental attitudes to education in Sunderland and juvenile delinquency in Cardiff respectively. A follow-up study by Herbert (1977) demonstrated that the spatial variations in delinquency could also be interpreted by means of factor scores. In this case a set of Varimax scores derived from a Principal Axes analysis of a set of census based variables were used as the independent variables to explain the variation in delinquency (measured by the numbers of juvenile offenses per area). Table 6.12 shows part of the results of a stepwise regression adding in the third, fifth and second factors. The first three factors explained 36% of the variation in delinquency rates in the city. Herbert (1977) emphasized that this approach cannot be thought of as being exclusive, in the sense of explaining all the variation. After all, any urban environment can be conceptualized as possessing a built-form, the quality of buildings and

Table 6.12

Stepwise regression results for delinquency study

Step	Factor Added	Highest Loadings* on Factor Added	S.E.	r2
1	3	+Overcrowding; unemployed males −Owner occupied; social class 1 + 2	1.72	0.15
2	5	+Unemployed males; foreign born −Sex ratio	1.60	0.27
3	2	+No amenities; no fixed bath; shared dwelling −Social class 1 + 2	1.51	0.36

* Varimax values   S.E. = standard error

Source: summarized from Herbert (1977 p.92)

spaces, a personal social environment linked to life styles, attitudes and behaviours, and an impersonal social environment, one that can be measured by its class or demographic structure, etc. Clearly a set of social indicators derived from the census only relate to this latter type of social environment. Such indicators provide a measurement of the local area or neighbourhood effect to set alongside the variation in delinquency associated with the behavioural variables. This is the contextual approach described in Chapter III. Given the variety of sources of variation, the level of explanation shown in Table 6.12 seems relatively high. It is possible, of course, that these built-form and impersonal social environments are intimately linked to variations in personal social environment, so the context of behaviour reinforces behavioural. After all, the factors, as described by the sample variables shown in Table 6.12, relate to poor physical conditions and larger numbers of low social class people. This means the dimensions of low Socio-Economic Status, Substandardness and perhaps Ethnicity are the ones linked to delinquent behaviour. Evidence of the importance of these dimensions of the social environment in accounting for other social variations, this time mental health has been provided by a series of studies (Bagley 1968; Giggs 1973; Bagley et al 1976). More recently, Dean and James (1981) working in Plymouth showed that the components of Accommodation, Quality and Single Person Migrants were most important in explaining variations in the incidence of male mental illness. However for females a Social Class construct became more important for high levels of re-admission to hospital although the Accommodation axis was important for lower admission levels. These results show the utility of factor methods in isolating variables of social area character to account for variations in personal behaviour.

6. SUMMARY

This chapter, like its two predecessors, has been concerned with some of the basic problems involved in the interpretation of factor solutions. Now that human geographers have begun to obtain experience in the application of factor methods many of the early problems revealed in the literature will be avoided. Nevertheless the need to routinely scrutinize alternative solutions and to search for a stable structure is a very necessary task for all factorial work if the results are to be used to build up a body of cumulative research. In this context it is fortunate that factorial ecologists can be guided by more advanced work in factor applications in psychology, although the areal concerns of the ecologist mean that different problems have to be faced. What is really worth emphasizing is that all applications of factor analysis will produce some type of results, so their utility will depend upon the care taken in problem definition, data screening, choice and justification of solutions. This may be particularly important in relation to the use of factor scores and explains why some investigators have preferred to go back to the original variables as indicators of sources of variation rather than depending upon some composite weighting process.

Before leaving this general topic of interpretation it is worthwhile re-emphasizing that factor analysis can, if applied with care, be used to:

(i) explore the interrelationships in a data set;

(ii) provide a measure of parsimony;

(iii) create new measurement scales for variables and areas;

(iv) transform a data set into orthogonal dimensions for use in other multivariate procedures;

(v) provide a means of formulating concepts or constructs; and

(vi) test the hypothesis that a certain number of common factors can be found in a data set, using the common factor model.

Such a range of objectives means that factor ecologists must be careful to specify in advance which goal is being pursued.

Before leaving this topic a final cautionary note about the utility of factor methods in causal analysis must be made. One of the best ecological examples in this context is Meyer's (1971) study of the relationships between correlation and factor analysis using a ten variable study of Detroit and Memphis. His results showed that Socio-Economic and Family Status axes could be identified in both cities using only the census tracts with larger numbers of blacks. In Table 6.13 part of the results for Detroit are shown. It can be seen that the correlations between the individual variables loading highly on the Socio-Economic Status axis are lower than the factor loadings. This is nearly always the case because it has been shown that the component is, after all, a new hypothetical 'variable' which can be viewed in a descriptive sense as 'averaging' the relationships between the variables.

Table 6.13
Socio-Economic Status variables, loadings, correlations: Detroit

Variables	Loadings *	Simple Correlations				
		1	2	3	4	5
1. Education	90	—	78	75	69	66
2. Family income	93		—	70	54	81
3. Sound housing (%)	77			—	51	59
4. Value of housing	66				—	40
5. Rent	80					—
6. % Females Labour Force	76					—

*Loadings > 0.3 on Socio-Economic Status only

Source: revised from Meyer (1971 p.337 and 338)

215

Meyer showed that in Detroit the simple correlation between the Education and Rent variables was quite high at 0.66. But when the variable Family Income was controlled for, by calculating the partial correlation between the two variables, its value was only 0.08. This implies that the Education-Rent association is a product of their mutual link with Family Income. By contrast, the Education and Sound Housing variables showed a simple correlation of 0.75, but the partial correlation between the two calculated by controlling for Family Income was 0.45. This means that Sound Housing is an important additional basis for the differentiation of black households by education in the black residential areas. So Education and Sound Housing can increase in black residential areas even in cases where the family income is constant. From a scrutiny of the factor loadings alone, it may be erroneously deduced that an increase in Education and Sound Housing is always linked to an increase in Family Income. Meyer concluded that factor analysis provided a mechanism for defining the general patterns of interrelationships in Detroit by means of a new construct or measurement. In this process some of the information about the variables is lost. This means that if a human ecologist wished to make statements about the relationships among specific variables of theoretical interest, correlation, not factorial methods, are the most appropriate. Now that the major basic pitfalls in the field have been identified, it is necessary to turn to the substantive results that have been obtained in three of the areas to which factor analysis has been extensively applied, namely the study of regions, inter-urban and intra-urban structures and patterns. As always in a developing field, some of these results can only be considered to be indicative given problems with the techniques used. Nevertheless by summarizing existing knowledge in certain fields a set of generalizations are created which can be more rigorously tested by future factorial work.

7 REGIONAL CHARACTER AND IDENTIFICATION

One of the most fundamental and enduring ideas in the literature of geography and human ecology is the regional concept. Regions are one of the ways researchers can organize, summarize or portray information on the spatial variations of the earth. Although a variety of alternative organizational procedures can also be used to describe the spatial character of areas and phenomena, regions still represent one of the most useful ways of fulfilling these objectives. Since both the creation of regions and the identification of regional character are primarily problems of synthesizing a variety of variables, factorial procedures can help in these tasks. This chapter demonstrates the utility of the factorial approach to the study of regions in two very different fields: first, the creation of regional generalizations in the study of formal regions; and second, the identification of functional regions. Before dealing with these specific issues in the second and third sections of this chapter, it is important to set the scene by summarizing the essential characteristics of regional study in the first section.

1. REGIONAL TYPES

Although many categories of types of region have been recognized it is traditional to draw a basic distinction between formal (homogeneous or uniform) regions and functional (or polarized) regions. The major difference between them comes from the type of information being dealt with: in the former case from the formal characteristics of places; in the latter from the flows, or functional relationships of places. Despite this difference of content both types of region are really similar in concept. Each can be defined as a homogeneous area of earth space composed of a series of contiguous locational entities having some similarity to one or more specified attributes or characteristics. Regions based on one attribute or characteristic are called single-feature regions; those based on several characteristics are called multi-feature regions. Where the complete set of characteristics of area are being dealt - as in most texts in regional geography - the term compage (Whittlesey 1956) represents a useful description, even though it is rarely used today.

In the case of the formal region the similarity or homogeneity that leads to the cohesion of the individual entities comes from some formal characteristic of the space being considered, such as a certain minimum level of ethnicity or an annual minimum amount of rainfall. Places that have less than these operationally defined amounts are not sufficiently similar to be considered to be part of the region. In the case of the functional region the homogeneity comes from some shared interaction or connectivity, producing an internal regional cohesion that is greater within than outside the area. Nodal regions, as Brown and Holmes (1971) have shown, are particular types of functional regions, since their cohesion comes from a common focussing upon one node or focal point.

All regional studies involve two very different tasks:

1) the characterization of the essential features of the region, the attributes that make it distinctive;

2) the identification of the area defined as the region, usually referred to as the regionalization or region-building process.

In many studies these two tasks are usually integrated within the confines of a single study. For example, in the qualitative approach to the study of regions the intuitive identification of the various regions is followed by a subjective interpretation of the character- istics of these areas. In the quantitative study of regions the tasks are usually separated. The result is that explicit attention is paid to the definition of the sub-areas upon which the characteristics of differentiation are being measured, as well as upon the choice of the variables or sources of variation being considered in the regional study.

In examples involving the definition of single feature regions, or regions based on small number of variables, the characterization of the region is a relatively simple matter since it is a function of the 'cut-off' decisions used for the variables. For example, in a study of the ethnic structure of cities, a black ghetto could be defined as the area in which more than half the people are non-white. An investigator would normally plot all the enumeration areas, city blocks, or whatever sub-unit is used to collect the information, with these character- istics. Those that are dominantly non-white and form a continuous area would be defined as a non-white region; those that are not contiguous with this area would presumably be outliers or form additional regions. This description of the simplest type of quantitative region-building process obviously focusses attention upon the decision rules used to define regional character. Why is fifty percent non-black used? Does the use of other areal units to collect or process the information produce different regions? Investigators have to justify these decisions if the generalization represented by the region is going to be accepted by other workers as a characteristic spatial pattern for the area. For those who prefer the qualitative approach, these decisions are made implicitly. An intuitive leap is needed to produce the regional definition and its characterization. This intuitive process is rarely exposed to review, although presumably some justification of the description is needed if the regional pattern is going to be accepted.

The problems of regional definition and characterization become much more difficult in the case of the multi-feature region since they involve the synthesis of different types of information. For those who favour the qualitative approach the integration of the spatial variables is an intuitive step by an investigator. It is linked to the degree of preparation of the researcher, as well as to the investiga-or's experience and synthesizing ability. For those who prefer the quantitative approach, the problem of synthesizing the data into component parts is one that can be solved by a variety of multivariate procedures. Certainly the family of factor analysis procedures provide some of the most useful ways of dealing with these problems of regional study. The next two sections provide case studies of the utility of factor analysis in characterizing the essential features of regional variation in formal regions, and the identification of functional regions. In both cases the individual factorial results are set within the context of alternative approaches to regional characterization and identification so that the methodological utility of the approach can be assessed.

2. GENERALIZATION IN REGIONAL GEOGRAPHY

(a) Difficulties of traditional regional description

The development of a quantitative approach to geography came under particularly strong attack in the early 1970s by traditionally minded geographers who condemned what was called the 'excesses' of quantifica-tion (Green 197), and the 'eclipse' of synthesis by analysis (Balchin 1972). Such critics pointed to what Fisher (1970) called the 'breadth of vision' which created regional generalizations from a concise inte-gration of a series of spatial distributions. In many ways these critics attempted to reestablish the traditional role of regional geography (usually expressed as the study of compages or the totality of phenomena in area). Yet such criticisms were inappropriate since they ignored the very real progress made by quantitatively minded geographers in the development of regional principles and techniques that began in the 1960s and continued to the present day. Examples can be found in the work of: Berry (1964) in relation to regional taxonomy; Grigg (1965) in terms of the logic of regional systems; Spence and Taylor (1970), as well as Johnston (1968, 1978) in the problems and approaches to regional classifications; Brown and Holmes (1971) and Davies (1980) in functional regionalization. These studies show that exponents of a quantitative approach to geography have begun to formalize an explicit body of regional principles and methods, features that were really only implicit in the work of previous generations. In addition, the point of this book that quantitative methods can be used to provide syntheses, as well as analytical studies, seems to have been ignored. This means that there is little point in challenging Fisher's view of the importance of the 'breadth of vision' represented by the regional generalizations produced by the 'new' geography of the early twentieth century. In any case there is a great deal of scientific support for the viewpoint that the creation of basic generalizations and concepts should always have primacy over the problems of measurement and empirical case studies. What is difficult to accept is the apparent belief that the quantitative approach cannot be used to help create and justify these generalizations. However stimulating the original ideas of geographers such as Mackinder and Vidal de la Blache proved to be, an excessive concern with the generalizations of the past masters is hardly a healthy sign in any discipline. It implies that the methodological approaches used by geographers working in the early twentieth century are still the major paths by which substantive generalizations can be achieved, and ignores the very real reservations that were expressed by some of these early geographers about the regional approach (Buttimer 1971).

Any review of the literature in regional geography will reveal that the field has always been <u>fact</u> rather than <u>concept</u> oriented. The result has been that the scope for the communication of regional principles or generalizations has been limited. Even when these principles were defined they tended to be overwhelmed by the wealth of factual material. This conclusion can be easily justified by reference to any standard regional text, for few of these studies differentiate between the various types of region, or try to justify the regional divisions used in the general description of spatial variations. The result is that most geographers have learned their regional principles inductively, rather than as part of a coherent methodology. Moreover, the specific spatial features of the region under study, whether Canadian Prairies or the Spanish Meseta, were always more important than the route by which these regional differences were derived. In many ways this is unfortunate for a discipline attempting to build up a corpus of concepts and principles, primarily because some of the greatest intellectual efforts in regional geography lie in the construction of the initial set of regions and the creation of the regional generalizations. Unfortunately, the logic of these procedures are rarely described; the organizational basis of the approach produces results that are presented primarily on a 'take it or leave it' basis.

The empirical emphasis of the traditional regional approach led to another problem in geography: the virtual abandonment of any attempt to build a theory of regionalization or even a series of distinctive regional concepts. Today the result is that in many areas other social scientists have pre-empted the field, and produced ideas such as 'growth poles' or 'theories of polarized development' (Friedmann 1973; Perroux 1955; Lausen 1969) giving an applied, or prescriptive value to their studies in addition to their ordinary, descriptive importance.

It must also be remembered that even the empirical study of regions is no longer one of exclusive geographical or even academic concern. Breakthroughs in mass media and communication, plus the increasing mobility of people, all mean that geographers only represent one group of individuals who are describing areas on a professional as well as a popular level. These trends, combined with the acceleration in the number of facts known and recorded about places, have broken down the virtual monopoly once possessed by geographers about the dissemination of information about areas. In addition, the rapidity of change in the socio-economic sphere, which alters the social patterns of areas, tends to provide additional support for the development of an analytical tradition that defines the role of geographers in terms of their success in the development of spatial <u>concepts</u> and <u>generalizations</u> dealing with distributions, organizations and change. Such an approach provides an additional focus to the traditional concern with the particulars of a place or region at any one time.

A final set of problems created by the need to provide spatial integrations of facts need to be considered. In the first place there is the difficulty of successfully coping with more and more pieces of information. Second, there is still a tendency for regional texts to be agricultural or landscape oriented, at a time when society is increasingly urban and mobile. If accurate generalizations are to be forthcoming these changing societal features need increasing emphasis. Third, even if one accepted the traditional argument that the region is the core of the discipline of geography, and tested this against the majority of textbooks, there can be little dissent with the view that concise and coherent generalizations are few and far between. The chapters of most texts contain a series of individual fragments, without developing an overall generalization or series of conceptual statements. Perhaps one can explain this by the fact that it takes a great mind to successfully integrate facts about area into an incisive regional generalization. The gift of generalizing in the Mackinder or Vidal de la Blache fashion is given only to a few, and it is carping to criticize the efforts of geographers with lesser ability. Yet this type of viewpoint lies dangerously near an acceptance of the role of vision or 'genius' in creating the vital regional generalizations that are seen as the ultimate justification for the discipline in some eyes. There is usually a great deal of subjectivity involved in the derivation of these generalizations, and these ideas are subsequently utilized by individuals of less vision. This would seem to confirm that geography, or at least the traditional approach to regional method, is to this particular school of thought, an art rather than a science.

These difficulties should not, however, only lead to the alternatives quoted in the introduction. There is no need to give up the hope of ever producing an objective regional generalization or to revert to the halycon days of yesterday. There are many paths to understanding and if human ecology or geography is going to claim sophistication then it must discuss alternative forms of investigating its problems. This does not imply that the mental process by which all individual concepts and generalizations are derived can ever be completely isolated. That is likely to remain an intractable philosophic problem. What one can see, however, is that any specific generalization is probably produced by individuals with sufficient knowledge of the relevant factual background, and with enough overall intellectual ability and determination to complete the task. Yet even this is not enough. Familiarity with the spatial approach, or with reasoning by analogy from other disciplines, is needed to complete the normal cycle involved in the creation of a generalization. In practice, most geographical concepts are still learned by rote, rather than being created anew each time. Dickinson (1970) has reminded us that it is likely to continue as a fundamental part of the learning process. However, the question must be posed as to whether individuals without a background in the spatial approach, or geography students without the very highest intellectual ability, can be taught how to derive generalizations with a spatial basis. In other words can the facts relevant to any situation be ordered or manipulated in such a way that their integration into simpler concepts is but a short step away?

Factor analysis would seem to provide a way by which spatial based generalizations can be derived. Obviously it is but one of the ways of

achieving this goal. But if it proves useful, then the integration of
spatial facts may cease to be only the preserve of a first class integrating mind. In other words factor analysis may provide a technical
aid in the process, and this might provide as big a breakthrough in
geographical simplification, concept formulation or even testing (by
the common factor approach), as radio, television and films have provided in the dissemination of knowledge about areas.

(b) A case study of French planning regions

So far, this study has concentrated upon illustrating the basics of
factor analysis by means of simple examples, as well as pointing out
the variety of alternative procedures and problems involved in the
application of factorial procedures. Given this fact, and the arguments developed above about the problems of traditional regional
geography, it is important to provide a case study which will illustrate the utility of the factorial approach in the creation of regional
generalizations. The data set used in the study (Table 7.1) came from
Hansen's study of French Regional Planning (1968) which used data on
the variations of 33 socio-economic variables over the 22 planning
regions in France in 1961. The information was deliberately obtained
from an already published source to allow future investigators the
opportunity to replicate the study. In addition, the example had the
advantage of going beyond the use of the data in the original source.
In Hansen's study the variables were described in a series of groups
which implied that there were different sources of variation in the
data set. No attempt was made in the original study to apply multivariate analysis to the data. Instead, the individual variables were
considered to be diagnostic indicators of the variations in the
socio-economic structure of France.

Table 7.1
Principal Axes component analysis: French regions

(a) Variables and communalities

Variables for each region	Final Communality
1. Total population	98*
2. Percentage population change 1963-65	97
3. Median age 1963	65
4. Percentage employed in agriculture	95
5. Average farm size	78
6. Value added per worker	77
7. Value added per hectare	91
8. Percentage housing of agricultural population with running water	79
9. Turnover tax on business	98
10. Industrial energy consumption	85
11. Important business headquarters	98
12. Commercial enterprises: net charge	80
13. Baccalaureats gained	99
14. Students preparing for Grandes Ecoles	99

Variables for each region	Final Communality
15. Students in secondary education	98
16. Gasoline consumption charge: July/August 1955-64	86
17. Hotels: tourist mean capacity	91
18. Campers	81
19. Average white collar salary	75
20. Average blue collar salary	93
21. Wage earners: percentage with under F5,000/year	63
22. Low tension electricity consumption: per capita	30
23. Wide roads: over 7 meters	88
24. Letters distributed	99
25. Telephone traffic	99
26. Phones installed	99
27. Bank deposits	99
28. Bank credit	98
29. Savings bank deposits	99
30. Fiscal receipts for local authorities	99
31. Local authority operating outlays: per capita	89
32. Local authority investment outlays: per capita	75
33. Caisse des Depots loans	97

* Decimal points are removed: 98 is 0.98

(b) Eigenvalues

Component	Eigenvalues (Unrotated Matrix)	Cumulative Proportion of Total Variance
1	18.04	54.65
2	4.63	68.69
3	2.93	77.56
4	1.82	83.07
5	1.58	87.84
6	1.05	91.04
7	0.89	93.72
8	0.61	95.57
9	0.32	96.54
10	0.29	97.41
11	0.27	98.23
12	0.19	98.81

Table 7.1 shows the results of a Principal Axes component analysis applied to the Pearson Product Moment Correlations obtained from the 22x33 data matrix. This was followed by a Bi-quartimin rotation of the axes using the UCLA Biomedical Series Package Programme (Dixon 1968). The method was chosen because of the advantages described in Chapter VII, although it must be noted that alternative procedures such as the Direct Oblimin rotation produced comparable results. Four components

with eigenvalues explaining more than 5% of the original variance were initially abstracted. However, interpretation of solutions abstracting four or more axes showed that the addition of a fifth axis clarified the loadings on the first four axes. This illustrated the utility of the method of abstracting and rotating one more component than the investigator wishes to really interpret, sometimes a useful procedure which avoids the last interpreted component becoming a 'catch-all' for redistributed variance.

The five axis solution explained 87.5% of the variance. This means that instead of having to individually describe a 33 variable x 22 region matrix, the factoring procedure has boiled the problem down to two smaller matrices:

1) a 33x5 component loading matrix, together with the correlations between those components; and

2) a 22 region x 5 component score matrix.

In doing so the study has only lost 12.5% of the variability contained in the original correlation matrix although this total value conceals some of the variations between variables. The list of communality values indicates the respective contribution of the 33 variables to the general patterns expressed in the factors. Only three of the variables have communalities under 0.75, meaning that 56% of the variance of these variables is accounted for by the solution. Only one variable, per capita consumption of low tension electricity (variable 22 in Table 8), which is primarily related to the domestic consuption of electric power, has less than half of its variability accounted for by the five axis model shown in Table 7.1.

In this particular study rotation did disaggregate the generality of the largest dimension by reducing it from 54.6% in the first component to 50.3% in the rotated solution. But this is a relatively minor amount, and shows that a strong axis is a function of the dimensional structure in the data set and is not consequence of rotation. Indeed, no less than 18 of the 33 variables have their highest, or first ranking loading on this first rotated component, with another five having loadings over 0.30. Scrutiny of this list of variables reveals that they relate to indices measuring size and prosperity, for instance, population (variable 1), amount of turnover tax (variable 9), number of students (variables 13, 14, 15), business headquarters (variable 11), communication indices (variables 24-26), etc.

In terms of their content the cluster of variables associated with this first axis represents a heterogeneous collection, but as a common characteristic is their relationship to overall Size and Prosperity, it is as well to name the axes accordingly. An obvious word of caution must be introduced at this stage, given the intuitive leap that is involved in naming the factors (see Chapter VI, Section 3). It must emphasized the title is simply a short-hand description that attempts to isolate the type of variation being abstracted by the first axis.

In spatial terms the distribution of component scores on this dimension in Table 7.2 and Figure 7.1a reveals the overwhelming dominance of the Paris region with a score of +4.16. No other region

Table 7.2
Loadings of the oblique components: French regions

Variable	Loadings on the Components					Number of Loadings over 0.3	Variable
	1	2	3	4	5		
1	97*	− 0	17	1	12	1	1
2	48†	11	16	74*	− 3	2	2
3	20	68*	− 9	−34†	− 9	2	3
4	−35†	15	− 8	−80*	− 9	2	4
5	35†	−62*	−15	2	−50†	3	5
6	20	−34†	−25	66*	23	2	6
7	17	− 1	−32†	28	94*	2	7
8	18	−20	− 9	·79*	8	1	8
9	97*	−14	− 3	3	2	1	9
10	21	−34†	70*	36†	6	3	10
11	98*	−12	− 7	− 1	1	1	11
12	53†	57*	5	33†	−14	3	12
13	97*	14	16	1	7	1	13
14	100*	− 3	− 2	− 4	− 1	1	14
15	96*	4	16	5	11	1	15
16	− 6	90*	−10	−22	18	1	16
17	77*	42†	31†	4	− 7	3	17
18	13	89*	8	− 9	20	1	18
19	49*	− 7	27	44†	−30†	3	19
20	66*	−15	3	53†	− 1	2	20
21	42†	− 2	−22	−72*	− 5	2	21
22	−12	20	3	− 4	50*	1	22
23	14	12	90*	6	−22	1	23
24	100*	− 3	− 0	− 2	1	1	24
25	100*	− 2	− 5	− 2	− 1	1	25
26	99*	0	− 5	− 0	− 4	1	26
27	99*	− 2	− 8	− 0	− 1	1	27
28	98*	−10	− 9	− 4	− 4	1	28
29	97*	14	11	7	0	1	29
30	100*	1	2	− 1	5	1	30
31	87*	21	−34†	6	− 6	2	31
32	−18	63*	−44†	45†	−15	3	32
33	98*	11	11	− 4	1	1	33

* First ranking loading.
† Other loadings over ± 0.30.

Decimal points are removed: 97 is +0.97.

Principal Axes component solution followed by
Biquartimin oblique rotation.

Table 7.3
Component score matrix: French regions

Regions	Case	Rank on Component 1	Rank on Population Size	Component 1	Component 2	Component 3	Component 4	Component 5
Paris	1	1	1	4.15879	-0.83508	-0.63446	0.51859	-0.72962
Champagne	2	17	18	-0.41975	-0.82063	-0.60747	0.60519	-0.80616
Picardy	3	11	12	-0.28976	-1.37582	-0.99905	0.88345	-0.41469
Upper Normandy	4	12	15	-0.30170	-0.46281	-0.99566	1.06017	-0.08512
Centre	5	8	10	-0.19646	0.09856	-0.60129	-0.40934	-0.84778
Nord	6	5	3	0.12414	-1.71147	1.02138	0.56959	3.19711
Lorraine	7	10	8	-0.22717	-0.94221	2.25338	1.21125	-0.37791
Alsace	8	16	16	-0.39033	-0.04678	-0.79872	1.01358	0.99733
French Comte	9	20	20	-0.49719	-0.50111	-0.23577	0.74330	-0.67738
L. Normandy	10	18	19	-0.11305	-0.34792	-0.84776	-0.83596	0.76570
Pays Loire	11	6	5	-0.20387	0.11973	-0.37695	-0.86578	-0.31440
Brittany	12	9	6	-0.62569	0.23459	0.79786	-1.77467	1.62906
Limousin	13	21	21	-0.35778	0.96352	-0.28769	-1.89480	-0.09382
Auvergne	14	15	17	-0.44845	-0.06207	-0.09815	-0.74626	-0.79306
Poitou-Charentes	15	19	13	-0.03017	-0.38568	-0.44595	-0.82906	-0.06458
Acquitaine	16	4	7	-0.17871	0.35602	0.26858	-0.99318	0.75330
Midi-Pyrenees	17	7	9	-0.34863	0.48767	1.19202	-0.69233	-0.82078
Burgandy	18	14	14	-0.34863	-0.09631	-0.50874	-0.38471	-0.89054
Rhone-Alps	19	2	2	0.57390	0.98961	2.12671	0.64006	-0.17808
Languedoc	20	13	11	-0.32230	1.72313	-1.10452	0.88518	-0.02303
Province-Rivera-Corsica	21	3	4	0.53719	2.61523	0.88202	1.29588	-0.22582

N.B. This shows a typical computer print-out. Values are usually adjusted to two decimal places.

Table 7.3
(continued)

Component Names	1 Size-Prosperity	2 Resort-(Small Farms?)	3 Industrial-Construction	4 Rural	5 Catch-All (Residual Densely Populated Areas)
Highest Scores*	Positive	Positive	Positive	Positive	Positive
	Paris 4.16	Provence etc. 2.61	Lorraine 2.25	Provence 1.29	Nord 3.20
		Languedoc 1.72	Rhone-Alps 2.13	Lorraine 1.21	Brittany 1.63
		Rhone-Alps 0.98	Midi-Pyrenees 1.19	U.Normandy 1.06	Alsace 0.99
		Limousin 0.96	Nord 1.02	Alsace 1.01	L.Normandy 0.76
			Provence etc. 0.88	Languedoc 0.8	Aquitaine 0.75
			Brittany 0.78	Picardy 0.8	
				Fr. Comte 0.74	
		Negative	Negative	Negative	Negative
		Nord -1.71	Languedoc -1.10	Limousin -1.89	Burgandy -0.89
		Picardy -1.38	Picardy -0.99	Brittany -1.77	Centre -0.84
		Lorraine -0.94	U.Normandy -0.99	Aquitaine -0.99	Midi-Pyrenees -0.82
		Paris -0.83	L.Normandy -0.85	Pays Loire -0.86	Champagne -0.80
		Champagne -0.82	Alsace -0.79	L.Normandy -0.84	Auvergne -0.79
				Poitou-C -0.83	Paris -0.73
				Auvergne -0.74	

* Only scores greater than ±0.75 are shown.

227

comes remotely near it in importance, although the South-East also has positive scores, albeit very small ones. The Massif Centrale displays a minor depression in the overall pattern of negative scores. Apart from these minor variations, the pattern of loadings on this dimension indicates the extent of socio-economic agglomeration of this Paris axis, or 'pole' of growth, within the French economy. Hence, despite a generation of planning efforts to counteract this concentration, J.F. Gravier's (1947) summary statement of the human geography of France as 'Paris et le desert francaise' is shown to remain a persistent feature of the country.

Table 7.3 also shows the rank order of scores on this size-prosperity scale compared to the rank of population size. The two scales are significantly related, with only four regions showing rank order differences of three or more. Given the 'core area' generalization already made, it is not so surprising to find some type of 'spill-over' effect. Upper Normandy is the only region with a higher rank on the comonent scores than on population, whilst it is the peripheral regions in the west (Brittany, Poitiou-Charentes, and Acquitaine) that have lower ranks in the table of component scores than their population size would suggest.

Component 4 in Table 7.2 accounts for 11.85% of the variance of the similarity matrix and is the second largest dimension with eleven variables having values of over $\pm$ 0.3. Five of these have first ranking associations. The most important of these loadings, whether negative or positive, portray various facets of agriculture and rural life. Hence, the naming of the component Rural proved a comparatively straightforward decision. The close association between the type of variables linked to this dimension produces a scale of agricultural prosperity, in which high percentages of population engaged in agriculture (variable 4) and with low wages (variable 21) are associated negatively with the component, whilst low or negative population increase (variable 2) and low percentage of value added per worker (variable 6) have high positive loadings. Expressed spatially, the most striking characteristic of the axis is a broad W.S.W. to E.N.E. slope of increasing component scores which testify to an easterly increase in agricultural prosperity (Figure 7.1b). However, the simplicity of such an explanation should not obscure the fact that some southern areas, Provence-Riveria-Corsica, and some northern areas, Upper Normandy and Picardy, also have relatively high scores. The greater capitalization of agriculture in the latter case, as well as the association with tourism in the former case probably account for this pattern. This could, therefore, be summarized as a truncated bowl with its open edge to the west.

The second axis in Table 7.2 accounts for 11.61% of the total variance. It represents a bi-polar dimension since it has high positive and negative scores. Nine variables have loadings over $\pm$ 0.3, and six of these have their highest correlations with this factor. The variables associated with the positive end of the vector form a heterogeneous set, but indicate a common reliance upon tourist and retirement functions. This can be summarized by the term resort. The negative end of this spectrum is less clear, and is dominated by the farm size variable (variable 5). Given the separate nature of the variables grouped at either end of this dimension, it proved difficult

228

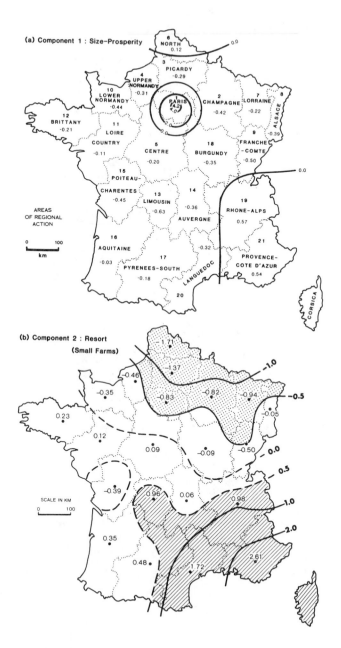

Figure 7.1 Pattern of scores for the French regions

229

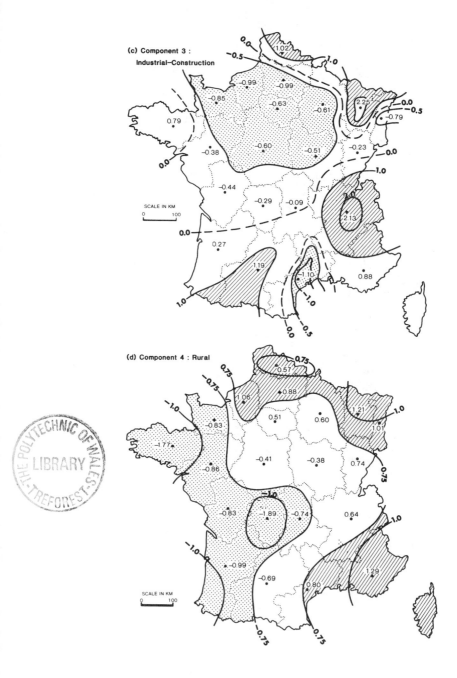

Figure 7.1 (continued)

to provide a generalization linked to a single concept. Hence the factor was labelled Resort—Small Farms.

The pattern of scores on the second component can be summarized in terms of a north–south surface with the highest positive scores on the Mediterranean coast, and the lowest negative scores in the Northeast (Figure 7.1c). Although this distribution reflects the orientation of the French holiday industry to the south coast, the association of these areas with high negative scores on the farm size and industrial energy variables must also be clarified. The small farms of Mediterranean agriculture, and the absence of heavy industry in the area, form a spatial association that parallels the presence of resort functions. Elsewhere, the high positive score of Limousin indicates another tourist area, but one in which small farms and low industrial activity are probably much more influential in differentiating this area from its nearest neighbours.

Table 7.2 shows that only two variables, consumption of industrial energy (variable 10) and the extent of wide roads in the region (variable 23) have their highest association with Component 3. This is the least distinct of the dimensions for it only accounts for 6.57% of the total variance explanation. Labelling this vector proved comparatively difficult. Eventually it was called an Industry—Construction factor because of the pattern of loadings. However it must be stressed that the vector does represent a rather subsidiary industrial influence, simply because the primary agglomerative effect has been removed with the first component. Spatially, the scores pick out areas such as Nord, Alsace-Lorraine, Rhone-Alps and Pyrennes–South. The broad pattern of scores (Figure 7.1d) could, therefore, be described in terms of a set of peripheral plateaux which are characterized by industrial areas. The primary socio-economic surface accounted for by the Paris node has already been removed in Component 1, but it must be emphasized that the loadings for per capita investment and operating outlays (variable 32 and 31) are highly negatively associated with this industrial vector. Perhaps this indicates the relative dimunition in the importance of these peripheral industrial areas during the early 1960s – despite the investment channelled to the regional equilibrium nodes of France (Rodwin 1970).

Component 5, explaining 5.34% of the variance, represents a 'catch-all' dimension in which the common variance redistributed in rotation is allowed to accumulate (Table 7.2). If more than five components are rotated many of variables that load highly on the vector become uniquely associated with additional components. In other words it tends to represent an arbitrary 'squashing together' of separate variables. It is not usual to interpret the last rotated axis when it is abstracted clarify the previous solution. The attempt was made here simply for the sake of completeness, since two variables had their highest association with the dimension. One of these, per capita low tension electricity consumption (variable 22), contributed very little to the overall explanation as its communality was very low. The presence of this variable and that of value added per hectare (variable 7) led to the decision to name the dimension Residual, Dense Population. The influence of Paris and other industrial regions have been removed by abstracting Components 1 and 4, so this axis picks out the other densely populated areas, that are high electricity consumers, such as urban areas, and agricultural regions with high value added per hectare. The pattern of scores shown in Table 7.3 reveal that Nord and Brittany represent the two most important regional planning areas on this dimension.

(c) Relevance of the results

In a review of the roles of regional geography Sugden and Hamilton (1971) pointed out that two distinct tasks could be recognized. The first lies in the identification and description of total spatial systems, the second lies in the development of theory about the behaviour of these systems. The application of factor analysis to the French data set has been primarily concerned with the first of the approaches, paralleling the pioneering examples of Berry (1960), Cole and King (1968) by illustrating a technique which can provide an additional aid to the creation of regional spatial generalizatons. In view of the scepticism with which quantification and other aspects of the research effect of modern geography has been viewed by defenders of the traditional approach to regional geography in the 1970s (Fisher 1970; Balchin 1972), it is worth emphasizing that this case study has made no claim to solve all the problems facing the student of regions. Rather the example stems from a belief that there are a series of alternative approaches to the creation of any generalization - as in all approaches to knowledge - and that the traditional qualitative approach to regional study is only one of the possible methods that may be employed. To continue to concentrate only upon the traditional approach underestimates the value of many of the technical and conceptual developments of the last two decades, providing an unconvincing support for the maintenance of a viable regional tradition in modern human geography and ecology.

Critics of this case study will argue that the results display little of substance, since they refer only to a set of socio-economic variables in very large planning regions. However, this restriction has proved to be an advantage, not a disadvantage in a college teaching context, rather than the research context of Berry (1960). When used in class as part of a survey of western European geography, the method usually produced a vigorous discussion, and series of question-answer sessions that seem to have stimulated rather than repressed further study, when compared to other approaches. This is true not only in relation to the results, to the crystallization of key elements in the human geography of France, but also to the problems of regional methodology in general. Four advantages in particular appear to have been gained.

(i) In the first place explicit recognition is given to the data upon which the generalizations are produced. Since the approach does not subjectively repress any of the information used to produce generalizations, greater confidence may be placed on the utility of the conclusions as applied to a specific data set. Additional gains would come from a follow-up study that objectively demonstrated the effect of adding new variables to the data set.

(ii) A second advantage is that the matrix of loadings and the communalities provide objective measures of the importance of each variable to each component and to the total study. This means that this type of research can present a more rigorous justification for the choice of any key variable, or set of variables, for a subsequent investigation. This might replace the subjectivity, although not the perspicacity of the key indicator approach used by Bowen (1959) or Carter and Thomas (1969) in creating regional generalizations.

(iii) Third, there is an explicit recognition of the ever-present 'scale' effect (Bird 1956). Since the results presented here are dependent upon the regional building blocks used to compile the original data, one can speculate, or more realistically, calculate, the effect of using different sized areas, or areas that may be more closely adjusted to the physical and historical 'pays' of France.

(iv) Finally, the gradual unfolding of the generalizations led to a much greater degree of class interest in the reasons that lay behind these spatial trends. Particularly noteworthy was the 'discovery' by the class of the utility of the 'growth pole' concept (Lausen 1969), and appreciation of the fact that this feature represented one of the most persuasive trends in the regional structure of modern society. Inevitably this led straight into the theoretical literature on polarization and into cross-national comparisons, providing a useful bridge between the descriptive and theoretical tasks of regional study as recognized by Sugden and Hamilton (1971).

It must be noted that this study could be extended by applying multivariate methods such as cluster analysis (Wishart 1978) to the scores to produce regional types from the planning regions. This goes beyond the objective of the study which was to use factor analysis to develop regional generalizations. Although the original objective has been fulfilled, it must be emphasized that the method described in this example will never replace either this standard approach to knowledge about France by the traditional study of individual trends or variables, as illustrated by Thompson's (1970) regional text, or the role of 'inspiration' or the 'flash of vision' in connecting separate pieces of information together into rigorous spatial generalizations. Rather, it provides a shorthand approach <u>to isolating, or synthesizing</u> <u>certain key elements in the spatial pattern of the country, and could</u> <u>be used to check the utility of generalizations produced by other means</u> <u>if the common factor model was used</u>. Moreover, as each stage of the generalizating process is explicitly laid out, it eases the ever-present problems involved in the replication of geographic work (Davies 1968). Also it enables students to arrive at a comparable set of results by their own efforts, and to question the conclusions reached by alternative routes. Hence it fulfills a vital educational role by throwing additional light on the reasoning process, upon <u>synthesis</u> in geography, and, in particular, on the <u>alternative</u> paths by which human ecologists or geographers produce generalizations.

3. SOCIAL DIMENSIONS AND TAXONOMIES

This emphasis upon the methodology of the example in the previous section means that no attempt was made to try and relate the substantive dimensions of this study to other regional analyses in which factor analysis has been employed to reduce the data set to a set of summary axes of differentiation. At this point in time the wide variations in both the data sets, and scale of regions used in such studies, make it difficult to rigorously interlink the various studies, although it is increasingly apparent that there are some common sources of differentiation such as: growth poles, regional economic disparities or modernization and ethnic disparities. In the context of more social sets of variables, however, evidence is emerging of some basic dimensional structures which can be used to evaluate the utility in regional description of the Shevky and Bell (1955) postulates described in Chapter I. The example used in this section is derived from Webber's (1977) classification of the 16,471 wards and parishes in England, Wales and Scotland using data from the 1971 census. This study represents a real breakthrough in its scope and attention to detail, one that ranks with Moser and Scott's (1962) trail blazing study of large urban administrative areas in Britain. Webber's primary task was to provide a national set of social area types against which local studies produced by individual planning authorities could be set in context. Since planners were likely to make extensive use of this national classification, as Webber's (1976) previous work had shown, an additional evaluation to the critique of Openshaw et al (1980) is needed. Of importance to the problem of this book is that the taxonomic approach rarely investigates the categories of social variation upon which the typology of area units are based. In other words the relationship of the taxonomic units to the variations in regional and urban dimensionality is lost. It can, of course, be argued that since the Webber's objective was to produce a typology of places any attempt to identify the dimensions of variation goes beyond the purpose of his study. Such an argument has merit, if used purely within the confines of the empirical context of the original study. Set more widely in any review of the relationship between urban and regional classification schemes and the bases of social differentiation the argument loses much of its force. By virtually ignoring the dimensionality issues much of the potential value of Webber's data set is lost, since the study only deals with <u>part</u> of the field of regional and urban classification.

Webber's (1977) study used forty variables derived from the 1971 census measured over all wards and parishes in England, Wales and Scotland. From this data base cluster analysis methods were used to derive thirty-six <u>clusters</u> or types of social areas. These were combined into seven <u>higher</u> order clusters called <u>families</u> of social area. This is not the place to comment at length (Openshaw 1980) on the problems of the initial analysis. But two comments must be made. The first concerns the variables. Webber attempted to obtain a balance between demographic, housing, social and economic criteria; no attempt was made to relate these variables to the extensive literature on urban and regional dimensionality that has emerged from the time of Shevky and Bell (1955) on. Also, several variables came from closed member sets so strong reinforcement effects in the clusters can be expected, whilst it was explicitly stated that no attempt was made to exclude correlated variables or to use normally distributed indicators. The second relates to techniques. The study did not explicitly identify the precise similarity measure and cluster technique applied to the data, although it can be inferred that a Euclidean Distance measure followed by a Hierarchical Clustering method was used. Moreover, the author admitted that 'it is difficult to explain just how the number thirty-six was chosen to constitute the number of clusters'

234

(Webber 1977 p.6). It is also worth stressing that the technical problems involved in handling such a large data set led to a three stage procedure, using 999 areas, then 3,996 and finally allocating all the observations.

In retrospect this study is a classic example of the Area Taxonomy approach in social area study for it classifies areas, but does not consistently provide any description of the interrelationships existing among the variables. Plots of the highest similarity between pairs of variables only provided a partial indication of the characteristics of the groups. In order to demonstrate the utility of factorial methods in dealing with this problem a replication based on Webber's published data was carried out, namely the average values for the forty variables over the thirty-six area types.

A Principal Axes Component analysis of the matrix of correlations for the forty variables based on the mean values for each of the thirty-six clusters was the chosen method, followed by a Direct Oblimin rotation of the preliminary solution. The use of Pearson Product Moment Correlations as the measure of similarity did differ from Webber's adoption of a distance similarity measure in his clustering algorithm. The decision to use correlations rather than similarities was based on the knowledge that distance measures exaggerate the effect of individual differences between any two values, even where variables are standardized to a common mean. By contrast the correlations repress the influence of individual size differences in favour of an overall measure of association. One obvious problem is that the values used are means of the thirty-six clusters. Since the clusters vary in size, in terms of the number of areas of the type, differential weighting is present. However, the similar problems apply to the original data, given the fact the values of each areal unit are based on units of different size. This vast data source was not available so the possible error associated with the data set had to be accepted.

The Principal Axes technique was used because previous urban research workers have shown that the procedure produces as good results as more sophisticated methods (Giggs and Mather 1975; Davies 1978). Incidentally, it also seems to produce acceptable results even when linear dependencies are found in the data set, a real problem here since so many of Webber's forty variables are related and total to 100%. These redundant variables should really be excluded given the mathematics of the approach. But since this replication is designed to evaluate Webber's results in the context of the existing literature on urban and regional dimensionality, on practical grounds it seemed more sensible to keep the same variable set. It is also worth noting that oblique rotation was preferred to an orthogonal solution because the method did not impose orthogonality on the final set of factor axes.

Scrutiny of a variety of alternative solutions led to the decision to use the seven axis results which accounted for 90.6% of the original variance in the 40x40 similarity matrix. A break of slope was found in the distribution of eigenvalues after the seventh axis and this accounted for all of the axes with eigenvalues greater than 1.0. Moreover, all the variables had communalities greater than 0.75. Table

7.4 shows the loadings for the seven axes. A summary title is given to
each factor (strictly speaking components) on the basis of the loadings
simply to ease the interpretation of the axes. The clearest vector is
Factor 3, called <u>Age</u> because the positive loadings measure high
proportions in the youthful age group, whereas the negative loadings
index old age characteristics. Factor 4 is the simplest axis. It is
called <u>Poor Quality Housing – Tenure</u> because the two largest positive
loadings are indicators of substandard housing and unfurnished rented
dwellings, whilst public housing is found on the negative side of the
axis.

Table 7.4
Loadings for the British wards – parishes study

(a) First order axes and loadings

Factor I – Degree of Impoverishment			Factor II – <u>Immigrant</u>		
			17)	85	New commonwealth born
			36)	84	Shared dwellings
Vari- able	Load- ing	Variable Title	25)	66	Single person h'holds (non pensioner)
33)	99	Serious overcrowding	2)	53	Female E.A. rate (2)
34)	95	Overcrowding	3)	48	Students
1)	85	Unemployment	31)	(46)*	Unfurnished rentals
40)	80	One/two room households	32)	(46)	Furnished rentals
			7)	(34)	Bus/train to work
7)	57	Bus/train to work	20)	(33)	15–24 years
15)	(50)	Unskilled			
38)	(50)	No bath	24)	(–40)	Marriage rate
30)	(35)	Council housing			
25)	(31)	Unfurnished	* Loadings in brackets are second ranking values or below.		
4)	(–39)	Cars per household			
24)	(–42)	Marriage rate			
29)	(–47)	Owner occupier			
35)	–71	Rooms per person			
28)	–80	Dwelling size			
Factor III Age			Factor IV Poor Quality Housing – Tenure		
18)	93	Aged 0–4 years			
21)	94	Aged 25–44 years	37)	86	No inside W.C.
27)	69	Five year migrants	38)	81	No bath
19)	65	Aged 5–14 years	31)	61	Unfurnished
24)	65	Marriage rate	6)	(37)	Walk to work
26)	57	Household size			
20)	(39)	Aged 15–24 years	12)	(–30)	Non manual
			3)	(–43)	Students
35)	(–36)	Rooms/person	30)	(–48)	Council housing
23)	–80	Aged 65 years +			
22)	–92	Aged 45–64 years +			

Factor V - Economy
          (Industrial-Rural)

2)	72	Female E.A. rate
8)	77	Manufacturing
13)	52	Skilled manual
7)	(42)	Bus/train to work
15)	(42)	Unskilled
31)	(-35)	Unfurnished
4)	-59	Cars/household
5)	-71	Two car household
39)	-73	Seven rooms
9)	-86	Agriculture

Factor VI - Mobility and Military

6)	-79	Walk to work
32)	-70	Furnished rentals
10)	-60	Services/dist'n/ defence
20)	-58	15-24 years
27)	-55	Five year migrants
5)	(32)	Two car households
8)	(37)	Manufacturing

Factor VII - Socio-Economic
            Status

14)	84	Semi-skilled
16)	78	Fertility
30)	51	Council housing
13)	(51)	Skilled manual
26)	(44)	Household size
19)	(38)	Aged 5-14 years

27)	(-30)	Five year migrants
40)	(-31)	One-two room households
3)	(-36)	Students
10)	(-46)	Services/dist/ defence
24)	-59	Owner occupier
11)	-65	Professional
12)	-75	Non manual

(b) Higher order axes and loadings:
     British wards and parishes

First Order Axes	Loadings for Second Order Axes				Communal-ities
	I	II	III	IV	
6. Military-Mobility	–	–	–	92	82
4. Poor Quality Housing	–	–	90	–	81
2. Immigrant	–	-56	–	–	68
3. Age	–	78	(40)	–	78
7. Socio-Econ. Status	–	59	–	–	73
5. Industrial-Rural	79	–	–	–	68
1. Impoverishment	78	–	–	–	72

Decimal points are removed: 79 is +0.79. The four axis higher
order solution is the first to have all communalities above 0.50.

Factor 2, the Immigrant vector, is also an axis with a clearly defined structure. The high positive loadings show its association with people born in the New Commonwealth and with inhabitants of shared dwellings. The low loadings confirm the character by the linkages with rented dwellings, high female activity rates, students and single person dwellings. Two of the other axes measure aspects of economic status. The easiest of these to interpret is Factor I, the Degree of Impoverishment axis. The indicators of overcrowding, unemployment, small households and unskilled employees all have positive loadings whilst the opposite characteristics, large dwellings and owner occupation, dominate the negative loadings. Factor VII is a more general Socio-Economic Status axis and is described as such; it separates the skilled, semi-skilled, high fertility and council housing indicators from the non manual, professional and owner occupation characteristics.

Less easy to describe are Factors V and VI. The former is called Industrial-Rural since it separates the characteristics found in industrial areas, namely high levels of manufacturing employment, females in paid jobs, skilled manual workers, commuting by public transit, from the higher levels of car ownership, larger dwellings, and particularly agricultural activity found in rural areas. The sixth factor is associated with rented accommodation, migrants, limited commuting, high proportions in the Service-Defence category and the young adult group. This suggests high levels of mobility. Since the only dominant factor score is linked to the Military cluster type, where armed force bases would have all these characteristics, the axis is given the title Mobility and Military.

The correlations between the seven axes were subjected to the same type of factor analyses at the higher order, and three axes accounted for 61.4% of the variance. The pattern of loadings shown in Table 7.4 is rather complex in the sense that the first order axes do not load exclusively on the higher order axes. Nevertheless, it is possible to give general titles to the higher order vectors. The first can be called Economy and Status since it indexes the Industrial-Rural axis and the two Socio-Economic Status axes: the 'General' and 'Impoverishment' status factors. The second vector is described as Ethnicity and Status because it is primarily linked to the Immigrant vector, but has minor associations with the socio-economic axes and the component indexing Mobility-Military characteristics. This is, however, a complex axis. The third vector is linked with the Age and Poor Quality Housing vector and is summarized as Age and Quality with the negative sign demonstrating that Poor Quality housing is linked to Old Age. The oblique rotation revealed that these axes were orthogonal to one another; the derivation of more axes in four and five factor solutions led to very correlated vectors produced by the split of factors that were distinct at the three axis level.

So far the description of the results has remained at the empirical level. No attempt has been made to compare the results with the dimensions produced in factorial ecologies of urban and regional systems. However, it is worth noting that this comparison does have problems. The ward-parish scale used by Webber cuts across the traditional division between the inter-urban and intra-urban literature on urban dimensionality. But since it is based on all wards and parishes in Britain it has a regional rather than an urban basis. Nevertheless, at the very minimum it is apparent that the seven dimensions that have been identified go far beyond the traditional explanations of urban social structure described in Chapter I, by Shevky and Bell (1955) or Timms (1971), and demonstrate the limitations of these explanations. Certainly the three basic axes related to Age and Family, Socio-Economic Status and Ethnicity can be seen in Factors 3, 7 and 2 in Table 7.4. Yet the four other axes indexed in this study are not unique. The Industrial-Rural axes represents one of the economic base dimensions picked out by Berry (1972) and Hodge (1968) in studies of inter-urban systems. The Mobility-Military axis can be linked to Webber's decision to add together the Service, Distribution, Administration and Defence employment types. This provides too general an indicator to separate out the nuances of urban variation in two types: one linked to Migration (McElrath 1968), the other to Economic Base variations. The Impoverishment axis is also one identified by Davies and Lewis in Leicester (1973) whilst the Poor Quality Housing – Tenure vector is closest to the Quality and Welfare dimensions proposed by Berry (1972) and Hodge (1968) in inter-urban analysis and identified by Davies (1977).

More complex relationships are found at the higher order level. The overlap between the first and second order axes may be a consequence of the scale of analysis employed; but since the overlapping structures have been found in other studies of British cities (Davies and Lewis 1975; Davies 1975), it could be a function of the way the basic axes have been heavily influenced by the presence of high levels of welfare. Only further research will reveal which explanation is the most appropriate. This is going beyond the scope of the case studies. What is more relevant here is the illustration of the way the factorial approach is able to uncover the dimensionality of urban social structure for the wards and parishes in Britain. It reveals the limitations not only of the scope of the generalizations provided by Social Area theories but also the failure of the Urban Taxonomy approach to provide a satisfactory description of the structure existing in the variable set.

So far the study has shown how the factorial method has compressed the pattern of relationships in the forty variable analysis into seven summary dimensions, which are, in turn, 'squeezed' into three higher order generalizations. A similar generalizing process can be applied to the thirty-six cluster types in such a way that the utility of Webber's grouping of these areas into seven families of clusters can be evaluated. Ward's Hierarchical and Wishart's Non-Hierarchical cluster analysis were applied to the matrix of Euclidean Distance calculated between all pairs of the 36x7 matrix of component scores (Wishart 1975). The major break of slope in the distribution of scores occurred after the tenth and ninth clusters respectively in the two methods. Since the initial groups formed by hierarchical clusters are perpetuated at the higher levels, the non-hierarchical method was preferred. Some minor differences between the non-hierarchical and hierarchical results were also seen, the most obvious being the discrepancy between the preference for the ten group 'cut off' of the former, on the basis of the first major break of slope, and the nine group interpretation of the latter. Two of the original clusters also changed positions:

'Villages with Non Agricultural Employment' was part of the hierarchical High Status group, and 'Local Authority Housing in Scotland' was part of the Older, Low Quality Settlement group in this same approach. Given the limitations of the hierarchical approach the non-hierarchical results were used in this comparison.

Figure 7.5 shows the comparison between the alternative descriptions of the clustering of the thirty-six types of settlement. Before looking at the degree of similarity between the results it is worth noting that the difference between Webber's seven clusters and the nine chosen here may not be so much of a variation as originally envisaged. Webber did not provide any real evidence for 'cutting off' at seven groups, and even maintained that a 'nine family solution' would also have been efficient (Webber 1977 p.67). In any case a plot of the loss of explanation in the fusion summary table (Webber 1977 Table 2 p.7) revealed that the biggest break of shape in the distribution came after the fifth grouping stage, with a minor break after the ninth grouping stage. At five groups Webber's Council Estates and Clydeside 'families' are merged, as are the Older Settlements and Rural Areas 'families'. At nine groups the Retirement Communities (35 and 36) separate from the High Status 'family' of the seven stage, whilst the High Status Student and Rooming Areas 'family' (28 and 29) break off from the Multi-Occupancy and the Immigrant 'family'.

Table 7.5 shows the differences between the two interpretations of the higher order clusters of 'families'. Given the differences in the statistical procedures used to process the original data it is rather encouraging to find such a high level of similarity between the two typologies. In terms of the composition of each of the higher order 'families' it is apparent that the core of each of the groups is similar in both cases; only six of the original cluster types move from one higher group to another, although the move does alter the balance of the characteristics. Those differences are more easily appreciated by summarizing the variations in terms of the labels given to the nine higher order clusters or 'families' of the replicated study.

(i) The addition of a separate Military group is the major difference at the higher order stage - demonstrating the very different type of community formed by these characteristics.

(ii) The Growth community is probably more precisely described as Growth and Good Housing since it is extended to include the Webber's Mock Tudor and Edwardian clusters.

(iii) The higher order group called High Status is more obviously a family of affluent communities of this type since it loses the 'Mock Tudor' cluster.

(iv) The retirement communities form part of a larger group in the replicated study. Since they now include 'Council Estates with high proportions of older people', and 'Market Towns', the 'family' is more coherently described as Settlements for Older People or Old Age and Retirement.

Table 7.5
Higher order groups (or families) of British wards and parishes

9 Groups	Original 36 Cluster Types	Webber's 'Families' Seven (and Nine)
3 Military	7 Military 7	

1 Growth and Good Housing	1 New Towns 1
	2 Planned Devlop't, Smaller 2
	  Towns
	3 New Council Housing 3
	4 Modern Low Cost Owner 4
	  Occupied Housing
	5 New Owner Occupied 5
	  Housing Growth Areas
	6 Modern High Status 6
	  Housing, Young Families
	31 Mock Tudor 31
	8 Edwardian *
		1
		Young Growing
		Population

2 High Status	33 Rural Established High 33
	  Status
	30 Modern High Status 30
	32 Established High Status 32
	  Suburban
	34 Very High Status 34
		7
		Established High
		Status and Resorts

5 Old Age and Retire- ment	35 Residential Retirement 36 36 Seaside and Retirement 35 22 Council Estates, Single * People 10 Market Towns 10	(8) (Retirement)
	* Edwardian 8	

4 Old, Low Quality Housing	9 Older Industrial, Low 9
	  Stress
	12 Poor Quality Housing, 12
	  Econ. Depressed
	11 Inner City, Low Quality 11
	  Housing
	19 Mining 19
		2
		Older Settlement

7 Council	17 L.A. Housing, Scotland 17
	  and & N. East
	* Council Estates, Single 22
	  People
	16 Overspill Estates 16
	20 Inter War Local Authority 20
	18 Council Estates, Good Job 18
	  Opportunities
	21 Inner City Council 21
	23 Clydeside Peripheral 23
		4
		Urban Council
		
		Part of 23/24

9 Groups	Original 36 Cluster Types		Webber's 'Families' Seven (and Nine)
8 Inner Clydeside	:24 Clydeside Inner	23 24	5 Acute Social Dis-advantage in Scotland
9a Immigrant/ Inner London	:25 Inner London :27 Immigrant, Multi-Occupied Inner City :26 Multi-Occupied Inner London	25 27 26	6 a. Areas of Multi-Occupancy and Immigrants
9b Young Adult/Room-ing Areas	:28 Student, High Status C. London :29 High Status, Rooming House	28 29	9 b. (Young Adult, Rooming House)
6 Rural	:14 Rural Areas Large Landholdings 15 Rural Areas Small Landholdings 13 Villages with Some Agric. Employment	14 15 13	3 Rural Areas

* Cluster type not in higher order group of this solution.

(v) The Old, Low Quality Housing Settlements include the Mining areas and lose the 'Edwardian' and 'Market Town' clusters – indicating a close association with the older, lowest quality housing areas.

(vi) A separate Council Housing 'family' testifies to the dominance of the public housing sector in the British urban fabric. Again the loss of the 'Mining' cluster from Webber's analysis reinforces the character of the group.

(vii) The impoverishment of the areas in Inner Clydeside is seen by the presence of a separate higher order 'family' or group composed only of this type.

(viii) The three Rural areas remain the same in the two studies, demonstrating that rural areas are still sufficiently distinctive in a social indicator sense to form separate higher order groups.

(ix) Finally, the areas of Multi-Occupancy, Rooming House, Immi-grants and Inner London form a final cluster in the replicated study. It has already been noted that this group splits into two at the next cluster level. Indeed, the ten cluster level is the stage at which the first major jump in loss of information occurs in the alternative non-hierarchical procedure. Given these facts the group was subdivided into (a) and (b) categories which separate Multi-Occupied, Inner London and Immigrant areas from the Young Adult, High Status Rooming areas.

These results provide a succinct summary of the basic social area types among the wards and parishes of England and Wales which provide a basis for comparing local planning results against the national standard. It cannot be denied that there are differences between Webber's original schema and the new results proposed here. Such variations might be expected given the differences in similarity method and the fact that the factor results based on the forty variable, thirty-six area types of Webber have filtered out the error variance associated with the data set. The cluster analysis used by Webber classifies all the data. Yet despite these differences in method, the core of the taxonomic results seem to be very similar, and give one confidence that a satisfactory classification has been produced from this data set. If anything, however, the replicated study clarifies the higher order groups proposed by Webber since the composition of each group appears to be more coherent in terms of the original variables.

Methodologically, of course, nothing prevents this approach being applied to Webber's original data set, although technically it becomes difficult to find computers big enough to handle such a large data set. Of greater methodological importance, however, is the fact that this example demonstrates how factorial ecologists can identify the dimensional structure (or at least that restricted to the census based variable set) for all areas of a country, and then apply another technical procedure such as cluster analysis to the scores to derive a summary classification of the units based on the dimensional structures previously established. In substantive terms, the results dispute the claim of Shevky and Bell (1955) that there are only three major axes of social differentiation, at least at the regional scale used here. Additional axes can be identified, a conclusion which will also apply to the inter and intra-urban structures dealt with in Chapters VIII and IX. Unfortunately the scale differences preclude a direct comparison.

So far this discussion has emphasized the utility of the factorial approach in the two studies of multivariate regional character using examples of economic and urban social character. In addition factorial techniques have also been used extensively by agricultural geographers in studies of agricultural crop systems (Aitchison 1974; Momsen 1966; Laut 1974). In most examples the standard R mode route in which many cases and several variables are measured at one time have been employed. Obviously the Q mode approach can reverse the study sequence. More importantly, alternative modes such as those shown in Figure 3.4 may be used to analyze temporal changes: for example, with many variables in relation to one case, or with many cases for one variable. These, however, do not exhaust the applicability of factor methods. Gould and White (1974) and others have shown how factor analysis can be used to summarize the variation produced in a data matrix consisting of people's attitudes or preferences for a set of places. In such studies the columns of the data matrix represent different people whilst the rows are the places being scaled by the people who are interviewed.

4. IDENTIFICATION OF FUNCTIONAL REGIONS

The study described in the previous section focussed upon the utility of the factorial approach in summarizing regional character by identifying the dimensionality of regions from a set of socio-economic variables. In the language of regional taxonomy the example involves the characterization of a multi-feature region. It did not deal with the related question of defining the areal extent of regions. In the case of flow data, however, factor analysis has proved to be useful in defining the structure of single feature functional regions. Factor analytical methods can be applied to such single feature matrices because the flows originate in one area and have a destination in others, as shown in Figure 3.3.

(a) Alternative approaches to the study of functional regions

During the last two decades increasing interest in quantitative regional taxonomy has led to the identification and solution of many of the problems associated with the definition of formal regions (Johnston 1968, 1972; Spence and Taylor 1970; Rees 1972; Cliff et al 1975) following on from Grigg's (1965) seminal work on the logic of regions. By contrast, the literature on the empirical approach to the definition of functional regions is far less cohesive. This is in contrast to the advances made by workers interested in the theoretical definitions of regions (Huff 1972). A consequence is that a substantive, and inter-related body of empirical literature on functional regions is still an elusive goal.

(i) Data. Part of the problem stems from the use of a variety of different indicators to define regions, whether journey-to-work, telephone calls, or shopping flows. Many of these indicators can be measured in different ways since flows vary at least by: mode of flow (e.g., shopping by car, train, bus, etc.); frequency of flow (e.g., the number of trips in a defined time period); size of flow (e.g., the crude numbers of travellers, money spent, etc.); purpose of flow (e.g., the difference between business and social telephone calls). Each of these examples could be broken down into even more detailed categories. The result is an array of possible measurements. Only rarely have students of functional regions chosen to define these flows in the same way, whilst the possibility of using surrogates of interaction, such as newspaper sales, to define community of interest areas, or transport types, such as airline movements, rather than people travelling on these aeroplanes, produce further elements of potential confusion.

As these various flows or interactions are usually used as single entities, to define what amount to single feature functional regions, it is difficult to know how many of these flows represent similar, or separate sources of spatial variation, since it is unlikely they represent completely different patterns. There is a pressing need to integrate these various indicators to identify the separate sources of variation. Factor analysis, using the dyadic mode, represents one way of resolving this problem as Davies and Thompson (1980) showed in their study of commodity flows in Western Canada, although this is between towns rather than regions (Chapter VIII).

Equally as important in accounting for the rather disparate nature of the literature, is the way in which a variety of different techniques have been applied to the data in the search for functional regions. These techniques have been applied to the sets on flows simply to identify the basic patterns of variation, because the number of interactions (even of one type) between places frequently make it difficult to determine the relationships inthe data set. In general, these techniques fall into two categories which Davies (1980) called the descriptive allocation and empirical derivation approaches.

(ii) <u>Descriptive allocation techniques</u>. The basis of this approach consists of the use of some specific 'cut-off' criterion applied to a data set of flows. This cut-off is assumed by the investigator to separate the major from the minor flows in such a way that these important patterns identify the basic structure of the data set. This approach has long been used by geographers interested in defining trade areas of towns. Unfortunately, from the viewpoint of consistency, the critical limits used by various investigators have often varied. Yet cut-offs based on the 50% value are typical. For example, in a study of trade areas around Swansea, Davies (1970) used cut-offs of 33-1/3 and 50% of the sampled households who visited Swansea at least once in the previous month to provide an approximate definition of the trade area of the city region, prior to subjecting the data matrix to a factor analysis.

Such inconsistencies need not matter if these cut-off values are justified as providing real or locally specific breaks in the distance-decay pattern of the flows. Unfortunately, the justification is rarely provided in the empirical studies. Hence, when such varied criteria are applied to the same area, very different results in the pattern of functional regions may be derived. This is especially true when varied data sources are used, in addition to the different cut-off criteria. For example, Davies and Musson (1978) used a case study of South Wales to show differences in sets of published results, such as Lawton's (196) use of 500 and 1,000 commuters to define the hinterlands of large British cities as compared to the Metropolitan Economic Labour Areas (MELA's) of Hall (1971). In the latter, several criteria (such as density, contiguity, size of work flow, size of community) were used to differentiate between the cores and the inner and outer rings of metropolitan areas. To these variations may be added the recent functional region definitions produced by the Centre for Urban and Regional Development Studies (CURDS) at the University of Newcastle-upon-Tyne (Coombes et al 1980). Although a series of criteria are used in the definition of areas, one of the critical pieces of information consists of the use of commuting flows up to 15% of the origin's employed residents. Workers in other countries, such as Berry's study of metropolitan areas in the U.S. (Berry and Horton 1970 pp.250-275) have used similar criteria to the MECA definitions, but with different figures for the final allocation of places to the metropolitan regions.

These alternative decisions may, if the cut-off decisions are justified, produce useful definitions of the functional regions in particular areas. But without an acceptable justification the results are hard to evaluate. More importantly these various decisions make it difficult to compare results between countries, although Hall's criteria applied to the 1971 census, and the CURDS criteria for 1981, were designed to eradicate these problems within one country. It is unfortunate that evaluations of the respective utility of these various methods has not been provided. After all, the descriptions of the functional areas depend upon the cut-off decisions made by those descriptive allocation procedures.

(iii) <u>Empirical derivation technique</u>. The dangers implicit in making arbitrary cut-off decisions about the importance of various flows, and the allocation of places to the functional regions, led many research workers to reject the allocation approach in favour of alternative procedures. The basic feature of these so-called empirical derivation procedures (Davies 1980) consists of the application of some statistical technique which is designed to uncover the essential structure of the pattern of connectivity in the data matrix. A variety of different techniques have been used in this context, and they have been applied to many different types of flow: for example, Graph Theory and Nodal Structure (Nystuen and Dacey 1961; Davies and Robinson 1968); Markov Chain Analysis (Brown and Holmes 1971); Transaction Flow Analysis (Soja 1968); Cluster Analysis (Slater 1976; Hirst 1977); and Factor Analysis (Pederson and Illeris 1968; Goddard 1970; Davies 1972). As yet there has been no formal evaluation of the utility of these various procedures in relation to a standard data set, whilst exponents of the various approaches have tended to use the methods without any explanation of the superiority of the particular procedure. The result is that the field is dominated by conflicting opinions about the appropriateness of particular procedures.

A major exception to this general characteristic is provided by Brown and Holmes' (1971) study, in which a set of desirable properties for functional regions were identified. They claimed that the Markov Chain approach fulfilled these properties and represented the most useful way of defining functional regions. Davies (1980), however, pointed out some of the limitations of the Markov Chain approach: in particular the imposition of a hierarchical structure upon the data, and the lack of real world measurements of the extent to which sub-areas are linked in the individual regions. Instead, Davies (1980) proposed that factor analysis, especially the higher order model (Chapter III, Section 3), represented a more appropriate way of solving the problem of defining functional regions. An important feature of the factorial approach was the way in which higher order regions, or successive levels of generalization, could be produced. This procedure highlighted the fact that functional regions are rarely self-contained, watertight areas. Instead they nest together, either in some hierarchical sequence or in an overlapping sequence, to produce higher level regions. Practically all the descriptive allocation and empirical derivation techniques seem to have ignored this fundamental property of functional regions since they deal with one level or scale of regionalization. The next section deals with the application of factorial procedures to some specific case studies.

(b) A case study of functional regionalization

The stages involved in using factor analysis to identify the structure of a connectivity matrix are shown in Figure 7.2. An origin-destination flow matrix is converted into a similarity matrix, usually of the Q mode type (Chapter III, Section 3) in which the similarities between a set of origin areas (rows) are measured over a set of destinations (columns). Application of an appropriate factor analysis routine to the similarity matrix produces a factor loading matrix which provides a parsimonious description of the pattern of relationships. Because a Q mode approach is used, the factors pick out clusters of areas with similar flow patterns. The loadings measure the importance of each origin to each factor or axis, whilst the factor scores for any axis measure the importance of each area as a destination for the cluster of origins identified by the axis. If there is a spatially structured pattern of flows, each factor has one dominant factor score, in other words, one important destination, and the areas linked to this destination have high factor loadings. In Figure 7.2 the places with high scores are shown as nodes, and desire lines can be used to link the places with important loadings on the factors to these nodes. In this way the structure of relationships between the areas is isolated.

In any real world example a number of problems have to be solved before satisfactory results are obtained. Indeed previous chapters have described how investigators must choose the factoring technique, and rotation type, as well as decide upon the various cut-off decisions in relation to the number of axes and importance of loadings. In flow matrices, however, particular attention has to be paid to the question of which form of the interaction matrix is the most appropriate to analyze. Six basic questions have been isolated by various researchers:

(i) Should the data be transformed to remove skewness, etc., (Clark 1973) or will transformations distort the distance:decay relations in the data as described in an example by Davies and Thompson (1980). Appendix B describes some of the procedures that can be adopted.

(ii) Should the matrix be adjusted to remove the differential effect of unequal sized units, a particular problem in flow matrices (Hirst 1977).

(iii) Is the problem best dealt with by altering the original data, such as converting an origin-destination matrix into an interaction matrix (Davies and Musson 1978), or using the ratio of interaction in each cell to the row totals, either as input to a direct factor analysis as in sociometry studies (MacRae 1960), or as input to a normal factor procedure.

(iv) Should indirect flows be calculated and added to the matrix (Nystuen and Dacey 1961; Tinkler 1972, 1975).

(v) Are similarity measures such as cos theta (Klovan 1966; Clark 1973) more appropriate for flow matrices than correlations because they incorporate size differences. Yet flow size may reflect area size.

(vi) Finally, there is the question of the need to choose a study area boundary that does not cut out the largest flows outside the area, perhaps to higher order towns or to adjacent regions. This type of a restriction frequently distorts the structure of relationships by imposing a closed system framework.

In practice all of these difficulties may be solved in different ways in individual studies, simply because of the specific problems of the study. Nevertheless they are issues that have to be faced.

In the study shown in Figure 7.3 Pearson Product Moment Correlations were used to measure the similarity of journey-to-work flows between 57

1. DATA MATRIX

Destinations

	1	2	....	N
Town Origins 1				
2				
3				
N				

2. SIMILARITY MATRIX (Q mode)

	1	2	....	N
1	1.00			
2		1.00		
·				
·				
N				

Q mode. Similarity between the rows (origins) over a set of destinations.

3. FACTOR LOADING MATRIX (usually rotated)

	I	II	III		N
Town Origins 1	0.9				
2	0.7	0.4	0.35		
3		0.8			
4		0.6			
5			0.9		
6			0.8		
·					
·					
N					

Factors pick out towns (origins) with similar patterns.

4. FACTOR SCORE MATRIX

	I	II	III	.... N
1	0.3	0.6	0.5	
2	1.2	1.0	0.1	
3	0.9	0.2	0.2	
4	10.3	0.6	0.3	
5		10.1	0.1	
6			11.2	
7				
N				

Scores measure the relative importance of each town as a destination. (Note the need for the extreme dominance of one town as a destination.)

5. DIAGRAMMATIC REPRESENTATION

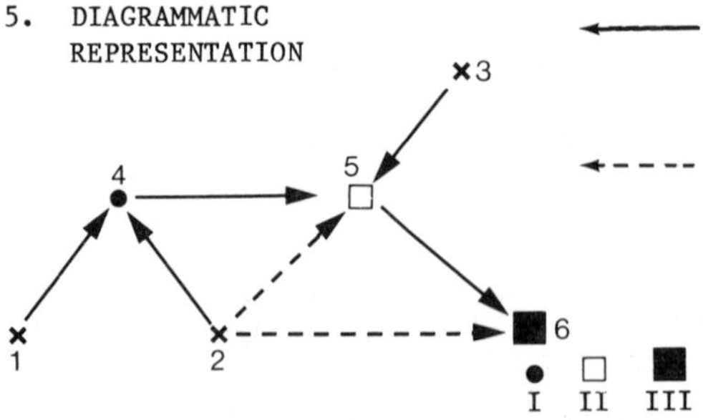

First ranking loadings on axis with high score destination.

Second ranking loadings.

(Rank differences are only made if the loadings are > 0.5 apart).

Possible sequence of nodes.

● □ ■
I   II   III

Figure 7.2 Factor analysis and single connectivity matrices

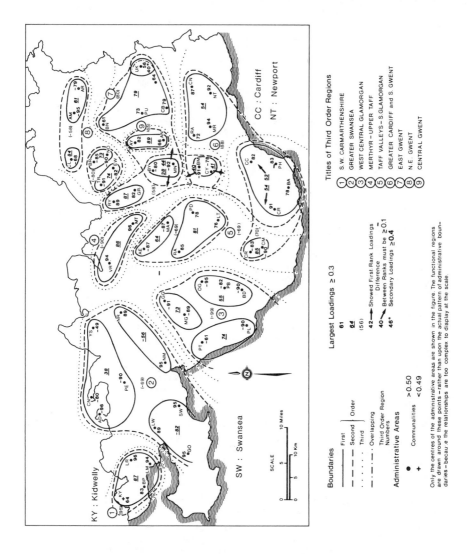

Only the centres of the administrative areas are shown in this figure. The functional regions are drawn around these points – rather than upon the actual pattern of the administrative boundaries because the relationships are too complex to display at this scale. The complete list of LW is Llwchwr, GO is Gower, NM is Neath Borough, NR is Neath Rural, YS is Ystradgynlais, PE is Pontardawe, CN is Cwmamman, AD is Ammanford.

Figure 7.3   Journey-to-work regions in Industrial South Wales, 1971

administrative areas of industrial South Wales in 1971. A matrix of
commuting interaction was used and the correlation measures were
justified by the need to identify similar patterns of connection, given
the differential size and shape of the study areas. The use of an
interaction measure means the matrix was symmetrical, so R and Q mode
results are similar. Experiments with the use of transformed data sets
did not substantially change the structure of the results. The 57
administrative areas 'collapsed' into 22 components produced from a
Principal Axes solution, which was followed by Direct Oblimin rotation
in which 80.2% of the variance was accounted for. A similar analysis
of the 22x22 factor correlation matrix led to the derivations of 12
second order axes accounting for 66.9% of the variance in this matrix.
A third order analysis following the same procedures led to eight third
order axes summarizing 72.3% of the second order matix. At this stage
the axes proved to be orthogonal to one another, so the factoring was
stopped.

It proved to be difficult to map what amounted to a complex interaction structure on the original administrative areas which were
variable in size and shape. So Figure 7.3 shows the functional
relationships in partial diagrammatic form. The loadings for each axis
at the first order of analysis are shown near the centre of the
administrative unit, and lines are drawn around these points if they
are part of the same component. Hence Gower R.D., Swansea C.B. and
Llwchwr R.D. are enclosed in one unit, since they are linked with
respective loadings of 0.95, 0.95 and 0.89 to the same axis. At the
second order this axis is part of a greater unit incorporating the
Neath Valley units and the Upper Tawe-Amman authorities, although the
association of the latter area is only a marginal one. At the third
order the unit is perpetuated. In the Cardiff area to the east it can
be seen that a much more complex pattern of overlapping relationships
is found. A particular feature of the analysis is that when communalities below 0.5 are found, local authority areas that are not part of
the overall pattern produced by the study are identified, since more of
their variance is associated with some other pattern. Kidwelli, in the
west of the area, provides such an example. This clearly demonstrates
that the method is sensitive to the fact that places included in the
study area may not have their major pattern of relationships with the
area. In other words the factorial approach produces a self-defining
areal property.

Figure 7.3 does display a rather complex pattern of relationships in
the eastern part of the study area, but this is a feature of the
complexity of interaction in the area. To impose a simplistic structure, such as a city-regional pattern based on the three largest
centres of Swansea, Newport and Cardiff upon the journey-to-work flows
(Hall 1971), does ignore the nature of interaction in South Wales
although the relationships around Swansea come close to this city-
regional structure. One of the problems of the study is, of course,
the use of journey-to-work flows upon administrative areas. But this
is the only published source of information on this flow in Britain.
Probably part of the complexity of the results is due to the unavoidable use of this areal basis. Nevertheless, such an information source
does provide a rather rigorous test of the utility of the factorial
method in uncovering the structure of relationships. Of particular
note are the ways in which the functional regions identify the

successive layers or scales of generalization from the similarity
patterns, and the degree of attachment of an area to the higher order
axes are shown by the size of the factor loadings. The method is also
capable of producing overlapping regional relationships, when this is a
function of the data. This can be seen by the differences between the
Swansea and Cardiff higher order axes. Once these general patterns
have been identified, it is possible to go back to the original data to
pick out the flows that do not correspond to these basic structures.
However, investigators must always realize that the analysis has been
based on a factor analysis procedure that identifies the linear pattern
of relationships in the data. As always this is a limitation, although
non-linear factor methods (McDonald 1967) have been invented. In this
study the fact that over four-fifths of the first order results can be
accommodated in such a model does demonstrate that a very useful
description of the similarity of interaction patterns has been provided. Davies (1980) showed that similar results were obtained when an
origin-destination matrix was factored. Yet it must be noted that any
similar analysis that wanted to incorporate the size differences
between the flows could use a cross-product matrix as input to the
factor analysis (Chapter IV, Section 4). Finally, the limitation of
this example to the interpretation of one type of flow, journey-to-
work, must be emphasized. Methods such as dyadic factor analysis, or
three mode analysis, exist for dealing simultaneously with several
flows (Chapter III, Section 3 and Chapter VIII, Section 3d).

(c) Utility of the factorial approach in flow studies

It is most unlikely that any single statistical procedure will be
satisfactory for all region-building purposes so it must be accepted
that alternative techniques may, in lots of cases, prove to be equally
useful in defining the structure of functional regions. Nevertheless,
factorial approaches - especially using higher order options and the
dyadic or three mode procedures - seem to have a series of advantages
over other techniques. These advantages were summarized by Davies
(1980) in terms of seven major points, many of which can be subdivided,
representing extensions to some of the basic ideas proposed by Brown
and Holmes (1971). Although these advantages are described in terms of
the factorial method, they obviously provide a checklist of desirable
properties for any methodology of functional regionalization, and as
such provide a means of evaluating the utility of other methods.

(i) The higher-order factorial approach identifies the latent
functional structure or degree of order present in any area for a
particular type of interaction, by defining functional regions at
successive scales or levels of generalization. The result is that
regions, hierarchies, and nodalities can be defined as 'related
entities', in the sense of Brown and Holmes (1971), since nodal and
regional structures as well as the relative ordering of areas or places
can be obtained. The fact that the method employs an open-ended
approach to higher-order factoring means that as many levels of
regional generalization as the data demands can be produced. No
artificial restriction of the regionalization at a particular scale of
investigation occurs.

(ii) The method has a very flexible data input so very different
information sets can be used. Since all the information is analyzed by

this method, there is no need to make a priori assumptions about the characteristics or form of the data set. Specific criteria for defining commuting or functional regions (Hall 1971) are not needed, and one does not have to make assumptions about the properties of the flows, such as dominance (Nystuen and Dacey 1961), rank order (Davies and Robinson 1969), transitivity (Soja 1968), or what constitutes a terminal node (Green 1950). In addition, the method does not depend on any comparison of observed with expected flows (Masser and Brown 1975). The approach can also use alternative similarity measures or different forms of data, such as nodality or directional interaction (Davies 1979), in place of total interaction, or even the adjusted matrices proposed by Hirst (1977). It can incorporate size, as well as the normal similarity differences, by using cross-product factoring in place of correlation measures.

(iii) Overlapping relationships are identified by the method, if these are characteristic features of the flow patterns. The method does not fall into the trap of exclusively classifying any place into one hierarchical unit, a problem of other procedures (Brown and Holmes 1971; Hirst 1977). Non-hierarchical cluster methods come close to this same property by not assuming hierarchical sequences. However the equivalences between the different orders or groups are difficult to make in cluster analysis.

(iv) The method can be used to provide standardized and real-world measurements of the relationships found within the functional systems being analyzed. These measurements are made in several different ways.

1) The amount of variance associated with each functional region can be obtained from the importance of the axis that defines each region. This represents a major advantage over alternative methods of defining regions, such as cluster analysis, where only total 'loss of information' measures are provided.

2) The degree of integration of every place in a region can be measured from the sizes of the factor loadings.

3) In origin-destination studies the importance of each destination to the cluster of origins that comprise a region can be obtained from the sizes of the scores in Q-mode analysis.

4) The amount of variance not described by the regional patterns can be identified. In other words this provides a measurement of the 'error' of the description represented by the factor solutions.

5) The low communalities pick out the areas that are not a major part of any of the axes or the functional regions of the study area. Hence the method can be seen to possess 'self-defining system' or 'areal closure' properties. The higher-order approach used in the example described above was based on the factoring of correlation matrices. Hence the higher-order loadings relate to the next lowest-order regions, not to the original administrative areas. Ways around this limitation are available, since methods have been worked out for directly calculating higher-order loadings from the original

units, or cases, if this is considered necessary (Cattell 1968; Royce 1973, 1976).

(v) The method is <u>multi-relational</u> in four different ways. First, different types of single data sets can be used, whether journey-to-work, journey-to-shop, or other types of flows. Second, different measurements of the relationships between places can be used, other than the interactional or origin-destination matrices used here. Third, alternative factorial methods can be used, with assumptions appropriate to different circumstances. Fourth, temporal changes can be dealt with, by methods such as three-mode factor analysis.

(vi) The approach is suitable for <u>data exploration</u> or <u>data testing,</u> via the component and common-factor models respectively. Specific hypotheses about the number of axes or regions postulated for the study area could be confirmed by the latter model. By means of the former procedure the factorial results can be used in exploratory or descriptive contexts. The component approach was adopted in the case study of South Wales primarily because there was no reliable way of estimating either the number of common factors or the communality of each area.

(vii) The method can cope with <u>multiple flows or interaction,</u> in the sense of integrating several different types of flows within one analysis by adopting the dyadic approach (Black 1973; Davies and Thompson 1980). This approach is shown in Chapter IX, in the case of inter-urban interactions. This means that multiple-feature regions could be defined if appropriate data are available. Unfortunately, it is very rare that comparable data for many different types of flow are available, so very few examples of such studies are found in the literature.

The objection can be raised that other and more simple methods for defining functional regions fulfill some of the properties noted above. This is undoubtedly true. However, it is suggested that few of these methods have so many of these advantages, whilst the cut-off decisions that have to be made can bias the final results. Whether the application of new statistical procedures such as Q analysis (Atkin 1974, 1975; Gould 1980; Gatrell 1982) represent advantages over the factorial approach in this area remains a moot point. Until an exhaustive evaluation of all existing methods have been made it is difficult to provide a final, definitive conclusion. What is apparent at this time is that factorial procedures possess many useful properties. Nevertheless, it is worth emphasizing that the approach is not without flaws. Many of these problems have been discussed in previous chapters so there is little point in repeating them at great length. The problems that seem most relevant are those relating to the linearity assumptions, and to the possibility of deriving invariant solutions by different methods unless a strong structure is found in the data set. In addition, care should be taken over the assumptions build into the analysis by the choice of data and especially similarity matrix. But in the last resort the fact that the unique, individual flows are filtered out of even a component model analysis must be noted. The concentration upon the general patterns by all generalizing statistical procedures does mean that the unique features may be ignored, unless the investigator goes back to the original data and compares these flows with the factor produced generalizations. This is almost always worth doing. It ensures the investigator does not fall into the trap of over-emphasizing the general patterns, at the expense of the ideosyncratic flows. If an investigator decides on this approach the researcher is really adopting a two stage methodology: first, the production of the factor results which pick out the general structure of the data by a sort of data screening or generalizing procedure; second, the identification of the features which do not fit into these structures, in other words, the highly localized patterns. The inferential approach of factor analysis, of course, is not really designed to deal with this latter issue since the whole methodology of the common factor approach is to eradicate these specific and unique patterns. By contrast, the use of the component model can pick up some of the specific patterns, whilst the variability not accommodated in the model is pinpointed by the communality values associated with individual variables. In most social sciences it is arguable whether these unique factors are of real interest; for human ecologists and geographers the region-particular patterns are often of great importance, at least in a descriptive sense so it does not seem sensible to completely ignore the local patterns or relationships. This type of two stage approach may help bridge the general-individual dichotomy that plagues so much of empirical research in the field.

5. SUMMARY

This chapter has provided three examples of the utility of factorial methods in helping students of regions to provide summaries of the variability existing in areas. The approach can be used to help generate regional generalizations and patterns of variations and provide a new data set derived from these generalizations for the construction of taxonomies or classifications. Obviously these generalizations are restricted to the original data sets, but there is no reason why this information cannot be extended. As yet the complexity of regional character has meant that the factorial results in regional study are still at the stage of helping to derive generalizations from the data rather than providing tests of theories of regional character or hypothesized structures.

8 STRUCTURES OF URBAN SYSTEMS

The term 'urban system' is frequently used to describe the field of study dealing with the formal characterisics of groups or sets of towns together with interactions between them. One of the major research efforts of the late 1960s and early 1970s consisted of attempts to identify the latent structure or underlying sources of variations in urban systems. Building upon previous work on economic classifications of cities these attempts were rarely related to theoretical considerations: researchers seemed content to simply identify the empirically derived categories of variables and associated groups of towns – without searching for explanations for these typologies. Such objectives are perfectly satisfactory in their own right but are rather restrictive. This may account for the decline of interest in the approach in the last decade. Yet the lefinition of structures, the collectivities of the data sets, represent the building blocks of theory (Blau 1976). So it is possible o look beyond these structures to the stages of understanding that provide the deep-seated causal explanations for urban variations, as anticipated by Gregory's (1978) discussion of the relevance of the structuralist literature to human geography. This chapter summarizes the available evidence on the character of urban systems as based on multivariate techniques, primarily factor analysis. The first three sections of the chapter deal successively with the formal characteristics, the perceptual or attitudinal character, and finally the functional or interactional features of urban systems. Following this description of the generalizations identified by factorial methods the last section makes an attempt to fit these variations within the context of a general conceptual model of urban system variation.

1. FORMAL STRUCTURES

(a) Early classification schemes

The first systematic of attempts to identify the sources of variation in urban systems can be seen in the earliest city classification studies. One of the first was carried out by the British Commission on the Health of Towns in 1840 where a five fold typology was proposed: metropolis, manufacturing centres, populous seaports, great watering places, country and other inland towns. By the 1920s the typology had become more comprehensive. Table 8.1 taken from the work of Aurosseau (1921) shows that the typical approach was to take a set of categories or types of urban characteristics and allocate individual towns into these categories. Such studies are certainly useful in pinpointing the differences between cities, but the subjective nature of both the categories used in the typologies, as well as the allocation decisions used to put cities in these groups represent real problems with this type of approach.

Table 8.1
Urban classification scheme of Aurosseau

Class I - <u>Administration</u>	Capital cities; revenue towns
Class II - <u>Defence</u>	Fortress towns; garrison towns; naval bases
Class III - <u>Culture</u>	University towns; cathedral towns; art centres; pilgrimage centres; religious centres
Class IV - <u>Production</u>	Manufacturing towns
Class V - <u>Communication</u>	(a) <u>Collection</u> - Mining towns; fishing towns; forest towns; depot towns
	(b) <u>Transfer</u> - Market towns; fall line towns; break-of-bulk towns; bridgehead towns; tidal limit towns; navigation head towns
	(c) <u>Distribution</u> - Export towns; import towns; supply towns
Class VI - <u>Recreation</u>	Health resorts; tourist resorts; holiday resorts

Source: Aurosseau (1921)

Recognition of these difficulties led to the application of increasingly sophisticated quantitative techniques designed to solve those problems and measure the character of towns (Carter 1975). By the 1950s, for example, Nelson (1955) used standard deviation measures to determine the degree of specialization in towns, whilst the minimum requirements method of Dacey and Ullman (1960) was popular before multivariate statistics came into widespread use. A wide variety of alternative methods have been applied, from Smith's (1965) use of cluster analysis, through factor methods to a current research frontier favouring Q analysis (Gatrell 1981; Bennett 1982). This progression in statistical procedures has added precision to our understanding, but Isserman (1977) has argued for the use of simple indices of local specialization such as the location quotient. These can be applied to various levels of economic categorization, from gross industrial types such as manufacturing to individual industries.

(b) Multivariate structures

By the 1950s the search for precision in description led investigators to the almost exclusive use of economic criteria in urban classification schemes, resulting in the virtual abandonment of the wide ranging data sources of the earliest classification schemes. As such, the economic taxonomies have limited value in the identification of urban system structures for they deal primarily with the economic differences between places. An exception is provided by Thorndike's (1938) pioneering study of the USA system. A wide range of economic, demographic and cultural data was collected, together with attitudinal data for the urban places. Unfortunately, Thorndike only used the data to derive a single scale, a G scale of 'quality-of-life', along which cities were arranged. Intuitively, it is much more likely that city systems will be composed of several dimensions or sources of variation. This adoption of a uni-dimensional approach to description is, of course, not unique to urban geography. Early personality theorists, such as Spearman (1904), began by assuming there was a single general factor of intelligence. The history of subsequent research has been the identification of more and more factors of intelligence. In the urban systems field parallel developments ocurred - albeit a generation later - with the gradual discovery of many different structures or dimensions of urban systems. In this work the application of multivariate methods has been particularly influential in isolating these structures.

(i) Origins. The origins of the application of multivariate techniques, such as factor analysis, to this field of enquiry can be traced back to Price's (1942) study of American metropolitan areas, and Hoffstaetter's follow-up analysis (1952). But the field did not come of age until the high speed computers became widely available in the 1960s. Stimulated by the pioneering work of Moser and Scott (1962) in Britain, Berry (1965) and Hadden and Borgatta (1965) in the USA, and King (1966) and Hodge (1968) in Canada, the late 1960s and 1970s saw widespread interest in the field (Berry 1972).

In part, the approach is similar to that of the earliest classification studies. The difference is that instead of subjectively isolating the categories (such as size, age, economy) the dimensions or sources of variation were quantitatively derived from the data by factorial techniques. Unfortunately for the development of substantive generalizations the use of very different data sets, factor methods, and decision criteria in urban studies produced a series of studies which have proved very difficult to rigorously compare. Nevertheless, Hodge (1968) postulated that there were a set of common independent dimensions which covered at least the following characteristics:

(a) size of population; (b) quality of physical development; (c) age structure of population; (d) education level of population; (e) economic base; (f) ethnic and or religious orientation; (g) welfare; and (h) geographical situation.

Table 8.2
City system dimensions from census based indicators: N. America

	Studies using census-based indicators						
Hodge 1968)	U.S.A. Hadden and Borgatta (1965)	U.S.A. Berry (1972)	Canada Ray and Murdie (1972)	Canada Davies (1977)	Canada Simmons (1978)	General-izations
1. Size	8. Total Population	1. Size	8. centre-periphery	1. Size	1. Size	Size (and Density?)
	6. Population Density					
2. Quality of Development				2. Substandardness		Substandardness: / Housing Quality
	1. Socio-economic Status	2. Socio-economic Status		2. Socio-Economic Status		Socio-Economic Status
5. Economic Base	9. Wholesale	6. College	4. Primary Manufacturing	3. Economy	2. Economy	Economic Types
	10. Retail	9. Manufacturing				
	11. Manufacturers: Concentrations	11. Special Service	3. Mining Service			
	12. Manufacturing (Durables)	12. Military				
	13. Communications	13. Mining				
	16. Transport	8. Recent Employment Experience				
	14. Public Administration					
4. Education	4. Educational Centre					Education
	5. High School Education					
3. Age Structure	3. Age Composition	3. Family		4. Life Cycle	3. Demographic	Age
6. Ethnicity-Religion	2. Non-white	4. Non-white	1. English-French	5. Prairie Ethnic	4. Cultural	Ethnic Types
	4. Foreign-born	7. Foreign-Born	2. Prairie-Type	6. Western Ethnic		
			3. B. C. Type	7. Bi-Cultural		
			7. Ethno-Metropolitan			
	5. Residential Mobility	5. Recent Growth	4. Post-war Growth			Growth Mobility
		10. Female Economic Participation				? Females
		11. Elderly Males Working/Inter-State Commuting				? Commuting
7. Welfare						? Welfare
8. Geographical Situation						? Situation

258

Table 8.3
City system dimensions from quality of life indicators: N. America

Census-based studies	Quality-of-Life Studies				
Generalizations	Coughlin* (1973)	Smith (1972)	Smith (1973)	Abrahamson* (1974)	Final Generalizations
Size	1. Size (Traditional Structure-Population)			1. Size-Density	1. Size-Density
				3. Regional centres (city dominance)	
Substandardness /Housing Quality	3. Social Conditions (Middle America)		1. Social Deprivation	2. Crime (Disorganization and Deviance)	2. Substandardness and Housing Quality
		5. Social Disorganizations	5. Social Problems	7. Isolation	
		2. Living Environment		8. Deviance (Eccentricity)	
				4. Modern Social (Specialized Goods and Services)	
	5. Social Facilities and Education (Cosmopolitan-Affluence)				
Socio-Economic Status		1. Income, Wealth Employment	2. Socio-Economic Status		4. Social Status
Economic Types					5. Economic Types
Education		4. Education			6. Education
	4. Retirement-Tourism	7. Recreation		6. Leisure Style	7. Leisure/Recreation
Age					8. Age-Life Style
Ethnic Types			3. Racial Segregation	5. Ethnic-Cultural (Population Diversity)	9. Ethnic Types
Growth Mobility					10. Growth Mobility
? Females					
? Commuting					
? Public Welfare		3. Health			11. Health and Welfare
? Situation					
	6. Environmental Quality (Clean Air)		4. Pollution		12. Environmental Conditions

* Factor labels have been altered to reflect relative size of loadings. Original titles are given in brackets.

In addition to the eight proposed dimensions Hodge (1968) went on to argue that the economic base of urban centres acted independently of the rest of the structural features of urban systems. Much more controversial was Hodge's assertion that these dimensions tended to be the same between regions regardless of the stage or type of development in the various regions. These ideas led Berry (1972) to suggest:

'in effect we are only at the beginning of what must be a long term effort to describe the processes giving rise to the structural organization and orderly functioning of urban activities and the innovation giving rise to periodic transformations in the structural arrangements. A first step is to determine whether urban systems have common latent structures.' (Berry 1972 p.50)

(ii) <u>North American cities</u>. The utility of Hodge's generalizations
for North America can be seen in Table 8.2 which compares five of the
most comprehensive studies of American and Canadian cities in which the
factorial methods went beyond the production of unrotated component
solutions. At first sight these sets of studies seem to have little in
common since they range from four to 15 factors. In practice, the re-
arrangement of the dimensions in Table 8.2 to focus upon the similarity
of the factors demonstrates a high degree of coincidence. It can be
seen that most of the variation is associated with the presence of sets
of Economic Base or Ethnic-Religious axes. The last column in Table
8.2 identifies the axes found in at least two of the studies. The
generalizations are similar to the Size, Quality, Economic Base,
Education, Age, and Ethnicity axes of Hodge (1968) but to this set must
be added a Growth-Mobility dimension. On the basis of this comparison
less weight can be placed on the existence of Welfare or Geographical
Situation axes, whilst the Commuting and Female axes of Berry (1972)
also seem to lack generality. It is worth noting that a number of
different Ethnic and Economic Base axes are likely to be found.

By the early 1970s criticisms were made about the use of data sets
composed only of 'objective' indicators derived exclusively from the
census. Coughlin (1973), Smith (1972, 1973) and Abrahamson (1974) used
sets of 'quality-of-life' and 'social condition' indicators in studies
of North American cities, and suggested that the dimensions produced
from these variables had little in common with the results obtained
from studies of census variables. Moreover Table 8.3 shows that they
appear to vary drastically from one another. But once the dimensions
of variation are arrayed against the dimensions identified in Figure
8.2 it is clear that there is a large degree of similarity. The
differences are primarily associated with a number of 'social
condition' axes and the failure of some of the 'quality-of-life'
studies to find some of the now traditional dimensions produced by the
census indicators. This could be explained either by the fact that the
'quality-of-life' dimensions really were different, or the studies
failed to incorporate enough variables to measure these characteristics, or inappropriate titles were applied as suggested in Table 6.3.

On intuitive grounds it might be expected that the 'quality' dimensions would reflect, or give an additional flavour to, the standard
dimensions produced from census indicators since much of the quality of
urban areas seemed to be linked to prosperity of places. Rather than
simply assuming that this was the fact Davies and Tapper (1979) studied
a set of 85 variables for the 125 SMSA's over 250,000 population in the
USA — the variables indexing one of the 12 dimensions that seemed
typical of the 'census' and 'quality' studies. After demonstrating the
stability of their results using a variety of alternative factoring
methods they concluded that the most satisfactory solution was a ten
axis Principal Axes component with Direct Oblimin rotation solution
accounting for 70.2% of the variance in the data set. The component
loadings in Table 8.4 show that the urban system of the USA can be
characterized by ten sources of differentiation.

1) The most obvious character revealed by the study is the difference in the size of places, a product of the differential success of
the cities in attracting growth throughout history. It is noteworthy
that several measures of social disorganization and quality (murder,

deaths and pollution) are linked to social condition in the USA. This means that the structural dimension is more properly labelled <u>Size and Quality</u>.

2) The literature on urban growth processes have revealed that cities are composed of bundles of different types of economic activities. This <u>Economic Base</u> variation is identified as Factor III in Table 8.4, specifically a white collar-manufacturing axis. Compared to other factorial studies of the urban system the economic differentials are rather limited. This is a product of the decision not to overload the analysis with a large number of variables of one type and the restriction of the study to large cities. Many of the smaller cities are usually more specialized economically so several economic base axes can be expected.

3) No urban system is ever static. Although large cities rarely decline some are growing apace, others only slowly. This fact is reflected in a <u>Growth</u> axis (Factor IV) which also incorporates the relative mobility of people in the USA urban system.

4) The presence of differential growth rates throughout time means that cities are going to show rather different age structures depending upon the timing of their growth or stagnation. Added to this is the highly selective nature of migration - especially with the development of retirement communities - whilst cities often display quite different fertility rates. The consequence of all these forces is a variation by age or <u>Life Cycle</u>, as shown by Factor II in Table 8.4.

5) Few societies or cities are egalitarian, having equal rewards for differential work. The consequence of an unequal occupational reward system, allied to the different economic composition of cities and their varied prosperity, is a differentiation based on status. Factor I shows <u>Socio-Economic</u> Status variations which includes ethnic, health and welfare variables as well as the standard income and possessions indicators.

6) Social heterogeneity is one of the more obvious sources of variation in cities. In the USA it has historically been important because the country was occupied by people of very different European backgrounds. In the last 50 years the migration of blacks to the cities, and, more recently, major immigration waves of Spanish speaking people, has given additional sources of ethnic differentiation. These processes are reflected in the <u>Ethnic-Suburban</u> axis (Factor IX) which separates centres of concentration of Spanish speaking groups from the more prosperous smaller SMSA's. Again it is worth stressing that only a limited number of ethnic indicators were used in the study to avoid results dominated by ethnic axes. The major variations in the black population were intimately linked to the Socio-Economic Status and Housing Condition axes.

7) Now that governments have taken a more active role in caring for the people in their jurisdiction it may be expected that welfare variations will emerge in urban systems, particularly in federal countries with a great deal of local initiative in government spending

Table 8.4
Component loadings for USA urban system study

Loadings*	(Var. No.)	Variables

Component I - Socio-Economic Status

75	(14)	Per capital income
74	(16)	Families income $15,000 plus
68	(80)	% Families with both parents
63	(19)	Housing with telephone
57	(07)	Median value of housing units
54	(57)	Av. monthly AFDC payments
43	(46)	% 4 yrs. high school or more
41	(58)	Pop. change 1960-70
40	(48)	Per capital local gov. expenditure on education
40	(44)	Median school years
39	(18)	Households with one or more cars
35	(41)	Housing segregation index
32	(12)	Housing % with 8 or more rooms
-34	(46)	Males <u>not</u> high school grads
-37	(39)	Negro families > 1.01 persons per room
-40	(36)	% pop. non-white
-44	(70)	Number of days above 90°F
-46	(11)	Families in households > 1.01 persons per room
-47	(54)	Number of hospital beds
-52	(51)	T.B. deaths
-66	(9)	Households lacking adequate plumbing
-67	(79)	% Families with female heads
-82	(42)	% Negro families below the poverty level
-87	(15)	Families below the poverty level

Largest Scores

2.2	San Jose, Calif.	-2.5	Charleston, S.C.
2.2	Anaheim, Calif.	-2.5	Shreveport, La.
1.6	Paterson/Clifton	-2.4	Mobile, Alabama
1.4	Oxnard/Ventura, Ca.	-2.3	Jackson, Miss.
1.3	Seattle	-2.0	Birmingham

Loadings* (Var. No.)   Variables

Component II - Life Cycle

88	(05)	Average number of persons per dwelling
86	(28)	% Pop. under 18
81	(31)	Birth rate
78	(32)	4 or more children present in family
51	(11)	Households > 1.01 persons per room
-75	(05)	Death rate
-87	(29)	% Pop. 65 years and over
-89	(30)	Medium age of pop.
-93	(33)	No. children present in family

Largest Scores

3.1	El Paso	-3.4	Tampa, St. Pet.
2.2	Honolulu	-3.3	Fort Lauderdale
2.0	Corpus Christi	-2.7	West Palm Beach
1.9	Salt Lake City	-2.2	Miami
1.9	Charleston, S.C.	-2.0	Wilke-Hazleton, Pa.
1.6	San Antonio, Tex.	-1.6	Jersey City
1.6	Flint	-1.6	Reading, Pa.

Component III - White Collar Status/Manufacturing

88	(21)	% Employed in managerial and professional occup.
87	(20)	% Employed in white collar occupations
79	(47)	% Married couples both graduates
72	(45)	% 4 years high school or more
68	(25)	% Government employees
65	(44)	Median school years
59	(53)	No. of physicians
44	(59)	% Moved residence after 1965
34	(23)	% Employed in wholesale and retail
32	(07)	Medium value owner occupied housing
31	(61)	Housing built after 1960

Largest Scores

3.0	Washington	-2.3	Johnstown, Pa.
2.9	Madison, Wis.	-2.0	Lancaster, Pa.
2.1	Austin, Tex.	-2.0	York, Pa.
1.8	Santa Barbara, Calif.	-1.9	Reading, Pa.
1.7	San Francisco	-1.8	Flint, Mich.
1.6	Albuquerque	-1.7	Greenville, S.C.
1.5	Honolulu	-1.7	Jersey City
		-1.7	Wilkes Barre

```
Loadings*   (Var. No.)    Variables
```

Component IV - Growth

Loadings	(Var. No.)	Variables
78	(63)	% Growth in MFDing
75	(64)	% Growth in selected services
68	(65)	% Growth in retail and wholesale
60	(62)	% Housing built after 1960
57	(77)	Incidence of V.D.
53	(46)	Males not high school grads
47	(58)	% Pop. change 1960-70
46	(39)	Negro families 1.01 persons per room
43	(59)	% Families moved residence after 1965
33	(66)	Net migration from central city
33	(74)	Violent crime rate
30	(36)	% Pop. non-white
-32	(27)	% Unemployed
-32	(60)	% Residing in state of birth
-33	(52)	Hypertensive heart deaths
-33	(57)	Av. monthly AFDC payments
-45	(54)	No. of hospital beds
-46	(71)	Number of days below 32°F
-60	(62)	Housing built before 1950

Largest Scores

3.2	Fort Lauderdale		-2.4	Duluth Superior
3.1	West Palm Beach		-1.9	Johnstown, Pa.
2.3	Greensboro, N.C.		-1.6	Spokane
2.2	Greenville, S.C.		-1.5	Albany-Sac-Troy
2.1	Dallas, Tex.		-1.5	Utica-Rome, N.Y.
1.9	Orlando, Florida		-1.3	York, Pa.
1.8	Miami			

```
Loadings*  (Var. No.)    Variables


Component V - Size/Quality

    81          (01)        Population total
    81          (02)        Total labour force
    77          (03)        Total manufacturing establishments
    67          (74)        Violent crime rate
    52          (69)        Water pollution index
    40          (76)        Incidence of forcible rape
    35          (04)        Population density
    33          (67)        Total suspended particulates
    32          (14)        Per capita income
    30          (26)        % Working outside country of residence

   -30          (73)        Motor vehicle deaths
   -39          (18)        Households with one or more cars
   -69          (66)        Net migration from city centre

Largest Scores

    5.8  New York                    -1.6  Appleton, Wis.
    3.9  Chicago                      -1.6  Madison, Wis.
    3.0  Detroit                      -1.4  Binghampton
    2.8  Los Angeles                  -1.3  York
    1.5  Baltimore                    -1.2  Lancaster, Pa.
    1.5  Philadelphia                 -1.2  Utica-Rome


Component VI - Contemporary Family Organization/Sun Belt, Service

   100          (34)        Marriage rate
   100          (72)        Areas of parks and recreation
    86          (81)        Divorce rate
    82          (24)        % Employed in services
    38          (73)        Motor vehicle deaths
    32          (70)        No. of days with temp. above 90°F

   -34          (60)        % Residency in state of birth

Largest Scores

    9.6  Las Vegas                    -1.0  Johnstown, Pa.
    1.2  Phoenix, Ariz.
    1.1  Fresno, Calif.
    1.0  Miami
    1.0  Tucson, Ariz.
```

```
Loadings*  (Var. No.)   Variables


Component VII - Housing Conditions

    63          (10)      % Housing delapidated
    58          (38)      % Foreign born
    58          (04)      Population density
    48          (07)      Median value owner occupied housing
    33          (79)      Families with female heads
    31          (13)      Housing with 4 or less rooms

   -51          (18)      Households with 1 or more cars
   -72          (08)      % Housing owner occupied
   -73          (06)      % Housing single detached

Largest Scores

    5.4  Jersey City              -1.8  Tulsa, Okla.
    3.6  Honolulu                 -1.3  South Bend, Ind.
    2.9  New York                 -1.1  Orlando, Fla.
    2.0  Newark                   -1.1  Spokane
    1.5  Boston                   -1.0  Peoria, Ill.


Component VIII - Welfare Expenditure

    80          (56)      Per capita local gov. expend. on
                             public welfare
    77          (55)      Per capita local gov. expend. on health
    70          (78)      % Families receiving AFDC
    53          (48)      Per capita local gov. expend. on education
    52          (27)      % Unemployed
    46          (17)      Savings per capita
    39          (75)      Property crime rate

Largest Scores

    4.8  Stockton, Calif.         -1.5  Honolulu
    3.9  Fresno, Calif.           -1.1  Greenville, S.C.
    2.9  Bakersfield, Calif.      -1.1  Wilkes-Hazleton, Pa.
    2.7  Los Angeles              -1.1  Allentown, Pa.
    2.4  San Francisco            -1.0  Fort Wayne, Ind.
    2.3  Sacramento
    1.9  New York
    1.9  San Bernardino
```

```
Loadings*  (Var. No.)   Variables
```

Component IX - Spanish Ethnicity/Suburban

Loadings	(Var. No.)	Variables
47	(12)	Households with 8 or more rooms
46	(26)	% Working outside country of residence
44	(43)	Medium income diff. between negro/white
38	(36)	% Non-white
40	(35)	Females over 15 in lab. force
-32	(27)	% Unemployed
-39	(23)	% Employed in retail and wholesale
-40	(11)	Families > 1.01 persons per room
-41	(13)	Households with 4 or less rooms
-46	(38)	% Foreign born
-67	(37)	% Spanish

Largest Scores

2.0	Washington	-3.8	El Paso
2.2	Richmond, Va.	-2.6	Miami, Fla.
1.8	Trenton	-2.4	Corpus Christi
1.6	Atlanta	-2.2	San Antonio
1.6	Baltimore	-2.1	Tucson, Ariz.
1.4	Augusta	-1.6	Salinas
1.4	Harrisburg	-1.6	Phoenix, Ariz.
		-1.6	Albuquerque

Component X - Traditional Entertainment Facilities

Loadings	(Var. No.)	Variables
77	(83)	Number of sports events
77	(84)	Number of dance and drama events
54	(85)	Cultural and entertainment facilities
34	(71)	Number of days below 32°F
-31	(40)	% Negro owner occupied housing

Largest Scores

3.0	Pittsburgh	-2.3	Jersey City
2.6	Appleton	-1.9	Salinas
2.5	Denver	-1.7	San Bernardino
2.2	Buffalo	-1.7	Augusta
2.0	Minneapolis	-1.7	Mobile
2.0	Baltimore	-1.6	Oxnard-Ventura
1.8	Boston	-1.5	Lorain-Elyna, Ohio

* Decimal points are removed: 75 is +0.75.

Source: revised from Davies and Tapper (1979)

patterns. The _Welfare_ dimension (Component VIII) reflects this source of variation.

8) Another consequence of the differential growth of cities through time is the fact that the building stock is likely to be of different ages. Added to the variations in the prosperity of places it is likely that different degrees of obsolescence will be found. Factor VII is a _Housing Condition_ vector which is linked to ethnicity and size variations. The most disadvantaged people are likely to live in these older areas and larger concentrations of these characteristics are found in the bigger cities.

9) A rather unexpected finding of the study was a separate _Entertainment_ axis (Factor X). Although only a limited number of entertainment or leisure indicators were included in the study the axis is strong enough to support the age-old contention that urban systems can be differentiated by what can be described as their cultural character.

10) The most complex axis shown in Table 8.4 is Factor VI, a _Social Character_ vector which is given the complex sub-title of 'contemporary family character, services, and sunbelt' to highlight its association with high divorce rates, a focus on a service economy and a location in the south and west. Although the Florida centres load on the axis the extreme high score of Las Vegas on this axis is the primary reason for its presence. It demonstrates how this unique centre has a sufficiently different character to other centres to produce a separate axis.

Although there is not an exact relationship between the variables in the postulated categories and the empirically derived axes the results demonstrate that nine out of the 12 hypothesized dimensions were identified; only one axis was an unexpected finding and this was linked primarily to the uniqueness of Las Vegas. Although the full rigour of common factor analysis was not applied because a complete population was used, the results confirm that there is a basic dimensional structure for the city system of the United States.

These results demonstrate that the census-based studies of urban dimensionality provide a more comprehensive view of the variations in the urban system than those based on the usual set of quality-of-life indicators. In practice, however, the two sources of information are not separate categories of variation producing independent axes. They overlap with one another. In terms of the indicators used in this study, social disorganization, education, and environmental conditions do not appear to be separate differentiating characteristics, although the presence of welfare or entertainment dimensions must be emphasized.

Most studies of urban dimensionality have artificially restricted their synthesis to first order axes of differentiation by using orthogonal rotation. In this study oblique rotation revealed that the ten axes were not orthogonal to one another. The application of the same factoring procedures used at the first order to the factor correlation matrix led to the identification of five second order axes shown in Table 8.5. More than half of the variation in the first order axes was accounted for by the model.

Table 8.5
Higher order loadings for USA urban system

| Components | | Second Order Axes and Loadings | | | | | Commun- |
Sec- ond	First Order	I	II	IV	III	V	alities
I	4 Growth	76					63
	3 White collar-manuf.	53	48				66
	6 Cont. family organiz. etc.	68					65
II	7 Housing conditions				78		64
	5 Size/quality				64		55
III	1 Socio-econ. status		62			-31	53
	10 Trad. entertainment		76				64
IV	8 Welfare	-70					68
	9 Spanish ethnicity/suburban	77					70
V	2 Life cycle					87	78

Source: Davies and Tapper 1979

The results show that:

Size and Quality is associated with housing conditions paralleling Davies's (1977) conclusions for Canadian metropolitan areas.

Life Cycle forms a separate higher order axis although there is a minor inverse link with the size-quality axis which shows some relationship between old age character and population size.

Welfare Expenditure is inversely related to the Ethnicity (Spanish) Suburban axis since the latter has the ethnic characteristics as negative loadings. This means the higher order axis relates Spanish and foreign born ethnic groups with higher welfare expenditures.

The Growth, Contemporary Family and Economic Base axes are all associated on one axis, although the latter has a lower loading (0.53) and is also linked to the Socio-Economic Status vector (0.48). Given the growth of the metropolitan areas in the south and west with their particular emphasis on service activity, the relationships confirm one of the major characteristics of US metropolitan places.

Socio-Economic Status, Entertainment and the Economic Base dimensions (with the proviso already made about the latter) are all found on a separate higher order vector.

In many ways these results parallel Davies's (1977) study of Canadian metropolitan areas, although the addition of quality-of-life variables extend the dimensional structures. This means that it can be suggested that the latent structure of USA cities can be described in terms of five higher order axes:

1. Socio-Economic Status, Entertainment and Economy;

2. Growth and Economy;

3. Ethnicity and Welfare;

4. Life Cycle;

5. Size and Quality, Housing.

It must be remembered that the structures that have been identified do not account for all the variation in the data set. Idiosyncracies are treated as error variance in this methodology. It is worth noting the unique features of places may be considered as the most important characteristics of places for those interested in the idiographic approach.

Many other multivariate methods can be used to produce taxonomies of towns but an advantage of the factorial approach is that structures and the way towns are scaled on these structures are produced in one analysis. Davies and Tapper (1979) showed the distributional pattern of factor scores associated with each of the ten sources of variation, but also demonstrated how these individual scales can be summarized using a non-hierarchical cluster analysis based on the matrix of Euclidean Distances between each pair of places (Wishart 1978). Eight groups or clusters of SMSA's were identified (Figure 8.1).

Group I consists of New York, Jersey City and Chicago. As such it clearly indexes the Larger, Older Northern Centres which are characterized by low growth ratios and high proportions of people living in the areas who are engaged in manufacturing. There is no doubt that Jersey City is an anomaly, given its size, but this is a function of the fact that SMSA's were chosen for analysis not the larger scale continuously urbanized called Megalopolis (Gottman 1964) for which data on the variable set used here is unavailable. As part of the Greater New York area Jersey City takes on the character of the larger entity. This issue of boundaries also accounts for the fact that the high concentrations of service and quaternary activities in central New York and Chicago were not emphasized. The census records comparable data primarily by residence, not the place of work. Since the reidence of many of these high status workers is often outside the SMSA the characteristic is downplayed. The results, therefore, testify to the importance of the definitions of metropolitan places, for they condition the results that are produced.

Group II is called Regional Service Centres because seven out of the largest ten SMSA's are part of the cluster. They are characterized by relatively high socio-economic status, high entertainment and welfare levels together with low growth and poorer housing. Their dispersal throughout the nation reflects their primary role as major regional nodes, but the inability of places in the south and southeast to join the group must be noted. The addition of some smaller centres such as Appleton to the larger nodes that characterize the group demonstrates that size alone is not the only differentiating feature.

Group III places are called Southwestern and Western Growth Centres, given their association with the emerging centrs of the region. The members of the group have high growth rates, are relatively youthful in demographic terms, have high white collar levels as well as concentrations of people of Hispanic extraction and high levels of welfare. As such they are closely associated with Zelinski's (1973) description of newly emerging post-industrial centres.

Group IV places are also very localized. This, combined with their characteristics, led to the label South Florida Retirement/Tourist centres. Miami, West Palm Beach, Fort Lauderdale and Tampa were all indexed by relatively old populations, growth and service economies in 1971, as well as by a high Hispanic element and lower socio-economic status. Again the similarity with Zelinski's (1973) description of the new retirement and pleasure centres must be noted.

Group V only contains one SMSA, Las Vegas. The high level of socio-economic status, a youthful population, and the uniqueness associated with the factor identifying high marriage and divorce rates combine to give the city its distinctive role as the 'pleasure' centre of the SMSA's in the United States.

Group VI consists of a set of 19 centres that are summarized by the title Small, Northeastern Low Growth Centres. These are the demographically older places in the Middle Atlantic and New England regions, with outliers in Duluth, Milwaukee and Huntingdon. They are characterized by low growth rates, high levels of manufacturing, relatively poor housing, as well as being smaller in size.

Group VII places consist of the 24 Low Status Southeastern SMSA's. Part of the traditional 'south', they have low socio-economic status, high social deprivation, high levels of welfare and relatively higher levels of recent growth after years of stagnation. Their lower levels of welfare and low Hispanic ethnicity also differentiates them from other US centres in the southwest and Florida.

Group VIII centres, 34 in all, are called Stable, Regional Centres. Their average factor scores do not exceed 1.0 which means that they do not have any extreme characteristics. In general they are the well-established, usually smaller regional nodes concentrated in the mid-west and in the industrial states that are slightly above average on socio-economic status and in the quality of their housing.

These results demonstrate that there is a high degree of geographical variation between the SMSA's of the United States based on the information used in this study. Obviously the association with geographical regions is not complete, unlike the comparable Canadian study carried out by Davies (1977). The largest centres and the regional nodes form major exceptions to the trend. However, the way in which SMSA's in the southeast, southwest and west, north, northeast and Florida areas

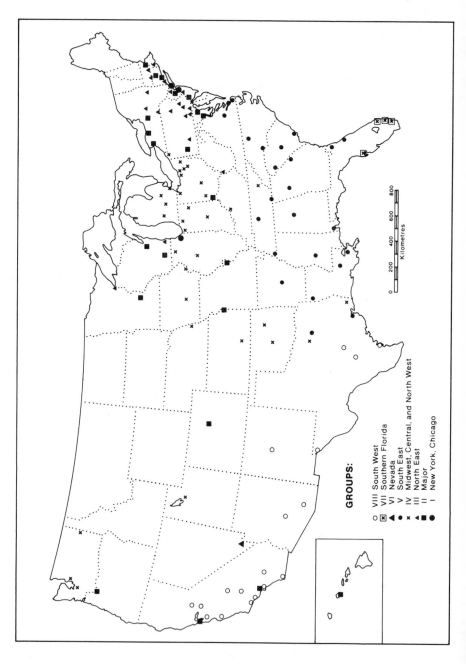

Figure 8.1   Groups of SMSA's in the United States

Key for Figure 8.1
List of the member SMSA's in each of the eight categories

I

'Large, Older Northern Centres'(3)

| Chicago | New York |
| Jersey City | |

III

'Small, North Eastern, Low Growth Manufacturing Centres'(19)

Albany	Paterson
Allentown	Providence
Binghampton	Reading
Bridgeport	Springfield
Duluth	Trenton
Harrisburg	Utica-Rome
Huntingdon	Wilkes-Barre
Johnstown	Worcester
Lancaster	York
Milwaukee	

V

'Low Status, South and South Eastern Centres'(24)

Atlanta	Jackson
Augusta	Jacksonville
Baton Rouge	Little Rock
Birmingham	Memphis
Charleston	Mobile
Charlotte	Nashville
Chattanooga	New Orleans
Columbia	Newport-News
Dallas	Norfolk
Greensboro	Orlando
Greenville	Richmond
Houston	Shreveport

VI

'Las Vegas'(1)

Las Vegas

VII

'Southern Florida Retirement and Tourist Centres'(4)

| Fort Lauderdale | Tampa |
| Miami | West Palm Beach |

II

'Mainly Major Regional Centres' (21)

Appleton	Newark
Baltimore	New Haven
Boston	Philadelphia
Buffalo	Pittsburg
Cincinnati	Portland
Denver	Rochester
Hartford	San Franciscio
Kansas City	St. Louis
Los Angeles	Syracuse
Madison	Washington
Minneapolis	

IV

'Midwest, Central, North West: Stable Regional and Industrial Centres'(34)

Akron	Lansing
Beaumont	Lorain
Cleveland	Louisville
Canton	Oklahoma City
Columbus	Omaha
Davenport	Peoria
Dayton	Rockford
Des Moines	South Bend
Detroit	Spokane
Erie	Salt Lake City
Flint	Seattle
Fort Wayne	Tacoma
Fort Worth	Toledo
Gary-Hammond	Wichita
Grand Rapids	Wilmington
Indianapolis	Youngstown
Knoxville	Tulsa

VIII

'Western and South Western Growth Centres'(19)

Anaheim	Sacramento
Austin	Salinas
Albuquerque	San Bernardino
Bakersfield	San Antonio
Corpus Christi	San Diego
El Paso	San Jose
Fresno	Santa Barbara
Honolulu	Stockton
Oxnard	Tucson
Phoenix	

display similar characteristics is worthy of note. Yet not all the centres in the 'traditional' textbook regions of the USA display homogeneous characteristics. For example, the Texas centres are not similar enough to form a separate group, whilst the New England metropolitan places are also members of several groups. Similarly, the places usually thought of as part of the manufacturing belt are split between the groups called Small, Northeast Centres and the Stable Regional Nodes.

These results demonstrate the way in which the multidimensional character of urban places can be summarized in terms of a set of basic dimensions or constructs. In addition, the results show how a typology or group of places can be derived from the respective importance of each place on these dimensions. This means the results show' how factor analysis methods provide a way of identifying the categories or types of towns by quantitative means, rather than subjectively defining the groups as in Aurosseau's (1921) study, as well as providing a mechanism for scaling towns along these categories. Cluster analysis, however, is needed to produce the classification or groups of centres that have spatial patterns. Yet, although regional character is obviously not a complete explanation for the differences, it still goes a long way towards summarizing the variations in the SMSA's of the USA.

(iii) National variations. So far the discussion on the dimensionality of the urban system has focussed only upon the North American case. Obviously an important question is whether different features characterize city systems in other nations in other parts of the world. Hodge (1968) after a series of Canadian studies seemed to dismiss the possibility of regional variation when he stated that the dimensions or the structural features of urban centres tended to be the same from region to region. He went on to emphasize this was regard- less of the stage or even the character of regional development.

Berry (1969), however, in a review of the evidence provided by factorial studies in which comprehensive data sets were employed, came to the conclusion that variations in the dimensional structures did occur, especially in developing or under-developed countries. The studies by Fisher in Yugoslavia (1966), Berry in Chile (1969), Mabogunje (1968) in Nigeria, and more recently Hirst (1975) in Tanzania, have all identified a large, yet rather general axis of differentiation indexing a traditional society-modernization split in which the agricultural, commercial and cultural urban centres of traditional society were separated from the growing, modernizing nodes. This type of axis combined the separate dimensions of Socio-Economic Status and Age found in most North American studies. Apart from this axis, a set of Economic Base, and Ethnic and Growth factors were also usually found in most studies of countries with developing economies. Although Size has also been identified in studies in the USSR (Harris 1970) and Chile (Berry 1969) it did not separate out in many of the other analyses. In some cases this is undoubtedly due to the proce- dures used in the study. For instance, in Nigeria (Mabogunje 1968), a size variable was included but it was virtually filtered out by using a common factor analysis apparently based on the communality esimates as diagonal entres to the correlation matrix. If size is a separate dimension the communality of a single variable will be very low (as illustrated in this study), so the variable proved to be virtually

Table 8.6
Welsh towns in 1971: axes and loadings

Title and Variables	Loadings for Different Factoring Methods (Oblique Rotation: Direct Oblimin 0.0)				

	Principal Axes				
1. AGE & MAJOR ECONOMY	Component*	Factor	Image	Rao	Alpha
Order of Axis	1(1)	1	2	1	1
4. Female Ratio	84(73)	74	73	78	+77
9. Old Age	82(81)	80	79	81	+79
15. Service-Distribution	73(80)	72	70	66	+75
27 Large Dwellings	67(79)	66	62	63	62
5. Welsh Speakers	49(56)	49	50	44	48
25. Young, Single Females	46(67)	51	54	40	50
23. Visitors	--(54)	--	34	--	--
7. Council Housing	-33(-49)	--	-31	30	--
11. Mining	-37(-38)	-32	--	-33	-31
10. Young Adults	-48(--)	-46	-33	-50	-46
12. Manufacturing	-60(-72)	-58	-55	-53	-59
8. Children	-71(-82)	-73	-75	-70	-66
3. Persons/room	-78(-88)	-81	-77	-78	-75

2. DEGREE OF RENEWAL (Substandardness, Growth)					
Order of Axis	2(2)	2	5	2	2
26. No Bath	-84(83)	-78	66	-77	-82
21. Born in Wales	-61(65)	-57	63	-58	-54
11. Mining	-60(64)	-54	65	-55	-55
19. Unemployed	-46(45)	-35	37	-33	-37
17. Local Movers	-38(38)	-34	30	-35	--
2. Population Change	60(-68)	62	-49	65	62
18. Distant Migrants	79(-80)	76	-67	75	73
15. Service Distribution	--(-33)	--	--	--	--

3. SIZE (and GROWTH)	Component*	Factor	Image	Rao	Alpha
Order of Axis	4	4	1	3	5
24. Retail Turnover	99(97)	100	100	95	100
1. Population Size	98(96)	98	99	94	96
2. Populaton Change	43(44)	35	36	38	34

4. COLLEGE-RESORT TOWNS (Young Adult-Visitors)					
Order of Axis	3	3	4	4	3
10. Young Adults	89(90)	93	83	−82	91
23. Visitors	79(79)	72	69	−76	73
25. Young, Single Females	70(58)	65	59	−71	64
4. Female Ratio	--(−36)	--	---	--	--

5. SPECIALIZED ECONOMIES					
Order of Axis	5(4)	5	7	6	4
13. Construction, Transport & Utilities	84(77)	77	80	71	71
22. Irish	71(66)	53	52	49	58
16. Government, Defence	56(72)	57	37	65	57
12. Manufacturing	−39(−50)	−54	−30	54	51
20. Economically Active Females	−39(--)	--	−30	--	--

6. TENURE					
Order of Axis	7	7	6	7	7
7. Council Housing	−80(78)	−87	79	−77	−85
17. Local Migrants	−47(48)	−34	33	−40	−34
8. Children	--(33)	--	--	--	−34
6. Owner Occupation	89(−87)	85	−80	84	87

7. WELSH CENTRES	Component*	Factor	Image	Rao	Alpha
Order of Axis	6(7)	6	3	5	6
5. Welsh Speaking	−46(51)	−37	−−	43	40
21. Welsh Born	−31(31)	−29	−−	30	28
12. Manufacturing	−31(−−)	−−	−57	−−	−−
13. Construction, Utilities	−−(31)	−−	−−	34	−−
17. Local Movers	−−(−34)	−−	−−	−−	−−
15. Service, Distribution	−−(−35)	−−	43	−−	−−
16. Government, Defence	59(−35)	30	49	−−	−30
20. Economically Active Females	60(−75)	59	−−	−63	−58

 * Figures in brackets represent loadings for varimax solution.
** Image analysis title should be 'major economy'.

 Source: Davies 1983.

excluded in the final analysis. This example demonstrates the problem of generalizing from the earliest studies in the field, where a wide variety of data and factorial procedures were used.   Indeed, it is possible that the generality of the major dimension found in some studies of underdeveloped countries could be a statistical artifact, produced by the extraction of unrotated principal components, or the rotation of too few axes.

Unfortunately interest in the dimensionality of urban systems seems to have been lost in the last five years for there are few published studies comparable to those in the late 1960s and early 1970s.   Yet some appropriate evidence for the utility of Berry's developmental sequence can be seen in Table 8.6, which describes the factorial dimensions extracted by Davies (1978, 1983) in a study of Welsh towns. It is argued that the presence of two general vectors, one indexing Age and Major Economy, the other Degree of Renewal (which summarizes substandardness and growth), is a function of the particular nature of the Welsh urban system derived from 50 years of economic depression, or at best halting growth, and a peripheral location in Britain.   This means that the combination of the axes represent a lapsed condition to a typology more typical of developing economies. Table 8.6 also shows that typical Size, Economic Base and Cultural (Welsh) axes are also identified, along with a Tenure axis which again can be linked to the high level of welfare in British society, as manifested by the council (or public) house programme.   Most of the axes are stable when different factoring methods, except for the Welsh dimension in the Image Analysis. This reflects the relatively minor nature of the axis, and the fact that the common factor basis of the procedure redistributes the variance associated with the vector. In general, therefore, the stability of these results give support for the idea that national variations in urban dimensionality can be expected.

Berry (1972) provided a summary of the dimensional structures of city systems and in general these studies reported here support his conclusions, subject to some modifications.

1) City systems are characterized by a hierarchy of centres based on aggregate economic power and can be measured by a Size dimension.

2) The Economic Base of centres frequently acts independently of the other dimensions producing specialized types of towns additional to the hierarchical pattern.

3) Social Status and Age or Life Cycle variations are usually dominant dimensions. However, these structures are frequently combined in less economically developed countries, sometimes with economic base features. This results in a general axis of Development or Modernization versus Underdevelopment, often expressed spatially as the core-periphery contrast (Friedmann 1973).

4) Culturally heterogeneous societies will provide separate ethnic axes of differentiation. If sufficiently different life styles can be found (as in the case of the Contemporary Family/Sunbelt axis in the USA study reported previously) a separate dimension may well emerge, thereby modifying Berry's (1972) ideas.

5) Growth will often be identified as a separate axis associated with particular economic sectors linked to specific centres.

6) Berry (1972) paid little attention to the post-industrial changes in society. They are likely to lead to Welfare (as expressed in financial support and housing) as well as Entertainment axes of differentiation. These are likely to provide an increasingly important source of variations in urban systems, although it could be argued that entertainment (in the sense of consumption of some product produced elsewhere) is really only another type of specialized economic base.

7) As yet there have been few factorial studies of urban systems in socialist countries using comprehensive data sets. Harris (1970) in the USSR identified Size, Growth and Density of Spacing components. Kansky (1976) in Czechoslovakia showed Size, Age and Housing, Education/Cultural an Young Family factors in his 1970 study with an additional set of factors measuring Apartment character in 1961. These results imply that there are few variations in the character of urban systems in different political systems.

(iv) _Temporal changes_. Although it is relatively easy to speculate about temporal changes in urban system structures we must be cautious. For example, only a developmental study through time will confirm whether the interpretation of the characteristics derived for Wales is appropriate. To date nobody has carried out a comprehensive temporal study of urban systems using a series of different times, so our knowledge is very limited. One example, King's (1966) comparison of Canadian cities, is restricted to 1951 and 1961, but here the evidence for stability is not very strong since he reported that there was:

'only weak support for the contention that the urban dimensions are stable over time.' (King 1966 p.161). At this time, therefore, the degree of stability in urban dimensions and the factors accounting for change are still largely unknown. No structural explanation similar to the Shevky-Bell (1955) model for the intra-urban case has been proposed. This is a major research task for the future, one that should be linked to the intra-urban differentiation case.

Despite this pessimistic conclusion, some exploratory studies of the spatial-temporal patterns of economic impulses in urban systems have been carried out using factor analysis techniques (King and Jeffrey 1972; Pigozzi 1975). Pigozzi's (1975) applied an S mode factor analysis to a data set consisting of 56 cities (arranged as the columns or variables), and bi-monthly unemployment rates for 10 years (as rows) from 1961 to 1970. Five factors with eigenvalues above 1.0 were extracted identifying groups of cities which were labelled:

1) National Urban Labour Market;
2) Midwest Fabricated Metals Labour Market;
3) Seasonal Labour Market;
4) Midwest Iron and Steel Labour Market;
5) Michigan Overspecialized Labour Market.

Obviously the S mode analysis does not provide information on the economic differentiation of cities; instead, it identifies the sets of cities with common behaviour patterns in rates of unemployment. In the case of the first factor, practically all the major national and regional cities were indexed, demonstrating the way the national economic trends are based on these cities. The other factors picked out groups of cities that either experienced the national trends at a later date or were associated with particular economies subject to very different unemployment regimes. Although this factorial approach adequately defines the structure of the impulses, Pigozzi concluded his study by pointing out that a spectral or cross-spectral analysis of the unemployment rate would provide more conclusive results of the nature of the temporal curves of unemployment.

2. PERCEPTION OF THE URBAN SYSTEM

Although the use of 'quality-of-life' measures has broadened our understanding of the nature of the urban systems, the indicators are still in the realm of the so-called 'objective' indicators. Since people act on their image of structures rather than upon the facts involved, a vital element in intra-urban differentiation is the perception of urban places. As yet most of the perceptual work in urban studies has been carried out in intra-urban space, as studies by Lynch (1960), Hall (1966) and Downs and Stea (1977) indicate. Gould, however, has carried out a number of studies which attempt to measure the attitudes, usually restricted to a question related to the desirability of living in towns and regions in the United States, Sweden and Tanzania (Gould 1974). When the results of interviews carried out in several locations are analyzed by location, a 'home town

Table 8.7
Loadings for perceptual axes in New Zealand

I. SIZE AND OPPORTUNITY
    (Socio-Economic Quality)*
    Loadings
    _____

28) 87 Many motorways
 7) 84 Large size
31) 84 Many cinemas
32) 82 Many theatres
38) 79 Variety of entertainment
18) 79 Fast growing
34) 76 Lively atmosphere
35) 74 Accesible to other cities
29) 74 Many schools
19) 74 Many job opportunities
20) 74 Wide variety of jobs
17) 67 Modern
26) 45 Fast moving
37) 60 Good sporting facilities
13) 52 Compact
33) 50 Many beaches

II. SOCIAL DESIRABILITY
    (Social and Housing
    Environments)
    _____

22) 84 Low building costs
23) 85 Cheap land
10) 81 Low crime rate
25) 84 Not congested
27) 85 Abundant parking

30) 80 No overcrowded schools
24) 68 Many flats-houses
11) 68 Spatious
 8) 57 Friendly
 9) 57 Few non-Europeans
16) 56 Clean
13) 51 Compact city

III. ENVIRONMENTAL QUALITY AND
     FACILITIES (Physical
     Structure)
     _____

14) 67 Many trees
15) 63 Many parks
36) 64 Easily accessible
       countryside
21) 54 No slums
 6) 52 Dry
 8) 50 Friendly
37) 51 Good sporting facilities

IV. CLIMATE
     (Climatic Scales)
     _____

 4) 71 Hot summers
 5) 67 Sunny
 3) 64 Mild winters
 2) 61 Calm

_____

* Titles in brackets are the originals from
  Jackson and Johnston (1972); the others are
  revised.

Decimal points have been removed: 87 is +0.87.

_____

effect', or local peak of desirability, is usually identified.   But
when all those cases are put together and factor analyzed by the Q mode
approach, a general axis identifying a ridge of desirability, focussing
on the core area of the country, is usually identified.   This means a
general desirability surface, as well as peaks of local preference need
to be identified.   Many of these images are persistent through time so
attempts to alter the regional balance within countries need to take
into account this stability if developmental strategies are to be
successful.

Questions dealing only with desirability obviously cannot encompass
the whole range of attitudes that people have towards sets of towns.

This led Jackson and Johnston (1972) to explore the images of New Zealand cities using 38 different questions dealing with eight major sets of character: population; economy; housing; traffic; education and culture; recreation; physical structure; climate. A Principal Axes, Varimax solution derived from the 38x38 similarity matrix based on the interview data led to the conclusion there were four axes or sources of differentiation. This accounted for 70% of the variance in the data set. A revised version of the results is shown in Table 8.7 where the axes are describing Size and Opportunity, Social Desirability, Environmental Quality and Facilities and Climate (Table 8.7). These results confirm that <u>perceptual</u> studies are likely to identify a different set of dimensions to those obtained from <u>objective</u> indicators, although size does parallel the results of previous work, and Social Desirability has some links with the Socio-Economic Status dimension described in the previous section. Studies with a much wider interview base, over many more towns than in this New Zealand study are needed in the future. Then, perhaps, perceptual indicators can be directly interrelated with the more objective census-type variables. Only in this way will the independence of these dimensions be established.

To some extent, of course, Thorndike (1938) anticipated this need in his social condition survey of US cities. He obtained ratings of American cities from civic leaders in education, business, religion and philanthropy, but the combination of this information into a single scale destroyed much of its utility. Despite this major problem Thorndike's results showed that civic leaders tended to over-emphasize the size, property value, and park acreage of cities, at the expense of education, morality levels and general living standards. This revealed a perceptual bias of the opinion of the elite in the cities, compared to the population at large. But the question of whether this is a bias is not as important as the recognition that differential attitudes can be measured and that these are likely to affect the future of cities. It is, after all, the managerial elite that usually control the growth of individual centres. Not surprisingly, it is possible to identify very different attitudes to growth as Adrian (1972) pointed out in his study of attitudes of government leaders. Although Adrian's study has a descriptive, not a quantitative basis, the results are worth summarizing since it is likely to set the standard for future studies in the field. He maintained there were four basic attitudes among city leaders in the USA which affected growth.

1) The first attitude was described as being dominated by economic growth considerations. Growing, optimistic centres in which producer interests were highly represented – such as company or resource towns – were examples of this <u>Boosterism</u> type. Artibise (1981) has isolated the characteristics of boosters in early Canadian cities.

2) The second were summarized as being <u>Amenity</u> oriented since they were concerned primarily with consumer interests and with the use of wealth to provide life's amenities. Upper middle class suburbs were deemed to be representative of the type.

3) The third were devoted to the maintenance of traditional services; they were so concerned with the status quo that government

was in the hands of the Caretaker types. Small towns, with a strong service centre base, with relatively slow growth were considered to be typical of the type, in which few regulations, freedom, and self-reliance were highly valued characteristics.

4) The fourth was described as the Arbiter type since the role of the governing body was seen primarily in terms of the arbitration of conflicting interests. In general the large, heterogeneous places were typical of centres possessing this attitude; the need for arbitration comes from the fact that every government action involves an advantage or disadvantage for somebody.

Adrian (1972) cautioned that his typology was likely to be culture-bound. Nevertheless, he provided a list of indicators suitable for measuring the various attitudes towards government, suggesting that these could be derived from content analysis of City Hall publications, speeches of city officials, etc. As yet, few examples of this type of comprehensive study using factor analysis have been completed. Hence, it is difficult to judge the utility of the hypothesized dimensions or their extension to deal with value differences between people and nations. Nevertheless, Adrian's four dimensions provide a useful first approximation to the question of the variations in government attitudes in a system of cities. If the area is developed in the future it provides two additional advantages for classificatory studies - apart from their value in avoiding the universalist fallacy that all city systems have similar structures. The first is that the study of urban government images takes one into questions of the management of urban growth. It means that much closer links with the prescriptive approach to urban studies could be forged. The identification of particular attitudes to urban growth may help predict or even modify the growth of towns in an urban system, ensuring that classificatory studies transcend their traditional descriptive role (Arnold 1972). The second advantage stems from the fact that each of the four categories of city government images were related to a set of socio-economic character-istics - such as size, economy and social status. This raises the intriguing possibility that one day it may be possible to interrelate all these dimensional structures and provide an integrated body of knowledge on the structure of the urban system. The opportunities are there. It is up to empirically minded investigators to deal with these problems and add to our overall comprehension of the urban system.

3.. THE STRUCTURE OF INTER-URBAN FLOWS

Although nodality is one of the basic characteristics of urban places and is likely to vary in importance from place to place much more attention has been paid to analyses of the formal characteristics of urban systems, whether economic base, commercial hierarchy or socio-economic dimensionality studies, than to studies of the flows or inter-actions for information, people, goods, and services between towns. Berry (1964) pointed to this imbalance almost 20 years ago, but apart from major developments in the modelling of flow patterns (Wilson et al 1977) it is still an unfortunate feature of the urban system litera-ture. Since it is obvious that the continued differentiation of the urban system depends upon these transfers of people, information and goods this imbalance in the study of inter-urban systems should be rectified.

(a) Problems of data

There seems little doubt that the limited number of inter-urban functional studies is very much a consequence of the dearth of data on flows between cities. Few census bodies collect information on inter-urban connections on any comprehensive basis - apart, perhaps, from migration or journey-to-work flow. This means that most investigators interested in flow patterns have to engage in costly and time consuming surveys and these are beyond the resources of most researchers. But of equal importance is the very complexity of inter-urban functional connections. Flows vary in many ways:

1) by type of flow, e.g., journey-to-shop, telephone calls, bank transfers;

2) by spatial direction, and field or area of coverage;

3) by mode or transport type, e.g., truck or car, airline or rail;

4) by size, e.g., volume or money costs;

5) by frequency in time, e.g., flows per week or year;

6) by duration or length of time taken to complete the flow.

As yet few of these sources of variation have been comprehensively evaluated or incorporated in individual studies. The literature, therefore, is dominated by studies that utilize only one or two of these parameters of variation. Moreover, for many years, investigators tended to use surrogates of flow, such as the number of bus services between towns (Green 1950) rather than actual flows. In those studies that have dealt with actual flows in an urban system the volume and direction of flows in a limited time period is the favoured data input. On a national scale the use of journey-to-work data or airline flows or telephone calls have proved to be the most popular indicators of movement. Inevitably the question must be raised of how representative these particular flows are of all movements between towns. In other words, even if a set of satisfactory results describing the pattern of flows between places has been derived, the degree of generality of this pattern must be queried. The familiar problem of the universal fallacy is present, as already critized in Thorndike's (1938) suggestion that a one dimensional 'goodness of life' scale for cities exists. Again this illustrates that investigators must avoid the ever present danger of assuming that there is only one pattern or model for inter-urban connectivities.

Confronted with the very size and complexity of inter-urban movements investigators have turned increasingly to multivariate techniques to help them solve their problems. Factor analysis has proved particularly useful in solving the two basic problems confronting any investigator of inter-urban flow patterns. The first problem consists of identifying the general patterns of interaction from within the mass of detailed flows. Factor analysis can be used as a data-reducing technique to summarize the key patterns. The second problem consists of the integration of several types of flows to produce a comprehensive description of the interaction patterns. This usually involves the alteration of the original data sets so as to provide one data matrix suitable for use in the so-called dyadic factor approach described in Chapter III. Examples of both these approaches will demonstrate the utility of the factorial approach in dealing with these problems as well as providing empirical evidence of the structures of urban system interactions.

(b) Analysis of single connectivity matrices

This is by far the most popular application of methods and has been applied to many different sets of data (Goddard 1968), telephone calls (Illeris and Pederson 1968; Clark 1973; Davies 1979), journey-to-work flows (Davies 1972) and trade flows (Russell 1968). Normally a Q mode component approach is used, which means that the similarity coefficients are calculated between the rows or origins of the origin-destination matrix. This ensures that the axes identify the cluster of origins, places with similar interaction flows, whilst the scores pick out the importance of the destinations to this group of origins.

Figure 8.2 shows the results of the application of the Principal Axes component approach followed by Direct Oblimin rotation to a study of the 1971 telephone calls between each of the 94 largest centres in Montana and to the 17 places immediately outside the state. The addition of the latter meant that the results were not artificially truncated by the use of state boundaries — ensuring that an open system approach was adopted. A total of 82.6 percent of the variance was accounted for by an 11 axis solution which proved to be stable when different factorial procedures were applied (Davies 1979). The results demonstrated the way in which the state of Montana could be broken down into separate areas dominated by the individual towns which acted as primary destinations. But, in addition, the result can be used to demonstrate that a rather simple structure characterizes this type of interaction. For example, a set of small towns Circle (59), Sydney (16) and Wibaux (79) have their largest connection with Component XI, the Glendive axis, but Glendive in turn has its largest loading on the Billings axis, Component I. If the structure of connections is organized in a non-spatial diagrammatic framework the results can be shown to conform to a hierarchical structure with upward or centripetal linkages, together with a rather complex series of reciprocal flows. These links are shown in Figure 8.3 by plotting the loadings between the largest places. Not all the places are hierarchically structured. In the northeast, a set of centres are linked to two nodes, rather than one major destination, namely Wolf Point and Williston. In the west, many of the larger places are linked directly to one another in a reciprocal fashion. These relationships can be explained by the rather specialized nature of many of the larger towns in the area: Butte's tradition as a mining centre, Helena's role as the political capital of the state, Missoula as the site of one of the major universities.

These first order factor results demonstrate the way in which the local features of Montana modify the simple hierarchical and reciprocal patterns of interchange between the towns in the state. But in

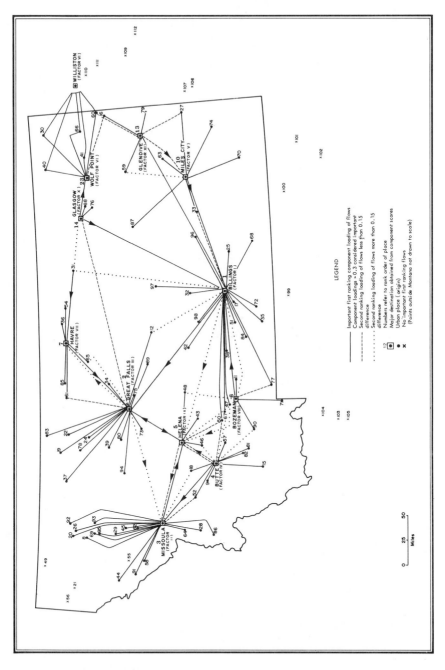

Figure 8.2   Structure of connectivity of the Montana urban system
Source: Davies (1978)

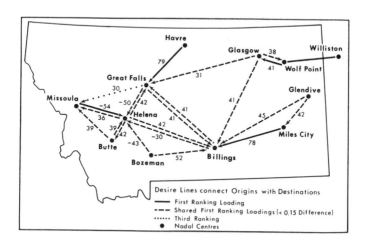

Figure 8.3 Connections between the largest Montana centres
Source: Davies (1978)

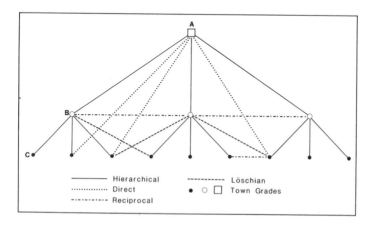

Figure 8.4 A basic model of urban connectivity

addition to these local structures identified by the factor analysis, one must remember all the individual flows which have been generalized out of the data set. If investigators want to isolate these flows a quick but non-statistical procedure would be to remove the general patterns displayed by the factor analysis and map all the remaining original linkages. In this way, the general and detailed local flows can be isolated. As yet, few methods exist for effectively separating the general and specific flows in the confines of the factor analysis, since most workers in the factorial field are not interested in the specific and unique parts of the variance. Gould (1980) suggested Q analysis may help in this regard.

After reviewing the evidence for inter-urban connections drawn from the type of empirical study described above, and the theoretical literature, Friedman (1973) and Pred (1977) have independently proposed that the basic structure of relationships between towns consisted of a hierarchical network, together with lateral or reciprocal flows between the larger regional centres. Pred's (1974) empirical work pointed to the need to incorporate Loschian patterns of dependency into models of inter-urban relationships. In addition, his studies of job control in large organizations demonstrated the importance of direct big-city to small centre linkages. A summary of these ideas is shown in the diagrammatic model shown in Figure 8.4. The postulated structure does conform to much of the available evidence of single flow structures. But on intuitive grounds it must be questioned whether any single model can effectively account for all patterns of flow, even when the minor or random connections are ignored. In addition, it must be pointed out that the problems of data sources mean that only a limited number of types of inter-urban flows have been investigated. Obviously, therefore, there is a pressing need to explore the utility of the model shown in Figure 8.4 by investigating the structures of many different data sets. Factor analysis, using the dyadic mode, has proved to be flexible enough to cope with this problem.

(c) Suites of networks

Friedmann and Pred's work on urban flows certainly takes our knowledge of urban flow structure several steps nearer to reality. Nevertheless the assumption that a single model can be identified must be questioned, since it runs the risk of the universal fallacy described in Chapter III. On intuitive grounds it is much more likely that several different types of structures will be found in the real world. This led Davies (1979) to propose a suite or set of alternative flow structures based on various combinations of several different types of linkage pattern.

Figure 8.5 (Davies 1979) summarizes the various types of linkages between any pair of towns. For the sake of simplicity, the linkages in the diagram (the flow or connection between the flow origin and the destination) only deal with two features: first the orientation or direction of the largest connection; second with the relative sizes of centres. These ideas could be extended to the second, third or subsequent most important flows. As far as possible the terms applied to the various relationships identified in Figure 8.5 follow those already in use in the urban system field, rather than using the analagous concepts developed by students of graph theory (Harary et al 1965).

The description of the relationships originally described by Davies (1979) can be quoted at length to identify the interactions involved:

'(i) Linkages. The simplest relationship shown is the <u>dependent</u> link in which the largest outflow from town A is received <u>by a</u> larger centre, town B. Normally the linkages are shown by desire lines in both the nodal structure (Nystuen and Dacey 1961) and the factorial literature (Davies and Musson 1978), and these dependent linkages join together towns of quite unequal size. Where towns of approximately equal size are involved the dependent linkage can be described as a <u>lateral</u> link (Pred 1977). Where the linkage is to a smaller centre the term <u>back flow</u> or link is probably the most suitable designation. Although only dominant flows are being considered here, these terms could be applied to three different types of flow. If B also sends its largest flow to town A a <u>reciprocal</u> linkage type can be identified. This type of association that has been pinpointed as a vital element in the interrelationships between the largest metropolitan areas (Pred 1973; Friedmann 1973). In many areas a sequence of linkages can be identified, for example, A to B and B to C. Although this type comes close to the idea of a network structure, the term <u>sequential</u> linkage can be used to emphasize that the linkage is part of a wider structure. Two very different patterns of flow can be envisaged. For example, if the direction of flow is from small to large the sequential linkage can be described as a <u>centrifugal</u> type. If the direction is from large to small, the term <u>centipetal</u> can be used.

Dunn (1970) and Pred (1973) have both drawn attention to the fact that any flow of a good or service is usually organized by an information flow in the opposite direction. The suggested terms, <u>dual</u> or <u>shadow</u> linkages, seem suitable to describe these patterns. However, it must be recognized that these are <u>not</u> really alternative linkage types as such; their difference stems from the fact that they are associated with separate types of flow.

(ii) <u>Networks or structures</u>. Theoretically, a variety of alternative networks can be constructed from the various directional components described in the previous section. Some of them are elementary networks composed of single linkage types. Others are more complex. Dunn (1970) attempted to summarize the range of network structures in terms of tree and circuit types by describing the functional characteristics they perform. His typology is not entirely appropriate for this study and alternative structures must be identified. Seven basic network structures are shown in Figure 8.5. Although some may also be relevant for identifying the flows between economic enterprises, they are primarily designed for inter-urban flow structures.

1) A <u>linear tree</u> structure consists of a linkage or connection between any two towns or centres, for example, coal carried from a mining plant in town A to a power plant or port in town B. This network is composed of a single dependent linkage relationship. Dunn (1970) suggested that a special purpose transformation of goods or services is involved in this type of network.

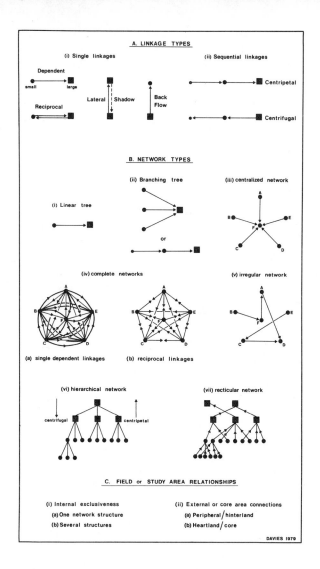

Figure 8.5   Alternative urban connectivity structures
Source: Davies (1979)

2) A branching tree is formed when the transport of goods is
followed by a transformation process that is multiple purpose,
for example, coal, iron ore and limestone moving from three
resource towns to a steel centre, or steel moving to a tin
plate town, shipyard or car fabrication plant.   Although Dunn
(1970) observed that limited attention has been paid to these
network patterns by students of inter-urban structure, the
ideas are obviously well understood in more descriptive studies
of economic processes.

3) The centralized network shown in Figure 8.5 could be considered as a variety of the branching tree but this time each town or centre, irrespective of size, has its linkage directly with one dominant centre. The principle of centralization is the vital element in its structure. Two different flow directions can be identified, one from the node to other centres, the other from the centres to the node. The terms centripetal or centrifugal centralized networks seem to be appropriate in these cases. Although the flows of manufactured goods and raw materials, respectively, may conform to these network types they seem particularly important for information flows in urban systems.

4) The term complete single circuit network can be used to describe the type of network illustrated in Figure 8.5 in which there is no single dominant flow; rather each place is converted to all other places by a single dependent linkage of equivalent size (Dunn 1970, graph A, p.242). A variety of this completely interconnected graph is shown by Dunn's (1970) type B, one in which reciprocal linkages replace the single linkages. The term complete reciprocal circuit conveniently summarizes the strucure.

5) Contrasting with these types is the specialized or irregular network, in which only some towns are linked together by an irregular pattern of linkages associated with the various specializations of the centres. This type may be characteristic of specialized industrial areas, where town A may be a mining centre with its dominant product flowing to the steel centre, B. But B is linked to C whose dominant role is the production of cars, as is D which produces rubber and plastics used for cars. Similarly, E is linked to F and this, in turn, is dominantly associated with B. Obviously, the irregular structure can have many different patterns.

6) The penultimate set of network structures proposed here may be more familiar to students of flow patterns. They involve a sequence of flows from small places to larger centres and from these large centres to the largest nodes of all. The hierarchical network is the most familiar since it was originally identified by Christaller (1932) and does seem to apply to the organization of service centres. However, there are several modifications of the type. Not only can there be centrifugal or centripetal structures, depending on the direction of the sequence, but variations obviously occur in the number of orders in the hierarchy and the number of dependent towns associated with each model. Such alternative structures were not clearly identified in models by Pred (1975) or Friedmann (1973).

7) Similar, but less regular than the rigid hierarchical type, are the so-called recticular networks. Figure 8.5 shows that distinctive levels or nodes at different orders are present but the simple rigidity of the hierarchical structure is missing: namely, lower-order places exclusively dependent on one higher-order centre, etc. Again varieties of the type can be

identified: first, by the direction of flow; and, second, by the number of orders involved. Since many functional regions have overlapping structures, as Davies and Musson (1978) have recently shown in their studies of commuting flows, the recticular model may be a more appropriate real world alternative to the rigidity of the theoretical hierarchical structure.

These seven network structures are postulated as providing an introductory set of models which will summarize the dominant structures of the majority of inter-urban flows. Additional types will also be recognized, not only as empirical studies of flow patterns grow in number, but with the use of analogies from graph theoretical ideas. However, two important limitations of this suite of models must be recognized. In the first place it is quite likely that the real world will provide evidence of combinations of these ideal types. In the second place there is no reason to assume that these structures are functionally dependent; for example, Friedmann (1973) has already emphasized that a town may be at one order for one type of hierarchical linkage and at another for different linkages.

(iii) Field or area of system. The linkage types and network models attempt to identify the general patterns involved in the functional organization of inter-urban connections but fail to provide measures of the internal cohesiveness of the connectivity structure and the external relationships of the study area.

1) Internal exclusiveness. The basic issue involved is the extent to which the network model encompasses all the towns in an area. It is possible that the structural model connects all towns, such that the centres can be considered as part of a single field or system or relationships. Alternatively, separate networks in different areas may exist and these networks need not necessarily be of the same structure. In other words, single or multiple linkage systems – or subgraphs to use the graph theory terminology – may be identified.

2) External connectivity. The degree of exclusiveness of the system has external ramifications since the linkages outside any study area also need to be considered. If the area is a dependent one, in the sense that most of the flows have external destinations, the system may be described as peripheral or hinterland area in the national or international economy. If the majority of flows have their destinations in the region under study then presumably a heartland or core region has been identified. There may, of course, be a series of alternative core regions for different types of function and there is no reason to assume that the patterns of flow have to coincide in any geographical space (Friedmann 1973).'

(d) Empirical evidence

As yet it is difficult to evaluate the utility of these various interaction structures, given the relative paucity of detailed studies dealing with more than one type of place. With the exception of studies by Berry (1968) and Black (1973) dyadic factor analysis has been used very rarely in geographical studies and there are few urban system analyses. In many ways this is because of the difficulty of obtaining comparable data on the same areal basis for many different types of flow. However, in a recent study of the prairie provinces Davies and Thompson (1980) illustrated the utility of the approach in interpreting the structure of the urban system. The study dealt with the 15 commodity flows (road and rail) between the 17 large places over 10,000 population in the Canadian prairies, and four destinations outside, namely the west coast of Canada, the Canadian east, the United States and northern Canada. In total the data set consisted of 15 matrices, each describing one of the commodity classes shown in Table 8.8. The rows of each matrix consisted of the flows from each one of the 17 towns, and the columns accounted for the volume of flow for the commodity having its destination in the centres. Although each of the matrices could be factor analyzed separately to identify and measure the structure of interaction, this approach could not be recommended because an investigator would still be confronted with the problem of integrating the 15 separate flow structures. As shown in Chapter III the dyadic approach solves this difficulty by providing one set of integrated results.

Application of a Principal Axes component model followed by Varimax rotation to the data set of road and rail flows within the Prairies only produced three axes accounting for 66.84 percent of the variation. Untransformed data and standard correlation measures were used because the influence of size needed to be repressed - given the differential sizes of the commodity flows - whilst transformations differentially compressed one flow compared to another (Davies and Thompson 1980). These results mean that two-thirds of the variability in the flows of 15 commodity types over 17 towns could be accommodated by the three axis <u>linear</u> model. Figure 8.6 shows that the loadings identified the importance of each commodity on the axes; the scores measured the respective importance of each dyad or link on the components.

The first axis or factor was called <u>Tertiary and Heavy Industry</u> because it was associated primarily with flows of retail goods, household goods and heavy manufacturing products. The flow pattern resembled the <u>downward hierarchical</u> type. Goods moved from the eastern manufacturing belt through Winnipeg to the other large metropolitan centres and down to the regional nodes. Some reverse and reciprocal cross links between the largest centres complicated the patterns.

The second factor was called <u>Raw and Semi-Processed Materials</u> because of the commodity types most closely linked to the area. The structure most closely resembles the <u>centralized network</u> in Figure 8.5 - with small towns connected directly to Winnipeg, the historic gateway city (Artibise 1972). However, the pattern is not a simple one. Again there are reciprocal and reverse flows, together with a rather complicated interchange between the Alberta towns.

The third axis was labelled a <u>Food Products and Light Manufacturing</u> factor because of the pattern of commodity loading. It is character-ized by an <u>irregular pattern</u> of interchange associated with the differental location of manufacturing in the cities of the province.

(a) Component I: Tertiary and Heavy Industry (29.4%)

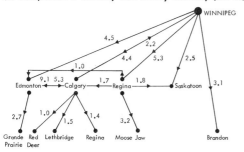

Commodities	Loadings
General Freight...	0.88
Metal Products	0.81
Machinery.........	0.79
Construction.......	0.77
Household	0.70
Others	0.62
Foodstuffs..........	0.49*
Petroleum..........	0.40

(b) Component II: Food Products and Light Manufacturing (22.3%)

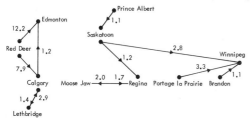

Commodities	Loadings
Trailers	0.91
Perishable Food...	0.87
Foodstuffs	0.68
Seed and Feed	0.60
Livestock	0.54*
Machinery.........	0.50*

(c) Component III: Raw and Semi-processed Materials (15.1%)

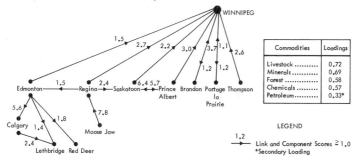

Commodities	Loadings
Livestock	0.72
Minerals	0.69
Forest	0.58
Chemicals	0.57
Petroleum..........	0.33*

LEGEND

1.2
⟶ Link and Component Scores ≧ 1.0
*Secondary Loading

Figure 8.6   Connectivity of the Prairie urban system
            (a) Dyadic analysis of individual flows within Prairies
            (b) Total flows within and outside the Prairies

Most of the flows are provincially based, although the Saskatchewan network does link into the Winnipeg based structure.

As in all multivariate studies in which the general patterns or structures are derived, it is important to remember that the solution does not deal with all the flows found in a data matrix. The idiosyncratic flows are considered as part of the error variance, or as specific axes in common factor studies. In addition, flow studies make it possible to deal with another level of analysis scale, namely the

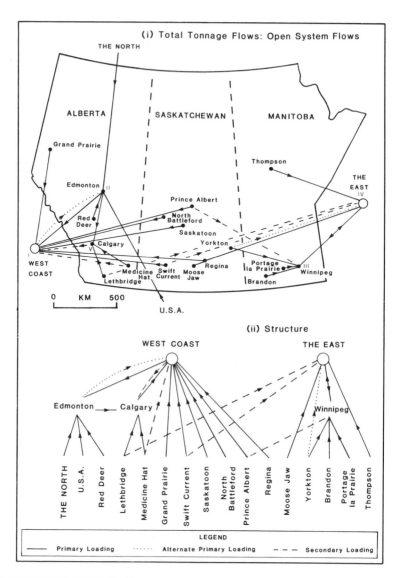

Largest Factor Scores		I	II	III	IV	V
City or Area	Edmonton	0.6	4.0	0.0	-0.0	1.4
	Calgary	-0.9	1.5	-0.2	0.2	-3.8
	Saskatoon	1.1	0.0	0.5	0.3	1.5
	Regina	0.2	0.1	-0.7	-1.1	0.3
	Winnipeg	0.0	-0.4	4.1	1.2	-0.3
	West Coast	4.0	-0.5	-0.6	0.2	-1.3
	The East	-0.1	-0.4	1.3	-4.2	-0.1

Figure 8.7 Total flows in Prairie urban system

Table 8.8
Commodity types and prairie centres

Commodity Class	City Name*	1971 Population
1. General freight	1. Grand Prairie	13,080
2. Non-perishable food stuffs	2. Edmonton	495,700
3. Perishable food stuffs	3. Red Deer	27,675
4. Heavy machinery	4. Calgary	403,320
5. Metal products	5. Lethbridge	41,220
6. Petroleum products	6. Medicine Hat	28,775
7. Bulk liquids and chemicals	7. Prince Albert	28,465
8. Bulk dry chemicals and minerals	8. N. Battleford	12,700
9. Forest products	9. Saskatoon	126,450
10. Livestock	10. Swift Current	15,415
11. Construction materials	11. Moose Jaw	31,855
12. Seed and feed products	12. Regina	140,735
13. Trailers - mobile homes	13. Yorkton	13,430
14. Household goods	14. Thompson	19,000
15. Miscellaneous goods; others	15. Brandon	31,150
	16. Portage La Prairie	12,950
	17. Winnipeg	540,265

* All centres over 12,500 in population

aggregation of all the commodities shown in Table 8.8 to a single matrix of tonnages moving between all 17 centres and the outside destinations of the west coast, USA and eastern Canada. In Figure 8.7 five axes account for 88.0% of the variance and display a centrifugal hierarchical pattern, focussed not on one centre but to the east and west destinations. The centripetal hierarchical structure revealed by the factor scores contains examples of recticular and reciprocal linkages between the largest places and recticular connections. As such, it is similar to Pred's model of interaction. It cannot, however, be assumed that these findings demonstrate the universal superiority of the centripetal hierarchical model with reciprocals. After all, the prairies are still dominated by an economy based on the exploitation of raw materials and agricultural products that have destinations outside the region. So in Friedmann's terms the area is still a periphery, a hinterland in Canadian terms, particularly if gross volumes of flow are considered. This means that the results cannot be considered to reflect a universal model of urban system character. They simply demonstrate the dominance of a single type of pattern in this particular area, a pattern that can be disaggregated into three very different structure associated with different sets of commodities. This implies that other flow structures are likely to be relevant for other types of interaction in other areas. The conclusion must be that just as the formal structure of the urban system can be disaggregated into a series of dimensions based on size, economic base, status, age, etc., (Table 8.3) so the pattern of flows in an area can be characterized by one or more alternative structures, such as centralized, irregular, hierarchical, etc., each associated with different sets of flows.

4. SUMMARY

These results show that recent work on the study of urban system structures has progressed beyond the stage of assuming that single set of models or relationships characterize the urban systems. Instead factorial methods have been useful in defining alternative structures in different domains (or content areas) of urban system character whilst a series of distinctive structures defining urban flows have also been identified. This means developments in this field have paralleled the work of psychologists in the study of human intelligence in moving from single to multiple dimensions of character. Now that some of these structures have been isolated, further work is needed to establish additional types, the conditions under which they emerge, and the areas in which they are found. In this way the structures of urban systems both in terms of dimensional character and flows can be determined, not as a set of place-particular relationships but as examples of a set of alternative system structures from which a theory of urban system character may emerge or can be tested.

9 SOCIAL DIFFERENCES WITHIN CITIES

This chapter demonstrates how the application of factor analysis methods has contributed to knowledge of the social differentiation of cities. With reference primarily to cities of the western world the contribution lies in integrating sets of variables and scaling areas to provide generalizations about the social dimensions and spatial patterns of cities, not in the discussion of individual variables. In the first section a brief introduction to the development of ideas in urban ecology provides the background to the introduction of factorial methods. This leads on to the second section, which attempts a reformation or re-arrangement of theoretical concepts of intra-urban structural differentiation. The third section describes the utility of these ideas by outlining the empirical evidence for variations in social structures and patterns. The fourth and fifth sections describe the contribution of factorial methods to the social region-alization and social typologies of western cities and explores the use of space:time methods.

1. THE DEVELOPMENT OF URBAN ECOLOGY

(a) Origins

Most textbooks in urban ecology claim that the origin of serious work in the study of social differentiation within cities lies in the work of the so-called Chicago School of Human Ecology (Park 1925; Bogue and Burgess 1964). Such an opinion over-emphasizes the North American contributions to the field and ignores the perceptive investigative reporting of Henry Mayhew (1862) and Frederick Engels (1845) in the mid 19th century, and Charles Booth (1902) at the turn of the century. Davies (1978) has reminded human geographers and ecologists that Booth cannot only be considered as the 'pioneer' of descriptive social surveys and mapping. After all, Booth (1893) attempted to go beyond the use of primary material in his studies of poverty to derive an 'instrument' for the measurement and classification of areas in London. Using six indicators from the British census, he averaged the values to produce the variations in what he called the 'social condition' of areas in London (Figure 9.1).

Table 9.1 shows a Principal Axes factor analysis confirmed Booth's assumption that a single dominant vector of variation was present in the data set. Similar results are found whether a component or common factor model is used, or orthogonal or oblique rotation is applied.

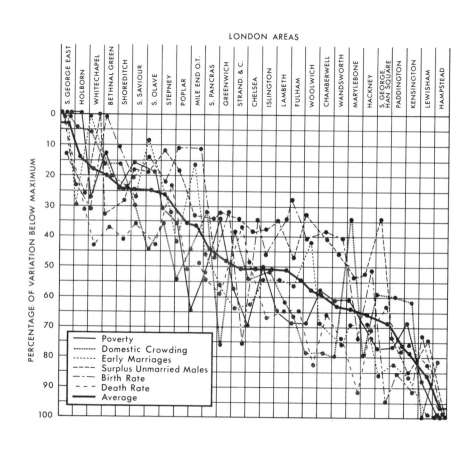

Figure 9.1   Variations in Booth's indicators
            Source: redrawn from Booth (1893)

Table 9.1
Component analysis of Booth's statistical indicators

(a) Principal Axes component loadings

Variable	Component 1 Loading	Communality
1. Poverty	0.93	0.86
2. Domestic crowding	0.83	0.68
3. Young married females	0.95	0.90
4. Birth rate	0.86	0.74
5. Death rate	0.90	0.82
6. Surplus unmarried females	−0.88	0.79

Eigenvalue   4.79 = 79.9% original variance

(b) Two axes solutions

Variables	Principal Axes Unrotated Loadings		Varimax Rotation		Commun-alities
	1	2	1	2	
1 Poverty	0.93	0.10	0.51 (62)*	0.53 (70)	0.87
2 Domestic Crowding	0.83	0.52	−0.52 (26)	0.99 (94)	0.96
5 Death Rate	0.91	0.11	0.49 (60)	0.52 (69)	0.83
6 Surplus Unmarried	0.95	−0.08	0.73 (76)	0.31 (58)	0.91
4 Birth Rate	0.86	−0.45	1.00 (94)	−0.18 (24)	0.94
	97.7%	9.0%			

*Oblique loadings (Direct Oblimin with gamma = 0.0) are given in brackets. The correlation between the oblique rotated axes is 0.63. Extraction of a third axis adds only 4.3% to the overall variance explanation.

Source: extension of Davies (1978). See also Table 2.12 for original correlation matrix and component scores.

If a two axis oblique solution is derived Table 9.1b shows that the additional axis is very oblique and contributes little additional variance. So it is not really worth extracting. The identification of one axis is similar to the extraction of a single axis of intelligence that characterized so much of the work of the early psychologists (Spearman 1904). But the subsequent breakthroughs in statistical manipulation made by the psychologists were not achieved by urban ecologists until almost a life-time later. So Booth's (1893) assumption of a single dimension was never checked or analyzed by multivariate methods.

Unfortunately for the development of quantitative urban ecology, this particular contribution by Booth was largely ignored by subsequent generations of human ecologists. Instead, the field is normally considered to have begun with a group of researchers at the University of Chicago in the period after World War I. There is no doubt that Park, McKenzie, Burgess et al (Park 1925) were responsible for producing some excellent behavioural studies of social groups and urban problems, and for formalizing and popularizing a field of endeavour. But two major limitations can be identified in the development of systematic measurements in the field of intra-urban differentiation: first, there was an apparent inability to link the study of spatial variation with the background changes in society; second, a measurement problem was present, namely that of depending upon single indicators to measure the social variations in cities. From the discussion in Chapters III and IV, it is apparent that at the very best this provides a one-to-one relationship between indicators and deep seated constructs. These problems meant that by the 1950s much of the discussion in the field hinged upon an often sterile discussion of the utility of three competing 'theories' of urban social variation to particular cities shown in Figure 9.2.

The first was the concentric zone model proposed by Burgess (1925); the second was the emphasis upon sectoral patterns identified by Hoyt (1939) after studies of changes in the residential structures of US cities; the third was the suggestion that cities displayed a series of separate land use and social variation clusters (Harris and Ullman 1945). Each nuclei was located in a particular area, attracted similar land uses, expanded and coalesced with other nuclei to produce a multiple-nuclei spatial pattern (Figure 9.2). Detailed descriptions of these so-called 'theories' are provided in the standard textbooks in the field (Timms 1971; Johnston 1971) and need not be repeated here. However, it is worth noting that they are not theories in the standard scientific sense; they are generalizations of social and land use patterns found in western cities. These generalizations were often assumed to be universal models by adherants of one or other of these types. Moreover, it was rarely possible to link the spatial patterns found in cities with particular processes of social or ecological change.

(b) From social areas to factorial ecology

In the mid 1950s two very different traditions were established. The Areal Taxonomy approach developed principally by Tryon (1955) applied cluster analysis procedures to a data set to derive a classification, grouping or taxonomy of areas, such that the areas most similar on a range of characteristics are grouped together. Through time more and more sophisticated cluster procedures (Wishart 1978) have been derived, whilst Multi-Dimensional Scaling methods (Golledge and Rushton 1972) and Q Analysis (Atkin 1974) have also been applied to the problem. Although these techniques have ensured the classifications have become more and more sophisticated, the approach simply represents an empirical methodology that is devoted to the grouping of areas. After all, there is no necessary link to wider societal issues, or to the definition of the collectivities of the variables as described in Chapter I, although the clustering can be applied to variables to produce a grouping of the variables in place of the areas.

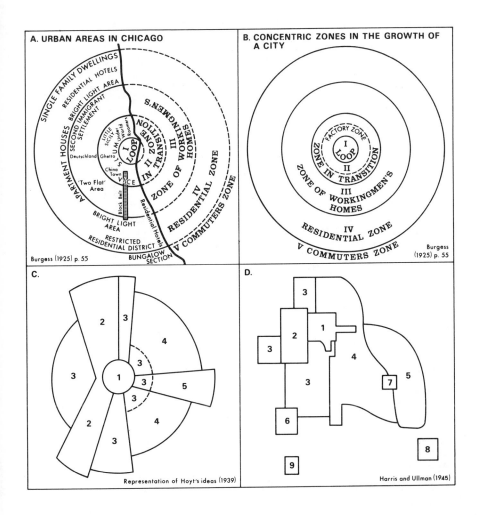

A. URBAN AREAS IN CHICAGO

SINGLE FAMILY DWELLINGS
RESIDENTIAL HOTELS
BRIGHT LIGHT AREA
SECOND IMMIGRANT SETTLEMENT
APARTMENT HOUSES
S.N.ERGMEN'S
I LOOP
II TRANSITION
III
ZONE OF WORKINGMEN'S
IV RESIDENTIAL ZONE
V COMMUTERS ZONE
Little Sicily
SLUM
Ghetto
Deutschland
China Town
VICE
'Two Flat' Area
Black Belt
Residential Hotels
BRIGHT LIGHT AREA
RESTRICTED RESIDENTIAL DISTRICT
BUNGALOW SECTION

Burgess (1925) p. 55

B. CONCENTRIC ZONES IN THE GROWTH OF A CITY

FACTORY ZONE
I LOOP
II ZONE IN TRANSITION
III ZONE OF WORKINGMEN'S HOMES
IV RESIDENTIAL ZONE
V COMMUTERS ZONE

Burgess (1925) p. 55

C.

Representation of Hoyt's ideas (1939)

D.

Harris and Ullman (1945)

Key for C and D

1. C.B.D.
2. Wholesale and light manufacturing
3. Low class residential
4. Middle class residential
5. High class residential
6. Heavy manufacturing
7. Outlying business
8. Dormitory suburb
9. Industrial suburb

Figure 9.2 Early models of urban ecological patterns

301

More comprehensive in scope, if less rigorous in measurement, was the second breakthrough in the 1950s. This was Shevky and Bell's (1955) Social Area approach, which has already been described in Chapter I. Based on a previous, largely empirical study of Los Angeles (Shevky and Williams 1949), the theoretical approach provided some really fresh insights into the study of intra-urban differentiation, particularly in its attempt to link the study of urban social differentiation within cities to wider societal changes. This provided a theoretical rationale that had been missing in previous urban ecological work. The identification of three separate sources of variation in western cities, based on Social Rank (Socio-Economic Status), Urbanization (Family Status), and Ethnicity, led to the recognition that these three social structures usually displayed very different spatial patterns, each of which could be linked to one of the 'classical' ecological models. In other words, Family Status was frequently distributed in a concentric pattern in western cities, Socio-Economic Status displayed sectoral variations, and Ethnicity was linked to small clumps or clusters of segregated groups. Murdie's (1969) valuable summary (Figure 9.3) of these patterns showed that when all three patterns were put together much of the complicated social mosaic of the western city could be reconciled. In many ways, therefore, Murdie's extension of the social area results of Shevky and Bell provided a reconciliation of what were previously considered to be competitive ecological patterns. His factorial work confirmed they were really separate parts of the complex process of differentiation that created the spatial mosaic of cities.

Despite the pioneering nature of Shevky and Bell's ideas, a number of criticisms of the social area approach have been made and reviewed by Timms (1971). In terms of measurement, the techniques used to combine the indicator variables into indexes of social variation (the axes of differentiation in factorial language) were very primitive. At the very minimum they prejudge the possible sources of variation, and ignore the fact that indicators defining any index have to co-vary before acceptable results are obtained. Such problems can be resolved by using more rigorous measurement problems such as factor analysis. Less easy to resolve is the fact that the various stages of the theory involve major inferential leaps from which other conclusions could be derived. Moreover, the fundamental mechanism of 'increasing scale' is too vague a concept upon which to base the postulated three sources of change. However, of crucial importance is the question of whether social area differentiation necessarily follows from changes in society as a whole. Udry (1964) observed that two theories must be involved, one providing the mechanisms for social change, the other for spatial differentiation. Unfortunately, his ideas were never fully articulated. Also, there is a very real problem in relation to the degree of universality of the theory. Shevky and Bell seemed to envisage the three axes as fundamental sources of variation in society that were applicable to all scales: regional, inter-urban, as well as intra-urban. Finally it is worth noting that Shevky and Bell restricted their ideas to industrial society, dealing only with the traditional: industrial society transition. Although it may be implied that non-western societies are likely to experience the same process of change as found in western societies, this represents an extension of the original formulation. Certainly the construction cannot be thought

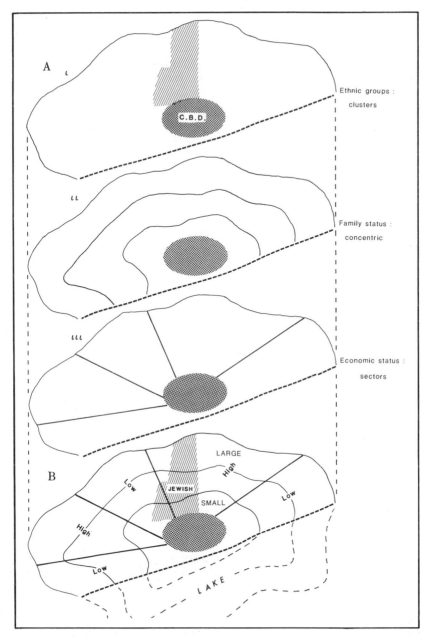

Figure 9.3 Murdie's (1969) summary of spatial patterns
of social dimensions

(a) Individual patterns
(b) Partial composite for Toronto (other ethnicities
excluded)

of as being relevant to pre-industrial, post-industrial or socialist societies.

The application of factor analysis methods to measurements of regional and intra-urban variation in the 1960s (Berry 1961; Robson 1969) removed many of the difficulties associated with the Shevky-Bell formulation. Of particular importance was the fact that the factors of differentiation were derived from the data. Also there was a clear separation of the problems of identifying the sources of social separation (the collectivities of the data set) from the measurement of the patterns of the areas. Nevertheless, most of the earliest applications of factorial methods were still based on the assumption that Shevky and Bell's three axes of differentiation represented the basic sources of variation in western cities.

(c) A paradigm of intra-urban social differentiation

At the same time that factorial methods were being applied to the Shevky-Bell indexes, several researchers attempted to clarify the various stages in any study by describing differences between the various objects of interest to students of intra-urban social differentiation. For example, Berry and Rees (1970) proposed a summary of the residential location decision process with differentiation between social, housing, community and physical space. Later on, Robson (1975) used Social Rank, Family Status and Housing axes to not only differentiate between categories of area, but to construct a model of spatial differentiation in British cities. Davies (1983) has described the difficulties in these approaches and attempted to clarify the methodological stages involved in describing the various components in the differentiation of urban social variation. In the elementary introduction to this problem in Chapter I, only two stages were identified: the first involved the measurement of the collectivities of the variables through factors; the second was the measurement of the importance of each unit on the factors by means of factor scores. Figure 9.4 (Davies 1983) shows that the measurement of urban social differentiation requires five stages if the various issues are to be adequately understood.

The first distinction that must be made by investigators is between the Content Area of the investigation and the level or Scale of Aggregation being used in the study. The content refers to the variables or indicators which identify and measure the phenomena being investigated; the aggregation relates to the size and type of collecting units used in the study. This means that the relative position of individuals, families or areas with respect to each variable can be identified by plotting one pair of social indicators at a time. The examples in Figure 9.4 show a bi-variate scatter, such as these produced by plotting the number of rooms in the house, against the number of years spent in high school: first, by showing how individuals can be located in this social space by their characteristics; second, how communities can be located by their average values on those indicators. These are called Social Indicator Spaces since they deal with separate indicators and are plotted without regard to geographical space.

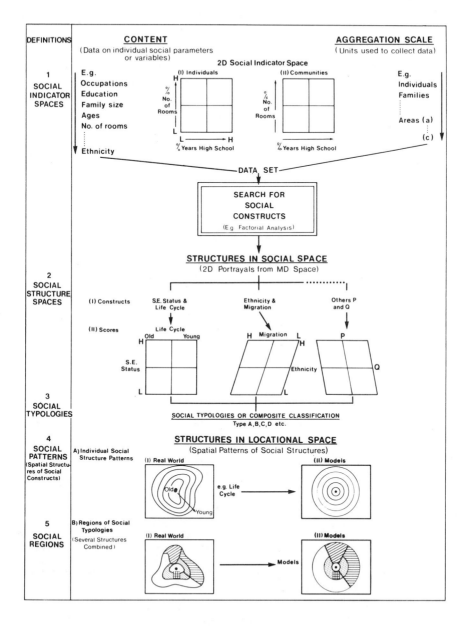

Figure 9.4   Social spaces and structures
            Source: Davies (1983)

These two dimensional portrayals are obviously limited because they only deal with two indicators at a time. Previous chapters have shown that methods such as factor analysis are available for defining the collectivities or latent structure of a data set of social indicators. The social structures, or Social Constructs in Figure 9.1, that lie behind the individual indicators are identified and measured in this way. In Figure 9.4 it can be seen that a series of alternative social structures are identified, such as Socio-Economic Status and Life Cycle, etc. Depending upon the scale of the aggregation – individuals or communities can be used – the units can be located with respect to one another by their scores on the individual factor structures. Several examples show how two-dimensional portrayals make it easier to conceptualize the relationships found in the multi-dimensional space. it is worth noting that some of these diagrams have angled axes. This simply draws attention to the possibility of oblique relationships occurring between the social constructs. Since the scores position the aggregation units with respect to one another on these social structures, they are called Social Structure Spaces. However, no geographical or real world connotation is implied by use of 'space' in this context.

The third stage identified in Figure 9.4 involves the construction of what can be described as Social Typologies. These describe the composite classification of the units on the basis of all the scores, and usually involve the application of cluster analysis methods to factor scores, although they could work directly on the data. Separate social area types can be identified. At the community area scale within cities, for example, a skid row type of area is often found in inner cities, one that is characterized by low or impoverished Social Status, Mobility, Family Disorganization and Substandard Housing Conditions. This means that in the same way that areas are given summary characteristics, so composite social types can be derived from combinations of social constructs.

The fourth stage deals with 'space' in a geographical sense, namely the structures that are found in locational space. As such, they describe the spatial patterns of the social structures. Each individual social structure or construct may have a separate type of spatial pattern and, for convenience, these variations are called Social Patterns, the spatial structures of social constructs. Figure 9.4 shows an example of a real world annular pattern, such as the distribution of the Life Cycle dimension in the city. This can be generalized into a simpler relationship in the concentric Social Pattern Model by assuming away the directional biases of the real world pattern. If all the social patterns are integrated, again perhaps by cluster analysis methods, a series of typologies of areas can be produced. If these different types of area are located on a map then similar contiguous units can be joined together. Usually these types cluster in particular areas so that a set of composite real world social patterns, or Social Regions, are identified. These can be generalized into Social Region Models as shown in Figure 9.4. To avoid confusion, it is worth noting that these analytical stages do not necessarily depend upon any particular set of multivariate techniques, whether factor or cluster analysis. The paridigm, therefore, should have an application beyond the use of such methods.

306

All the various stages given precise terms in Figure 9.4 represent examples of what Blau (1976 p.3) has described as social structures: the regularities, configurations and patterns discernable in social life that have empirical meaning. It is the objective of factorial methods to identify these studies from the various data sets. But lying behind these structures are the deep seated mechanisms, the study of which is the objective of the so-called 'structuralist' school of social science (Levi-Straus 1952). This additional approach to social structure has been described as:

'a system of logical relationships among general principles which is not designed as a conceptual framework to reflect empirical conditions but as a theoretical interpretation of social life.' (Blau 1974 p.615)

Despite its flaws it must be accepted that the Shevky-Bell theory of social areas was structuralist in intent since it provided a theoretical basis for the various 'structures', constructs, or axes of differentiation that can be isolated by factorial methods. The difference between these structures must be contrasted with Rumley's (1979) concept of 'structural effects', which describe the way the attributes of space, or spatial patterns, influence or affect individual behaviour. This has been described as the contextual approach in Chapter III. Although such approaches have a long tradition in urban ecology, they are not dealt with here. Instead, the rest of the chapter will deal primarily with studies using the concept of structure as isolating social constructs or patterns.

2. THEORETICAL CONSIDERATIONS

(a) Societal variations

As with all theories the specific details of the Shevky and Bell's (1955) Social Area Analysis can be criticized, but its methodological importance in the history of ideas remains undiminished, for it has linked societal differentiation with social variation in cities. Through time, a number of theoretical modifications of these ideas have been made. These refine and extend the theoretical basis for urban social differentiation. One of the first of these modifications was provided by McElrath (1966). He suggested that an additional major construct should be identified, namely Migration, one which followed logically from the original theory of Shevky and Bell (see Figure 1.5). He also suggested that two mechanisms underlay social differentiation. Social Rank (more usually called Socio-Economic Status), as well as Family Status, were postulated as being products of the societal changes described as Industrialization. However, Ethnicity and Migration were linked to Urbanization, in other words to the process of settlement agglomeration which brought in new residents from other areas and cultures.

Modifications of the Shevky-Bell thesis were also provided by students of cities in developing countries such as McElrath (1968), who found rather different structures to those predicted by Shevky and Bell. Elsewhere, Abu-Lughod (1968) demonstrated that the Family Status and Social Rank axes were combined in Cairo (Egypt). Berry (1972) and his associates in several studies of Indian cities also confirmed the fact that cities in under-developed countries did not show the typical separation of Social Status and Life Cycle, and emphasized the importance of Male dominated axes linked to sex-selective migration and Communal or cultural axes of differentiation in Indian cities. Generalizing from all these results, Berry (1971) suggested that intra-urban social structures varied in terms of three types of society: western; non-western; and socialist. In the latter case high levels of welfare would modify the standard western pattern. In addition, a descriptive review of intra-urban residential patterns in several countries led Johnston (1972) to propose a three-stage model of development for the process, based on stages called: pre-industrial, industrial take-off, and contemporary modernization. However, it was Timms (1971) who produced the most comprehensive extension of the Shevky-Bell typology. He proposed that a family of social area structures (social structures in the terminology of Figure 9.4) should be recognized. The four axes of McElrath (1966) were used as the basic constructs and were combined in different ways to produce six types called: feudal, pre-industrial, colonial, immigrant, industrializing, and industrial.

Figure 9.5 interrelates Timms' (1971) family of social area types into a developmental model of change between them. The model is based on the same assumption that McElrath's (1968) four major dimensions of social differentiation are found in cities, and these are combined in different ways in various societies to produce varying social structures. Unlike Timms (1971), however, the structures are not considered simply as ideal types: they are related to three categories of society to which a fourth, or post-industrial type has been added to take contemporary conditions into account.

(i) Traditional. In traditional or feudal societies, most cities were small and consisted of a small, dominant elite and a commercial-artisan group, with the bulk of the population in the servant or worker class. Family-related considerations dominated the social structure, since prestige and status were based primarily upon kinship. In spatial terms the elite usually occupied the city centre, with segregated clusters of other groups in the rest of the city. These segregations were based on occupation or family. In the so-called Feudal city, therefore, a single axis of differentiation can be expected, combining Social Rank and Family Status as well as the limited amount of ethnic and migrant variation.

(ii) External contracts. The addition of external contracts through trading or political conquest is assumed to lead to the growth and differentiation of cities. Division of labour increased and a merchant class is added to the political elite, thereby beginning the process of differentiation of the traditional city structure. Am important change was the fact that economic wealth becomes an index of prestige, and this wealth can be derived from trade or industry or through political or family-related domination. Selective migration streams added to the social complexity of cities. It can be postulated that these changes led to the creation of three very different social structures in cities, each composed of two dominant axes of differentiation that combined the four basic constructs in various ways.

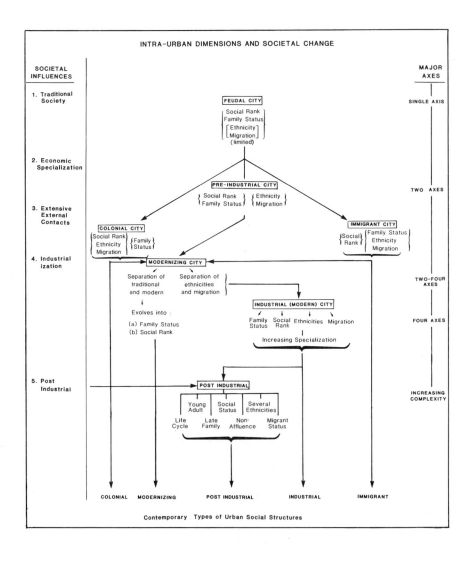

Figure 9.5 A developmental sequence of social structures

In Colonial cities that were located in previously populated areas, the immigrants are politically dominant in the society and are ethnicity differentiated from the 'traditional' population. The result is the creation of one axis of differentiation combining Social Rank, Ethnicity and Migration, with another axis indexing Family Status. In the Immigrant city, the indigenous political elite remain dominant, thereby producing Social Rank axis based on this characteristic. But given age and sex selection in the migration process, which is usually large enough to overwhelm the other groups in numerical terms, the Ethnic and Migrant constructs are intermingled with family variations. This creates the second dominant axis combining Family, Ethnic and Migrant characteristics, which are associated with the major immigrant streams. In the Pre-Industrial city type external contract can be assumed to lead to the addition of ethnic and migrant groups which are separated on a single axis of differentiation. By contrast, the perpetuation of the older indigenous social elite, and the continued importance of family kinship patterns, ensure that social rank and family variables are combined on a single axis of differentiation called Social Rank-Family.

(iii) Industrialization. The third phase identified in Figure 9.5 is associated with the consequences of mass industrialization on the various urban types. Cities grew apace and experienced great division of labour and specialization, whilst the growing importance of service or tertiary occupational groups in mediating between producer and consumer created very different economic structures. The new source of wealth, and the ability of those engaged in the productive process to keep an increasing share of this surplus, meant that wealth became an important new index of prestige in society. The rigid class boundaries of kinship-based pre-industrial cities become more porous, leading to differentiation by wealth and occupation. The addition of rapid transport routes enabled first the upper, then the middle classes to escape from the inner city and build their own exclusive areas, occupied by one type of class, often located in a particular sector of the city.

In the earliest stages of industrialization these changes produced a transformation in the characteristics of social prestige and family relationships providing a contrast between modernizing and traditional groups. Eventually this led to the separation of quite distinct Social Rank and Family Status sources of differentiation. Similarly, the processes of segregation led to the separation of various ethnic and migrant groups in different parts of cities. In this process of modernization of the typical industrial city, the central area changed from being the abode of the social elite, to the place of occupation of the new commercial groups, creating the distinctive central city business area. This was usually surrounded by a zone in transition, an area of mixed uses produced by the expansion of commercial uses from the business core. All three previous types, Colonial, Pre-Industrial and Immigrant, are likely to experience the onset of industrialization and hence modernization, so it is very likely that a variety of transitional forms will be identified which can loosely be called the Modern city type. This type is eventually transformed into the typical industrial city in which separate Family Status, Social Rank, Ethnic and Migrant constructs can be recognized.

(iv) Post-Industrial. The fourth societal type postulated is a product of the changes that have occurred in western countries since World War II. Usually called the post-industrial society (Bell 1974), the changes are based on a series of interrelated characteristics that make it difficult to reproduce the simple hierarchical linkages between change and construct found in the Shevky-Bell formulation. Moreover, the increasing complexity of society means that there are likely to be many more axes of differentiation, some of which are derived from the fission of the four basic constructs identified in previous societies.

The first of the new processes of change that produce a societal separation is linked to technological advances. These have led to more rapid and easier transport and communication linkages, removing many of the older frictions of space. The increasing dispersion of contact in social relationships has led to an increase in mobility, producing higher levels of mobility, as well as the dispersion of cultural origins, before immigration controls were imposed in many countries in the late 1970s. One consequence has been the possible re-emergence of the importance of Ethnicity as a source of societal variation. Many, but not all, of the identifiable ethnic groups of the early 20th century may now be indistinguishable from the general population through acculturation. But the new immigration of the post-war period has led to new sources of differentiation based on the new ethnic groups, rather than a single axis. These transportation changes have also seen an increasing degree of mobility in the general population, ensuring that Migrant status becomes a more potent source of differentiation.

The second set of changes creating post-industrial society are associated with an increasing control and intervention by organizations, both corporate and governmental. This has led to a rapid expansion in the tertiary and quaternary sectors of the economy. The partial replacement of free-enterprise societies by mixed economies, composed of government and private organizations, has perpetuated this trend. But the acceptance of the principles of 'equality' and 'welfare' mean that they now rank with economic growth as some of the primary objectives of government. The resultant general commitment to egalitarian policies have only been modified by a neo-conservative trend in the last few years. Such changes have led to the possibility of new sources of differentiation based on the Service sector and differences in the degree of Welfare, the latter as expressed either in the receipt of direct income, such as through pensions, or through indirect benefits, such as housing.

The third set of changes is really a continuation of the forces at work in industrial society, namely the increasing specialization of the economy. New skills and jobs have been created relatively quickly, as a consequence of an increase in the rate and scale of innovation and change, thereby leading to Economic Base variations. Differential access to wealth based on the various occupational wage levels perpetuates the differences in social rank in the population, although the impact of egalitarian policies may have removed some of the societal extremes. Until the last decade full employment government policies seemed to be on the verge of eradicating the pool of unemployed. However, the effects of automation, competition from the industrializing countries, and the slow-down or reversal in the growth rate of the western economies in the last decade have led to the creation of new levels of long-term unemployed. This ensures that the 'trailing edge' of society, as reflected in Non-affluent Groups and Sub-Standard housing may emerge as a special axis of differentiation. This will be additional to the economic base and economic status variations. Such disadvantaged groups seem to lag further and further behind the general populace, and may be a growing entity if long-term structural unemployment persists.

Fourth, a post-industrial society has been characterized by an increase in the degree of personal individualism and choice in social relationships. This has led to a greater separation in societal roles. For example, teenagers and young adults often socialize in their own groups, rather than in the context of families. Not úntil these younger groups achieve financial independence, or separate support, can they leave home to establish their own residences. In addition, the separation in values and the possibility of very different life-styles, (linked to the new leisure time created by both economic growth, preference, and government regulation) provide more diverse sources of social variation. The family:non-family divisions are likely to be perpetuated, since the decision not to have children is aided by the widespread adoption of contraceptive devices. But more detailed variations in age groups are likely to be found. One example is the separation of Young Adults who have access to apartments built specially for them, and to jobs in the new service economies in the large cities. Another is the creation of Old Age or retirement communities. Life style choice now means that special interest groups can be perpetuated, thereby providing additional sources of possible variation. Yet, when combined with the decrease in local contacts, it is likely to create the possibility of more isolation or disorganization in social relationships.

Finally, a set of potential sources of variation produced by the combination of some of these changes can be recognized. The first means stems from the ability of many people to use the personal transportation revolution brought about by almost universal car ownership to perpetuate a home:work difference in roles. This is frequently seen in an extensive Urban Fringe or a commuter village or area type, in which commuters, usually of middle and higher income, live in agricultural areas. In its more extreme form the counter-urbanization process identified by Berry (1976) and others in the USA provides a long distance culmination of the same process. Many small settlements have grown because people prefer to live in smaller settlements. This process has been lubricated by government support to decentralization schemes, and to technological changes which enable workers in such areas to maintain contracts with other areas.

(b) Ecological processes

None of the societal changes can be translated into sources of variation within cities if the changes lead to a uniform dispersal of people throughout the city. Sub-area differentiation will only occur if there are concentrations of the various groups. This means that the ecological processes leading to spatial differentiation have to be in operation before these societal variations are translated into spatial variations, and hence into axes of intra-urban differentiation.

312

Although it is rare to find this two-stage explanation in the factorial ecology literature, it must be emphasized that there is no need to invent a new set of theories. Many of the causes for spatial separation can be found in the traditional ecological literature that has been carefully reviewed by Johnston (1971) and Timms (1971).

The main reason for the spatial separation of the high socio-economic classes in cities can be found in the operation of the bid-rent mechanism (Muth 1969) that underlies the ownership of land in western cities. This implies that land will always be occupied by uses that command the highest return, subject to political constraints. In the specific context of the residential land market, higher income groups can always pay more for what is considered to be a more desirable piece of land for residential use, whilst Firey's (1945) 'symbolic' associations can be used to explain the perpetuation of high income groups in a status area. Such areas will only change if higher paying uses are allowed by the higher income groups to intrude; the term 'allowed' is used deliberately. The sectoral patterns stem from the fact that higher income groups represent a small part of the city population and have often deliberately segregated themselves in distinctive communities of high priced housing. Moreover, once a sector is established, the buying power of the group, or developers acting on their behalf, frequently enable the cluster to be perpetuated outwards to the city boundary. The reverse arguments apply to lowest income groups who can only afford cheaper land. Moreover, there is the additional association of low income housing with the industrial areas or with transportation corridors. These frequently occupy low-lying river valleys because of the initial ease of building in such areas. However, the hazards from flooding, or pollution, ensure that the land is less desirable to the higher income groups, and is cheaper for low income residences. Sectorial expansion, therefore, is perpetuated.

In the specific context of ethnicity some additional familiar ecological forces can be recognized. The most obvious is that of involuntary segregation through prejudice from the host population which creates a clustering of specific ethnic groups. But in all these ethnic clusters there is an element of voluntary segregation through sentimental ethnic associations (Firey 1945). Boal (1978) has identified a series of additional functions provided by segregated residence patterns such as defence, avoidance, preservation and attack. These provide more specific explanations for the maintenance of segregated clusters. Once society has become specialized enough to allow role and value separation to occur, voluntary concentrations of old-aged or special interest groups are increasingly likely to be found.

The concentric zonation associated with the Family Status dimension stems from quite different processes. In part it is a product of the age of the city since the suburban areas built in the post-war period are primarily for families. However, a more potent explanation comes from the contrast between access to 'space' and 'accessibility' in various parts of the city for various stages in residential locational requirements (Foote et al 1960). Other things being equal, such as the context of a standard income, individuals and families have had to 'trade-off' accessibility to the central city or business area with its jobs, entertainment, and shopping, yet high cost of a unit of land, with space in the suburbs. Hence low accessibility is compensated by the opportunity to buy more land per unit of money. In North American society families with small children seem to prefer their own yard and larger homes rather than high-rise or walk-up apartments, so many young families have chosen space, not accessibility. This produces high levels of family status in the suburbs and low levels in the inner city. However, the addition of suburban shopping centres, places of work, and entertainment, have modified the simple city-centre and suburban contrast in terms of space versus accessibility. Moreover, in the last generation, both the expansion of the black population and the physical deterioration in the housing stock in the inner cities, have led many white families to move into the suburbs. This has created another reason for the differential concentrations of family types. If distortions in the transportation surfaces, and specific topographic variations are assumed away, a simple pattern of concentric zonation for Family Status usually emerges, unless the city is an irregular shape, such as when a city borders a lake or ocean.

Within recent generations this simple concentric pattern has been modified. The most obvious is that concentrations of black families in inner city ghettos ensure that the family:non-family differentiation is modified in those cities with high populations of ethnically different groups. More generally, however, the late family groups, residents whose children have grown up and who have high levels of immobility, are primarily found in the older suburbs. These are now located in the middle city since new suburban areas have been added. This creates a donut-shaped concentration of <u>late family groups</u> and low mobility in the middle city, due to the process of insitu aging. Not all the middle or high income groups have been prepared to trade space for accessibility. The so-called '<u>gentrification</u>' process, by which middle income groups have re-populated the edge of the inner city by refurbishing older houses, has complicated the simple patterning. At the same time private developers have created new luxurious apartment blocks for high income residents who wish to live near the city centre. Added to this are concentrations of <u>young adults</u> found on the edge of the inner city, in areas close to jobs in offices, shops and institutions such as hospitals and colleges. Such groups have left home and have high levels of mobility due to the transitory nature of their jobs and residences. These groups prefer high accessibility due to the location of jobs, entertainment, or training facilities. Since their income levels are low, their ability to 'consume' space is limited. The groups are frequently concentrated in older areas of high status housing, that have been split up into one or two-room apartments, as in European cities, or in the North American case, in specifically designed apartment blocks.

All these processes are ecological since they provide explanations for the specific concentration of social groups in particular areas. Nevertheless they can only operate if the requisite social differentiation produced by societal characteristics has led to the creation of a pool of available people. It is now necessary to turn to the empirical evidence on urban social differentiation to determine the utility of these theoretical ideas of societal and ecological change.

3. VARIATIONS IN SOCIAL STRUCTURES AND PATTERN

So far, relatively few rigorous tests of the utility of Timms' (1971) family of social area types, or the extended form shown in Figure 9.5, have been carried out. Several problems are present. A major limitation in historical or cross cultural studies comes from deficiencies in the available data base, ensuring that the studies are rarely comprehensive. In a technical context many of the early studies used unrotated Principal Axes solutions using the component model or paid lip service to the need to justify the stability of solutions, thereby casting doubt upon the utility of the results. Yet of equal importance is the fact that individual cities have been studied by reason of their intrinsic interest to the investigator, not because they are hypothesized as being good examples of any of the types identified in Figure 9.5. Since these types are ideal constructions, and real world cities are likely to deviate from the typology in some measure, this makes the task of confirming the developmental model even more difficult. Finally it is to be expected that the four basic constructs identified in Figure 9.5 are only the major sources of differentiation. Additional minor axes may be expected. In this brief review of the empirical evidence of urban social structures, six sources of variation are identified: historical, cross-cultural, the industrial-post industrial transitions, city size effects, place-particular features, and the relationship between the intra-urban, regional and urban system scales of analysis.

(a) Historical city structures

Figure 9.5 proposed that urban social structures will vary according to the historical stage of the city. Unfortunately, relatively few rigorous factorial studies of cities in the Traditional or External Contact categories of Figure 9.5 have been carried out. Part of the explanation is that the surviving data base for Traditional or Feudal cities is very limited in scope and restricted to a limited number of indicators, so factorial methods are rarely appropriate. Whether there is such a city type as a feudal category common to all pre-industrial types as proposed by Sjoberg (1960) is now open to serious doubt. For example, Langton (1977) has shown that occupations were loosely patterned along the main streets in late medieval Gloucester and there are suggestions that spatial differences in social status occurred. So although factorial methods have not really contributed much to this literature, it is likely the proposition of a single feudal type is far too simplistic an interpretation to remain.

Within the last decade factorial methods using rotation techniques have been applied to census information on early industrial cities. One of the most useful is Shaw's (1977) factorial study of the emerging industrial centre of Wolverhampton (England) between 1851 and 1871. This has provided important evidence of the process of change. However modifications of the author's original factor labels are used here since they appear to improve comprehension of the results. The two largest axes in 1851 shown in Table 9.2 can be called a Social Status, Family and Economy vector and an Immigrant-Local vector on the basis of the loadings shown in Table 9.2. The results imply that the dimensional structure of Wolverhampton in 1851 lay between the Immigrant and Modernizing city. However, the presence of three additional familial

Table 9.2
Dimensions in Wolverhampton 1851 and 1871

I SOCIAL STATUS, FAMILY, ECONOMY
(Social Status)

No.	Variable	1851 Loadings	1871 (V)	(III)
03	Servants per family	82	(55)	
01	Socio-Econ. I & II	79	(37)	(−67)
23	Loners	74		
25	Dealing	74	(49)	
22	Lodgers	54		(−68)
05	Y. adult working	−37	(−50)	(54)
19	Males/females	−39		(86)
20	Males employed	−49		
04	Children working	−53		(65)
11	Children in family	−57		
10	Fertility	−66		48
02	Socio-Econ. IV & V	−68		51
24	Mining	−70	(−86)	−
26	Manufacturing	−		72

II IMMIGRANT − LOCAL
(Newly Immigrant Community)

No.	Variable	Loadings	(II)
21	Persons/h'hold	88	(76)
18	Family nuclei	89	(78)
16	Irish, Scottish	87	(78)
17	Density	61	(78)
22	Non kin h/hold	74	−
19	Males/females	52	−
02	Socio-Econ. IV & V	44	(68)
01	Socio-Econ. I & II	−	(−39)
10	Fertility	−49	−
23	Loners	−42	(−51)
14	Local born	−74	−

III FEMALES   FAMILY (I)
(No Title)   STATUS

No.	Variable	Loadings	(I)
12	Old	34	(84)
25	Dealing	−	(53)
07	Female heads	82	(78)
06	Females working	77	−
13	Age of head	58	(66)
8	Working wives	35	
	Servants		(42)
11	Children in family	−	(−76)
10	Fertility	−40	(−59

<table>
<tr><td colspan="3">IV AGE
    (No Title)</td><td>MALE
EMPLOYMENT</td></tr>
</table>

IV AGE (No Title)		MALE EMPLOYMENT	
No. Variable	Loadings	(VI)	
12 Old	81	–	
13 Age of head	50	–	
10 Fertility	–	(–44)	
18 Family nuclei	–	(41)	
20 Males employed	–63	(79)	
05 Y. adult working	–67	–	

V EMPLOYMENT (No Title)	
No. Variable	Loadings
08 Working wives	75
14 Local born	50
26 Manuf.	66
06 Females working	31
04 Children working	–45
24 Mining	–49

VI MIGRANT/DENSITY (No Title)		
No. Variable	Loadings	(IV)
15 Non local Eng./Welsh	–81	(–65)
14 Locals	32	(62)
19 Males/females	32	
26 Manuf.	35	(41)
04 Children working	47	(37)
17 Density	55	(49)
06 Females working		(82)
08 Working wives		(82)

VII FAMILY FERTILITY (No Title)		
No. Variable	Loadings	(V)
09 High fertility households	82	(–90)
11 Children in family	–46	

Decimal points are removed.

Original titles are given in brackets.

1851 loadings are given first. Those for 1871 are shown in brackets.

Source: revised from Shaw (1977)

axes, identifying Females, Age and Family Fertility, must be noted, demonstrating that there was some separation of the Family Status variables even in 1851. In addition separate Employment and a Migrant-Density axes complicate the pattern. Whether the separation of these axes is a real one, or a statistical artefact produced for the use of varimax rotation is unknown. Oblique rotation showing the correlations between the axes would be needed to determine this fact. What is worth noting is the presence of separate economic base or economy axes that are not postulated by the model in Figure 9.5.

By 1871 Shaw's results show that the Social Status, Economy, Family axis had split into two, one indexing commerce and servants as opposed to mining (V), another a status and industrial split (III). However, the Working Females (III) and Age (IV) axes merged in a modern looking Family Status axis (I), leaving only minor Male Employment and Family Fertility axes. The Immigrant axis is found again in 1871, with more conditions of overcrowding, whilst a Migrant/Working Females (IV) axis completes the analysis. Even without the benefit of oblique rotations these results show that Wolverhampton between 1851 and 1871 had progressed almost to the stage of the standard Industrial City type. Yet the presence of additional dimensions to these postulated by the model shown in Figure 9.5 must be noted.

Evidence for the presence of a Colonial City type is far more sketchy since it requires the separation of high Status, Ethnic, Migrant groups from the Family status dimension. Carter and Wheatley's (1979) study of Methyr Tydfil in Wales in 1851 provides some glimpses of the type. Again, modifications to the original factor labels have been made and it must be noted that the authors only interpreted seven out of the ten axes abstracted and omitted evidence on the factor correlation matrix. However, the fact that several different solutions were produced should be emphasized. Table 9.3 shows the Principal Axes Oblique rotation results with the new titles applied.

The largest axis shown in Table 9.3 is a high Social Class, Non-Welsh – Mining vector, whilst there is clear evidence of the separation of Family Status variations with the Non-Family and Mature-Young Family vectors. Three separate economy vectors, Employment, Quarrying and a Mining-Manufcturing split provide evidence of Economic Base specialization in different residential areas. In addition, three primary Migrant axes, associated with people from West Wales, Mid Wales and Ireland respectively imply that the social areas were also structured on regional migration lines. Finally a Low to Medium Social Class axis shows that status is not only linked to ethnicity. Carter and Wheatley were rather equivocal in their conclusions, finding relevance for the Colonial, Immigrant and Industrial types. However, the renaming of the vectors suggested in Table 9.3 suggests that Merthyr in 1851 lay somewhere between the Colonial city and Industrial type as identified in Figure 9.5. Nevertheless the presence of more than one family axis and the addition of economic base vectors demonstrates that the simplistic model of Figure 9.5 (with its focus on only four basic constructs) does need modification. Both Merthyr and Wolverhampton show that differentiation by economy as well as by ethnicity is found in these 19th century cities. Such a characteristic seems likely to be a product of the limited transportation surface in these centres, which meant that the early industrial workers had to live close to the work

Table 9.3
Dimensions in Merthyr Tydfil 1851

I HIGH SOCIAL CLASS/NON WELSH (High Social Class – Retailing – Professional – Non-Welsh)		II WEST WALES (Elderly – Large Family – Local Born – West Wales)	
Professional	94	West Wales	91
Servants	90	Density	43
Socio-Econ. I & II	88	Young heads	26
Dealing	84	Large family	-31
Unmarried heads	82	Glam/Mon/Brecon	-33
S.W. English	76	Local born	-37
English/Scottish	61	Elderly	-39
Other manuf.	60		
Domestic	44	IV NON FAMILY (Rural/Urban)	
Foreign born	41	Widowed heads	-85
Builders	30	Elderly	-44
		Domestic service	-44
Local	-38	Paupers	-38
Large family	-42	Independent	-33
Mining	-45		
		Large family	35
III LOW-MEDIUM SOCIAL CLASS (Low Social Class – Irish/Skilled)		Agriculture	35
		VI MID-WALES, LODGERS, TRANS- PORT (Mid Wales Born)	
Socio-Econ. IV & V	82	Mid Wales	78
Labourers	71	Lodgers	75
Irish	33	Transport	52
Socio-Econ. III	-69	English/Scottish	-32
		Elderly	-33
V EMPLOYMENT (Employment)		Other manuf.	-33
Not employed	-89		
		VIII IRISH, SHARED ACCOM. (Not Interpreted)	
Independent	32	Irish	82
Paupers	63	Shared accom.	76
		Lodgers	40
VII MATURE-YOUNG FAMILY (Age – Family Size)		Building	33
Mature heads	93	Local	-41
Large family	42	Old	-42
N. Welsh	-48	X MINING/METALS (Not Interpreted)	
Young heads	-77	Mining	-72
IX QUARRYING (Not Interpreted)			
Quarrying	69	Iron Manuf.	76
Glam/Mon/Brecon	36		
Independent	35		

New titles are provided. Original labels are in brackets.
Decimal points are removed: 91 is +0.91.

Source: revised from Carter and Wheatley (1979)

place. Only later, with improved transportation technology, could the industrial worker live at distance from the plant, allowing the city to become one residential housing market. This means that many of the earlier industrial centres may have dimensional structures closer to the current urban system types since the separate areas of the contemporary city functioned, in part at least, as individual industrial camps.

(b) Cross-cultural variations

A second major source of variation associated with social differentiation must be attributed to cultural differences, variations that can be identified by cross-national studies of individual cities and societies. Timms (1971), in a study of the Cook Islands, identified three axes, measuring a Traditional versus Modern way of life and a Migration axis in 1966. He observed:

'those who accepted modern values will pre-empt the higher social ranks in island society ... so ... the modernization factor will thus tend to become a social rank factor. Conversely, the factor relating to the traditional way of life, with its emphasis upon family variables, may give both to a separate family status dimension.' (Timms 1971 p.176)

Berry and Spodek's (1971) studies of a set of large Indian cities identified axes measuring Socio-Economic Status, various Communal types and a Familism dimension, usually working males located in the new industrial areas created by recent migration. This contrast suggests the evolution of the traditional/modern city structure dichotomy will take various forms in different areas, making it difficult to identify any universal model of evolutionary change as apparently predicted by Shevky and Bell (1955).

Further evidence of cross-cultural differences comes from factorial ecologies carried out by Abu-Lughod (1968) and Latif (1971) in Egypt. Three axis factorial solutions using the Principal Axes procedure and Varimax rotation were reported. Table 9.4 shows that in 1947 the largest axis from the 13 variable data base was a so-called Social Organization vector interpreted as a traditional-modern life style axis of differentiation, since it separated the variables measuring Muslims, fertility ratio, unemployed males, persons per room, from persons who were not married, literate or in schools. The second axis was called Social Disorganization and was clearest in Cairo where the density and handicapped persons indicators loaded strongly. In Alexandria divorced females, unemployed males, Muslims and density were linked to the axis. The third axis was only associated with two variables in Cairo, with the sex ratio and females never married variables, although males never married was added in Alexandria. This was interpreted as a type of Familism axis, the sex imbalances reflecting the differential migration of males to the city. This type of imbalance was previously reported by McElrath (1968) and in the studies described previously. Rather similar results were shown for analyses carried out in 1960 in the two cities, with the Social Disorganization vector showing the most changes, and some additional place-particular differences. Although these studies are restricted in terms of the limited data base and the use of only orthogonal rotation, the results seem closest to the

Modernizing city category of Figure 9.5, since the separation of Social Rank and Family Status does not seem to have occurred. However, these conclusions are only tentative given the limitations of the data base, always a problem outside western countries. Abu-Lughod (1968) also described the conditions under which the separations should occur.

Table 9.4
Dimensions in Alexandria and Cairo

Variables	Factor I		II		III		h2
	A	C	A	C	A	C	A
1. Sex ratio	03	−00	−19	−05	75	96	60
2. Fertility ratio	−82	−81	30	20	−07	−20	76
3. % Males never married	31	67	35	01	71	12	74
4. % Females never married	68	91	17	−07	47	62	72
5. % females divorced	01	42	47	31	11	19	24
6. Persons per room	−91	−80	17	10	−13	15	87
7. % Males literate	91	90	−02	12	02	06	82
8. % Females literate	91	95	−08	−03	01	17	83
9. % Females in schools or employed	78	72	11	−07	21	13	66
10. % Handicapped	−76	−08	−32	68	−21	−04	72
11. % Male unemployed	−26	−68	64	−38	05	−28	49
12. % Muslims in tracts	−81	−65	42	16	02	02	84
13. Density (sq. km.)	05	−18	62	62	−28	−16	47
% Variance explained	45.5	49.2	13.5	10.8	8.7	9.7	

A − Alexandria   C − Cairo   h2 − communalities for Alexandria only

Source: abstracted from Abu-Lughod (1968) and Latif (1971)

(c) Industrial and post-industrial transitions

Many of the earliest factorial studies of urban social differentiation in modern western cities were based on the premise that only three axes of differentiation were present. By selecting variables to define the three Shevky-Bell axes of constructs and then confirming that a three factor model was an appropriate description, such studies became part of a self-fulfilling prophecy. Yet although these studies failed to answer the question of whether the Shevky-Bell model was a complete description of urban social character they certainly provided evidence that the axes were an important part of urban differentiation.

One of the most comprehensive studies in this regard has been provided by Hunter (1974) in Chicago. His study of the factorial dimensions in Chicago from 1930 to 1960 using a standard variable set confirmed that the three Shevky and Bell axes were important sources of differences throughout the 30 year period. However, the degree of explanation in the model decreased through time, suggesting that the level of ecological complexity increased because the simple three axis model accounted for less of the variance. In addition, the Social Rank and Ethnicity axes became <u>more</u> important between 1930 and 1960, implying that socio-economic and ethnic variations increased in importance, presumably a product of the black immigration to the city, and to the increasing affluence of most of its residents in the post depression period. The fact that family differences were less important is surprising given the onset of post-industrial changes. In part the study deals with a period before these changes had taken effect, but they are also a product of the limited variable set used by Hunter, which was primarily designed to deal with the Shevky and Bell sources of variation, rather than with a more comprehensive set of constructs.

Throughout the late 1960s and 1970s larger and larger variable sets were used by investigators in factor ecologies of western cities. The results indicated that the three axis model of urban social differentiation had limited applicability even for modern western cities. For example, Rees's (1970) study of Chicago showed that several additional axes could be identified. When a city had several ethnic or cultural groups it was quite typical to find a series of ethnic axes of differentiation, most of which had a clustered pattern in the inner city areas. Of the other axes that have been identified, six dimensions seem the most ubiquitous in western cities and transcend place-particular associations: Migration Status, Non-Affluence; Pre-Family; Late or Established Family, Tenure, Urban Fringe. Also a group of other dimensions are identified. Since so many different variable sets have been used by various investigators it is difficult to make exact equivalences. Indeed, some of these dimensions described by an alternative title to the original prefered one may eventually be shown to represent separate axes. Moreover, the degree of separation of these axes from the three already described remains a moot question; few studies have used oblique rotation or higher order analysis. In Figure 9.6 each dimension has a separate spatial pattern which extend the list of types originally proposed by Murdie (1969).

(i) <u>Migration Status or Mobility</u> (Schmid 1960; Rees 1970; Jenson 1963; Davies and Lewis 1973, 1974; Davies 1975). This axis differentiates between areas of high and low mobility and migration in cities. Three interlinked spatial patterns can be found: the high migration associated with the new growth areas of single family dwellings on the edge of cities, sometimes producing a separate 'growth' axis (Murdie 1969); the high levels of migration linked to the transient or bed-sitter areas in part of the inner city; the low levels of migration found in the middle zones of the city dominated by late family characteristics. The result is to produce a spatial pattern that is best described as a roundal, with the low migration zone separating the central and suburban areas of high mobility. This pattern is very markedly influenced by variations in the transportation surface of the city, whilst high inner city mobility is usually only associated with one of the inner city sectors.

(ii) <u>Residential Quality, Substandardness or Skid Row</u> (Schmid and Tagashira 1964; Robson 1968; Herbert 1970; Evans 1973; Davies and

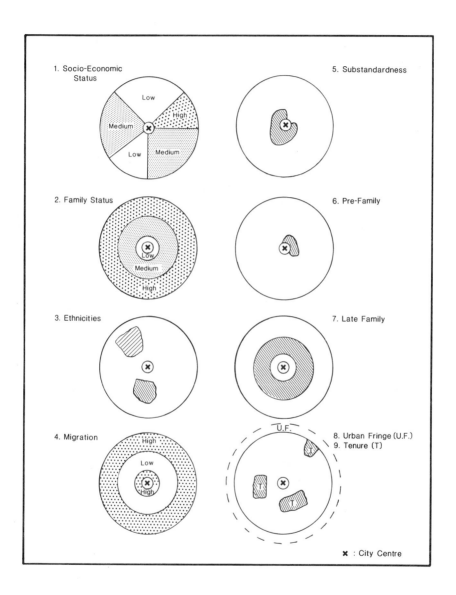

Figure 9.6 Spatial patterns for dimensions

323

Barrow 1974; Davies and Lewis 1973; Davies 1975). Whatever the title employed there is a very wide acceptance of at least one other axis to add to the Social Rank dimension. This picks out the poorest city areas, those associated with the worst housing stock and often with an aging population. Sometimes the axis is less obviously a scale measuring the poor quality housing stock and is linked to behavioural characteristics of the most substandard area, the home of the 'down and out', or alcoholism, even vice. Identified in this way, the axis has been described as the 'skid row' area in North American cities. It could be argued that 'skid row' is really a particular part of the substandard housing area, and, as such, may have a separate existence. Such a distinction has yet to be made in factorial studies. What does seem clear, however, is an inner city location, either concentrated in one area or partially surrounding the inner city core in a U shape. Since the axis is usually only found as a distinct vector when larger variable sets are employed, it has been suggested that the axis really is a variant of the Social Rank construct that identifies the seemingly irradicable 'trailing edge' of society, the home of the poorest, or least affluent in urban society.

(iii) Pre-Family (Young Adult) or Working Women (Davies and Lewis 1973; Davies 1975; Sweetser's 'Urbanism' Factor in Boston 1964; Davies and Barrow 1974; Herbert's 'Urbanization Factor' 1973; Robson's 'Young Adult Factor' 1968). Several factorial studies have shown that the inner areas of most larger cities are characterized by higher concentrations of young adult groups, or families without children in which high proportions of working women are found. To some extent these inner city features represent the low family status character already identified in the stage in family cycle dimension. However, the higher concentrations of young adults and working females frequently mean that a separate axis of Pre-Family or Young Adult character is found, one that can be linked to the pre-child stage in Foote et al's (1960) model of residential requirements and to the typology of areas in Figure 7.5.

(iv) Late or Established Family, or Postgenitive, or High Economic Participation (Sweetser 1964; Jansen's 'Middle Age' Factor 1968; Davies and Lewis 1973; Davies and Barrow 1974). In the same way that the 'Pre-Family' characteristics may separate out from the general Life Cycle dimension so the concentration of middle aged, late family groups with high work participation levels produced by women going back to work, or children joining the work force often leads to the identification of a third family axis called Late Family. It is usually found in the middle urban areas. Its doughnut shape is usually a discontinuous one broken by physical barriers such as rivers or the major radial links of transportation as shown in Leicester (Davies and Lewis 1975).

(v) Tenure (Robson 'Factor 2' 1968; Herbert 'Factor 1' 1970; Herbert and Evans 'Tenure and Life Cycle' 1973; Davies and Lewis 1973). The high levels of public authority housing in Britain has been shown to exert a significant effect upon the dimensionality of British towns. The more recent rotated solutions have shown that this characteristic does not appear as a separate axis, but is one that is linked to Life Cycle or family character. In other words, there is a societal modification of the Life Cycle or Family axes in Britain linked to the operation of public housing policy, since it is only old age, or large family size, that guarantees access to old people's flats or new council estates, respectively. The distribution is normally a series of clusters, a location depending on the particular city circumstances.

(vi) Urban Fringe (Davies and Lewis 1973; Davies 1975; Murdie 1968). The growing outer edge of the city has often been identified as a separate axis in studies investigating changes through time. In part this type of suburban area is probably best indexed by the migration dimension. The 'urban fringe' area is rather different, being the areas around the continuous built-up area containing a mixture of people with urban and rural occupations in which the former commute to the city to work. Obviously the axis can only be found if areas outside the political boundary are used by the analyst. The inclusion of these areas provides one advantage, namely that of relating the factorial evidence to established generalizations on the nature of the urban:rural zone of contact (Pryor 1968).

(vii) Other dimensions. Among the other axes that have been identified in more than one factorial study the following sets of dimensions may prove to have significance for particular types of western cities although more empirical evidence is needed.

1) Additional Social Status axes. Brady and Parker in Dublin (1974) and Davies in Southampton (1975) both identified social status axes which index the general range of social rank as well as an additional axis picking out the higher levels of status, rather than only the poorest groups of all. Similarly, Tyron in San Francisco (1955) and Timms in Brisbane (1971 p.177) showed axes related to self-employed professionals, whilst Davies and Barrow (1974) noted that the least affluent socio-economic groups in Prairie Cities formed the opposite side of a Service Economy axis. It is possible that the identification of these separate social rank axes relate to particular cities or individual urban economies, in the sense that high service oriented places, or those with greater degrees of prosperity may have more people in these higher status categories. This fact is reflected in the dimensionality of the places if these individuals are spatially clustered rather than being spread through the city.

2) Additional Family axes. Timms, in Brisbane (1971) and Borgatta and Hadden (1964) pointed to the presence of Family Dissolution and a Family Disorganization axes respectively, which imply that family status can split even further than we have described above – this time associated with the differentiation of groups with high levels of family breakdown. In most cases, the association is often with new high rise apartment areas of the inner city, or, if children are involved, with public housing projects. However, the axis would only seem to be partially identified in bigger cities by the usual range of census based variables. It is likely that the increased use of social welfare indicators or mental illness variables, such as those used by Bagley (1968) or Giggs (1970, 1974), will clarify the presence of such an additional axis. The partial evidence we have suggests that the 'disorganization' axis may be one of wider relevance than originally expected.

3) Size and density. Rees in Chicago (1963) and Janson in Newark (1968) both identified a Size and Density axis related to the size and

Table 9.5
Axes of intra-metropolitan differentiation: Prairie communities 1971

Second Order	Dimensions	First Order	Variables (Examples
1. Family	−78	1. Family Status	+Apartments; Old aged −Children; Single detached dwellings
	−81	9. Young Adults	+Young, single adults
2. Migrant Status	−70	3. Migrant	+Low mobility; Middle aged −High mobility; Immigrants
	−78	7. Housing	+New houses −Post war houses
		2. Socio-Economic Status	+Graduates; White collar workers −Manuf. workers; Limited education
3. Social Status	+79 +68 +40	5. British-East European Ethnic	+British ethnicity −Ukrainian and Polish ethnics
		4. Participation-Non-Affluence	+Economic participation; High school graduates −Substandard housing; native ethnic
4. Ethnicity	+37 +73 −67	6. Canadian-Immigrant	+Immigrants; Italians −Canadian stock
		8. French-German Ethnic	−German ethnic +French ethnic

−67
Size of loading
at second order

Source: Davies (1978)

density of population in the areas used in the analysis. Comparatively few other recent intra-city studies have used variables indexing these characteristics, so the general relevance of a separate axis of this type is unknown. If it is a general dimension it does seem to be restricted to the largest centres.

Many of these social structures can be linked to the postulates of post-industrial society whilst the different spatial patterns associated with the constructs extends Murdie's (1969) model of urban social differentiation. Yet at this stage it is difficult to estimate the extent to which all these additional axes of differentiation are typical of all western cities since there are so few comparative studies of several places. Moreover, many of the studies from which they have been derived have used different variable sets or component models in single city analyses. So there is always the possibility that the axes are city or society-specific. Davies (1978) in a comparative study of the five largest Western Canadian metropolitan areas (Winnipeg, Edmonton, Calgary, Saskatoon and Regina) in 1971 used a Joint Analysis of 358 community areas to identify the basic axes. Table 9.5 shows that among the nine first order axes accounting for 75.6% of the variance are the four typical McElrath axes with two extra ethnic vectors together with young adult, participation-non affluence and housing factors. However, these axes are all correlated since they collapse to four higher order vectors that confirm the McElrath model of differentiation.

Davies and Lewis's (1973) previous study of Leicester (summarized in Table 5.4) displayed a similar generalization of the first order vectors but in this case the three axis Shevky-Bell constructs provided a closer fit. The difference between Britain and Canada reflected in these and other results suggests that the North American case displays a more complex ecological structure, probably due to more advanced levels of post-industrial society. These tentatative results should not be taken as implying that there is a single industrial to post-industrial transition, providing an inevitable temporal sequence. Society-specific variations are likely to occur as demonstrated by the Canadian-British differences. Within Britain Davies (1978, 1983) has shown that the dimensions of the Welsh urban system display more overlap between the largest axes found in other western countries, for Age and Major Economy as well as Degree of Renewal provide the major sources of differentiation. Within the largest city, Cardiff, a medium sized British city yet comparable to others that have been studied (Davies 1975), similar types of overlap have been identified. The results for Cardiff (Davies (1983) showed more general axes of differentiation than expected. Separate Substandardness and Migration axes were not found. At the higher order level a general Family-Social Status second order vector was identified as well as Non-Urban and Ethnicity-Non Family vectors. Like the Leicester study a simple hierarchical pattern was not found between the first and second order vectors. There were minor associations with other second order vectors but Table 9.6 shows the Non-Urban higher order vector was only marginally associated with the first order Life Cycle vector. Certainly the size of the Life-Cycle linkage to the higher order Non-Urban vector means that it is difficult to use the dimension as part of the second order label. Only a rigorous temporal and cross-national study will show whether the argument that post-industrial changes have been slower

Table 9.6
Dimensions in Cardiff (Wales): 1971

Factorial Axes		Postulated
Higher Order	First Order	
	N.F.(part of 4)	1. Substandard-ness
I Non-Urban 87	5. Urban Fringe	2. Urban Fringe
37	3. Life Cycle and Tenure	3. Socio-Economic Status
II Family and Social Status −59	3. Life Cycle and Tenure	4. Life Cycle
+62	5. Late Family/ Mobility	5. Late Family
+71	1. Socio-Economic Status	
III Ethnicity- Non Family 60	2. Young Adult/ Non Family	6. Young Adult/ Pre-Family
77	4. Housing and Ethnicity	7. Ethnicity
−37	N.F. (Part of 7.)	8. Mobility
	(Life Cycle and Tenure)	9. Females
		10. Tenure
		11. Economy

Source: Revised from Davies (1983 p.94)

in arriving in Britain, or even more so in South Wales, compared to the midlands of England, is valid. Such studies are difficult to complete, not only because of the problem of obtaining comparative data but because of the variations in the meanings of indicators, as described in Chapter IV.

The differences that have been described in terms of cities in North America and Britain will undoubtedly be complemented by studies of variations between European countries, especially between east and west. Indeed a series of essays on urbanization in socialist Eastern Europe led French and Hamilton (1979) to conclude that socialist cities did differ from their capitalist counterparts in internal structure after thirty years of central planning, lower levels of affluence and the decision of planners to ignore the land value surface of western cities although modifications occurred because of the heritage of the pre-socialist past. Wedawowicz's (1979) factorial ecology of Warsaw in 1970 (Table 9.7) did not display completely different axes of differentiation to western cities in his study of 41 variables over 923 spatial units using the traditional Principal Axes component approach

Table 9.7
Dimensions in Warsaw 1970

I Socio-Occupational Position		II Housing and Social Situation	
Journalists/writers/artists	94	Dwellings 1945-60	87
State owned housing	88	Dwellings before 1944	86
Higher educated	83	Manual workers	82
Aged 25-64	83	Independent workers	81
Women in single households	80	White collar employees	79
Private dwellings	77	Scientists	77
Young adult 15-24 yrs.	74	Dwellings 1961-70	73
Middle school education	69	Dwellings with piped water	47
Single person households	68	County born population	41
Independent employees	56	Young adult (15-24 years)	40
Self-employed	55	Population 25-64	34
Female population	55	Single person h'holds	31
Basic education	52	Two person h'holds	31
Agricultural workers	48		
Doctors	42	Two + h'holds/dwelling	-32
		Population density	-43
Persons per room	-43	Persons/room	-48
Three, four person h'holds	-46	Workers	-67
Two person h'holds	-58	Living space/person	-86
Dwellings with gas	-72		
Co-operative housing	-81		
Living 5-9 years in city	-81	III Economic Position	
Children 0-14 years	-83		
Office workers	-88	Managers	89
		Population density	72
		Large households	67
IV Family Status		Self-employed	45
Population over 65 years	47	Dwelling with w.c.	42
Three, four person h'holds	45	Private dwellings	36
Two h'holds/dwelling	39	Population 25-64	34
Children (0-14 years)	35	Under 5 years in Warsaw	31
Five person h'holds	31		
		Workers	-48
Persons/room	-37	Young adults (15-24 years)	-53
Self-employed	-43	Doctors	-60

Decimal points are removed.

Source: Wedawowicz (1979)

followed by Varimax rotation. Yet there are important modifications to the standard western model. The largest axis, Socio-Occupational Position, separated variables indexing higher educated people, journalists, state-owned housing, from those measuring overcrowding. This was interpreted as categorizing 'ability to work and labour value' rather than the income-class differentiation at the basis of western Socio-Economic axes or one found in Warsaw in 1931. The Housing and Social Position vector was described as being related to differential housing conditions, which also indexed a contrast between new housing, white collar workers and scientists with overcrowding, although manual workers were linked to the better housing side of the axis presumably due to egalitarian housing policies. The third component, called Economic Position, was linked to managers and high density housing and was interpreted as identifying the groups having the greatest freedom to locate. The very weak fourth axis was called Family Status and had few large loadings but separated old age from large households. These results demonstrated the dominance of economic and housing factors in the study. If this is typical of Eastern Europe, it shows the failure of the socialist city to display the strong Family Status axis so typical of western cities, and the lack of separation of family and economic variables. Just as economic depression or lack of modernization has inhibited this separation in some cities in the western world, so the socialist planning process seems to have created similar conditions in Eastern Europe and has emphasized a distinct housing dimension. In addition, the variations within countries need to be investigated. To some extent the differences between Cardiff and Leicester already described, or Swansea and Southampton (Davies 1975) represent regional variations. In the context of the Shevky-Bell model Van Ardsol's (1958) work showed that southern US cities had different social structures; the fertility variable was linked to Social Rank alongside occupation and education, not the Urbanization (Family) axis.

All this evidence means that it is no longer possible to accept the concept of a universal model of urban social structure even for modern cities in the developed world. Moreover, the accumulation of information for the presence of many more additional axes of differentiation than predicted in the simple model of Figure 9.5 demonstrates that investigators are still some way from providing a comprehensive statement on the urban social differentiation of cities. Yet the fact that the first order axes can be generalized means that the Shevky-Bell or McElrath models still have relevance in western cities, except that they are second order rather than first order generalizations and imply that the post-industrial changes have still not worked themselves through to the level of complete axial independence at this community area scale. This means that the question of how many dimensions are needed to account for the complex social differentiation of cities has re-emerged as an important research question in its own right. So empirical explorations of the dimensionality of data sets may be as important as tests of existing theories, which have been shown to be limited in their degree of comprehensiveness. In this context it is quite possible that additional sources of variation may be identified, such as a variation based on economic types. As yet the factorial evidence is not clear on this point. What is apparent, however, is a variation in urban social differentiation associated with city-size.

(d) City size effects

Increasing interest in the imbalances caused by the accelerated growth of large cities in the last 30 years, and concern about the quality of life in these centres, has re-opened the old debate about the effect of size upon the differentiation of cities (Hoch 1973; Richardson 1973). Unfortunately, comparatively few attempts have been made to specifically investigate the city size or scale effect in factorial ecologies, although enough evidence has accumulated to demonstrate that such an influence can produce a fundamental effect on the patterns produced in any city. For example: Berry (1971) pointed to the 'gentrification' effect in the larger metropolitan areas, where middle and upper class enclaves have been re-established or have expanded in the inner areas of the larger cities; Rees (1970), showed that the dominant pattern was zonal for Family Status but although sectoral patterns were related to Social Rank zonal variations became more and more important as the size of city increased; Timms (1971) suggested that in the bigger cities of New Zealand, the familism axis split into two axes, one indexing Age Structure, the other a Housing-Way-of-Life axis that is close to the original concept of urbanization in the Shevky-Bell schema; Davies (Table 5.1) showed in a set of small British towns the level of overall explanation and the number of distinct axes were lower than in larger cities. Yet all these points are side issues in the various studies; the individual studies were not designed to specifically test the influence of size. Only in studies of small Ontario and Quebec centres by Bourne and Barber (1971), Prairie cities (Healey 1977), and Puerto Rican cities by Schwirian and Smith (1973) can be seen specific attempts to test the influence of the city size variation.

In the Canadian study Bourne and Barber were more concerned with spatial patterns than social dimensions, since only a restricted range of variables could be obtained for the cities in the range from Peterborough (56,177) to St. Catharine's (109,418). Basically they showed that the smaller cities displayed lesser order and hence less inner differentiation in the structure. These features were attributed to: the lower competition for land; the fact that the linkages between functions was less intense, causing lower concentration; and the lower barriers produced by such spatial restrictions as zoning and density requirements. Such factors are likely to provide the basic explanation for small city variation in most countries, but to this list we can add four other features. First of all there is the fact that in larger cities there may be a much greater cultural variation placed on certain land areas, providing an extra source of differentiation. Second, even where a small town shows the beginning of a separate type of pattern, such as the 'skid row' block near the transportation terminals, the area is unlikely to be large enough to be picked out by a factorial analysis at the normal scale of census tract or enumeration area. Hence relatively minor patterns can be obscured by the size of observation unit effect. Third, smaller cities are likely to display more specialized economic bases or certainly a lower economic mix of functions, so the range of possible variation is undoubtedly smaller. Finally, one must take into account the effect of what has come to be known as the urban size rachet (Thompson 1965), which in diffusion terms means that smaller centres are likely to display a lag in the impact of modernizing changes. So far Schwirian and Smith (1974) are the only ecologists to have investigated this point in factorial terms. Their analysis of Puerto Rican towns is shown in Table 9.8. It can be seen that in San Juan (the largest city) Social Rank, Housing, Fertility and Ethnicity are separated on four axes, whereas in a small centre, Mayaguez, Social Rank, Fertility and Ethnicity are combined

Table 9.8
Axes in Puerto Rican cities (1960)

Variables	San Juan (432,377)				Mayaguez (50,147)		
	I Social Rank	II Hous- ing	III Fer- tility	IV Eth- nic	I General Axis	II Hous- ing	III
Occupation	88	62	01	20	94	15	-17
Education	91	-09	-01	14	99	-10	-03
Infertility	25	10	83	12	88	-05	-41
Women at work	82	-14	27	-15	97	-04	13
Dwellings	-08	99	05	02	02	99	-03
Foreign born	35	11	14	76	94	04	37

Decimal points are removed: 88 is +0.88.

Source: K.P. Schwirian (1974 p.329)

into one axis. The explanation is that the modernizing process begins at primate city level and percolates down the city size hierarchy. At the time of analysis in 1960 these trends had not reached Mayaguez, which showed a traditional city pattern, compared to the western, or modern orientation of San Juan. However, it is possible the results may be exaggerated by overfactoring in the primate city: the correlations between the axes are not shown.

Similar reasoning can be applied to the split in the Familism axis reported by Timms in the large New Zealand cities. Only in the largest places is there the range of jobs likely to attract the young, mobile and more highly educated workforce, as well as the opportunities and social acceptance to indulge in alternative life styles.

All these studies, and the suggested explanations, support the idea that some kind of city size scale effect is present in the social differentiation of cities. But some researchers go further. Schwirian and Smith (1974) have suggested that ecological patterning is, to a large extent, a function of two independent variables: city size and level of economic development of the society in which the city is located. The authors emphasize the effects of the independent variables are not 'additive' for in large scale (modern) and low scale (traditional) societies there are very high and very low levels of differentiation respectively so city size has no effect. This means the city size effect should be only found in traditional societies such as Puerto Rico, where the impact of modernizing trends filter down through the hierarchy. Despite the value of Schwirian and Smith's study of the city size scale in modernization, and their attempt to rationalize the process by distinguishing ecologically differentiated from undifferentiated cities, the evidence already quoted would suggest they have overstated their case. It may be true that the major effect of city size lies in the degree of separation between Family Status and Socio-Economic Status in transitional societies. However, the work of Timms (1971), Norman (1971), and Bourne and Barber (1971) all show that city size effects are present in western, modern countries. Additional evidence on this issue can be found in Table 5.1. In a comparative study of three medium sized British cities and three small centres, Davies (1975), using the Variance Allocation method described in Chapter VI, showed that the bigger centres had more general axes, in the sense that the common variable variance contributed a greater amount of variability to the axes than the specific variable variance. In the smaller centres the minor axes, although similar in terms of the largest loadings, were more general than expected with the addition of a number of extra variables, as shown in Table 5.1. More specifically in Australia Houghton (1975) has also shown that spatial differentiation varies with size. These various trends have still to be documented and integrated thoroughly into the literature.

(e) Place-particular features

Advocates of the common factor model are only interested in the common, or as Cattell (1978) noted, the general sources of variation. The individual sources of variation found in the individuals in psychology, individual cities in ecology are usually condemned as specific or error variance. Most urban ecologists, however, are interested in the individuals, the city-specific sources of variation, so it is important to take these characteristics into account. Unfortunately, most of the earlier factorial ecologists tended to ignore these city-specific effects. One advantage of the factorial approach is that the city-specific characteristics can be teased out from the overall generalizations. The danger here is that it is very easy to attribute to individual cities characteristics that may be a result of the society in which the city is found, or the effect of city size, or even of the specific variable set or technique. Only by a comparative analysis of sets of cities can local patterns be separated from the general ones and even here what may seem to be a city-specific association could be a regional trend.

So far there are few comprehensive studies of several cities in a country (Jansen 1971) but even fewer have effectively separated out all the various relationships in such a way that place-particular characteristics can be unequivocally identified. In any case, reliance upon comparative measures such as congruence coefficients (which provide an overall measure of similarity between two studies) fails to help the investigator since it provides an overall measure of similarity. Several ways of resolving the problem were suggested in Chapter V but the Variance Allocation method proposed by Davies (1975) does seem to provide a relatively simple approach. Table 5.1 has already summarized the results of a comparative study of three medium sized British cities and three small towns: Colchester, Llanelli and Pontypridd.

(a) Common Variables	Average Loadings			
Car ratio	76	(72)		
High social status	75	(81)		
Employers, managers	73	(79)		
Service, distribution	71	(71)	SO	
Two car households	71	(73)		
Professionals	65	(68)		
Car commuters	57			(SO)
Non residents	–	(38)		(LE)
Pedestrian commuters	–42		ST	(SO)
Industrial workers	–48	(–69)	SO	
Personal service	–57			
Foreman – skilled workers	–58	(–80)		
Unskilled manuals	–64			(SO)
Non car households	–69			(SO)
Low social class	–76			(SO)

Figures in brackets are the loadings from the Medium:Upper Class
status construct in Southampton. (SO) Variable missing in
Southampton, (LE) in Leicester, (ST) in small towns,
(SW) Swansea.

(b)   Specific variables on Socio-Economic Status axis

Leicester	Intermediate workers	71	Unfurnished tenancy	–59
			Irish	–40
Southampton (General Status)	New residents	47	Local movers	–54
			Pensioners	–41
Southampton (Medium- Upper Class)			Medium status	–83
			Married females	–49
			Works transport	–47
			Mature adults	–45
Swansea	Rooms per person	55	Employed	–55
	New residents	52	Economically active	–51
	Fertile females	41	Council tenants	–49
Small Towns	Intermediate workers	78		

Decimal points are removed: 76 is +0.76.
Source: Davies (1975)

Table 9.10
Axes of differentiation at three scales in Wales (1971)

NATIONAL SCALE	CITY-REGIONAL SCALE	INTRA-URBAN SCALE
3. SIZE AND GROWTH		
5. SPECIALIZED ECONOMIES		
4. COLLEGE-RESORT		
1. AGE & MAJOR ECONOMY — Old, Service Children, Manuf.	II/I 4. RURAL-INDUSTRIAL — Manuf. Agric.	II 6. URBAN-FRINGE — Agricultural Commuters
6. TENURE — Council Owner Occup.	II 2. TENURE-AGE — Old, Owner Occup. Council, Y. Adult 0.67	I 3. LIFE CYCLE — Children Old
2. DEGREE OF RENEWAL — Migrants, & Growth — Mining, Welsh Substandard	I 5. ECONOMY AND STATUS — Service, High Status — Mining, Sub-Standard 0.63	I 5. LATE FAMILY-MOBILITY — Middle Aged — Local Movers
	I 1. ECONOMY AND ETHNICITY — Foreign Born — Welsh, Industrial 0.62	I 1. SOCIO-ECONOMIC — Status, English — Manuf., Welsh
		II 4. HOUSING AND ETHNICITY — Substandard, Foreign Born — Welsh, Council
7. WELSH — Welsh — Econ. Active Females	II 3. NON FAMILY — Single, Old — Children	II 2. YOUNG ADULT — Young Adult — Owner Occup.

First order axes are shown by capital letters and numbers. Examples of the original variables are in brackets.
Second order relationships are shown by Roman numerals. Source: Davies (1983 Figure 32)

The example of the Socio-Economic Status axis in Table 9.9 shows a clear high and low status association in the pattern of the average loadings. However, in Swansea the variables rooms-per-person, new residents and fertile females are closely linked to the high status side; the employed persons, economically active and council tenants indicators are on the low status or negative side. This suggests that the higher status characteristics in Swansea are primarily associated with the influx of new residents and young, adult females in new houses, the 'spiralist' groups described by Bell (1968) in his study of middle income families. By contrast, low status is specifically linked to the degree of employment and public housing in Swansea, a reflection both of the historically high levels of unemployment in the city and intervention by municipal action in the public housing sector.

(f) Urban, regional and national differences

The basic premise of the Shevky and Bell (1955) model, and of several empirical studies such as that by Carter (1974), is that there is scale invariance in the dimensions of urban social differentiation. Davies (1983) in a study of Welsh towns at three scales (urban system, regional and intra-urban) rejected this argument. The results shown in Table 9.9 indicate there were differences in the factorial dimensions at the various scales, although some comparable features could be identified. If these results have general applicability it provides additional evidence for the need for a two stage theory of the ecological differentiation of social characteristics. Thus, if societal processes alone explained the spatial differentiation in society there should be invariance in the dimensional stuctures, since these societal forces apply relatively uniformly in modern western society: unlike the 19th century pattern. The fact that variations occur means that other processes must be at work to provide differential spatial concentrations of social character.

In general it is likely that family variations will be more obvious within cities, given the different requirements of various age groups in cities already summarized in terms of space and accessibility. The large number of very different employment possibilities in cities, and standardization of wage rates across jobs, mean it is unlikely that economic base variations will be identified in sub-area differences, for the city usually forms one job-market. The exception seems to be in economic associations with ethnic groups that are themselves localized, or in concentrations of service related jobs in the immediate vicinity. In the past, economic base differences were present in intra-urban differentiation because of specific concentrations of industries in particular areas and the inability of workers to live outside walking distance of the plants. In contrast to these features the dimensional structures of urban systems will show more economic base variations because of the differential concentrations of particular industries in specific towns or sets of towns, due to the localization requirements of these industries. Regional studies will probably show an intermediate pattern between the city system and intra-urban dimensional structures. So far no satisfactory integrated two stage theory of social differentiation has been produced in which these differences can be satisfactorily explained.

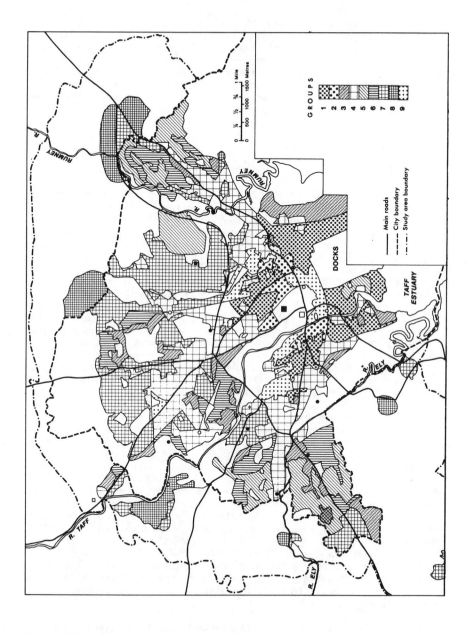

Figure 9.7   Social regions in Cardiff 1971
            Source: Davies (1973 Figure 21)

This brief description of the major sources of variation in studies of intra-urban social differentiation demonstrates that the universalistic assumptions of the classical urban ecology models or the theory of Shevky and Bell can no longer be upheld. There are a variety of alternative city structures which are still not clearly identified and are certainly not explained in a satisfactory manner. Although the developmental model of social area types in Figure 9.5 has been shown to possess limitations and needs modification, it provides a useful framework upon which to hang ideas or to base criticisms. Yet what is important in the context of this study is that the evidence from factorial studies has played an important part in expanding our knowledge of intra-uban social differentiation, or urban social structures.

4. SOCIAL REGIONS AND MODELS

(a) Empirical and theoretical social regions

Most factorial studies of individual cases plot the distribution of factor scores to identify the spatial patterns associated with the various dimensions. However, by applying other multivariate techniques such as cluster analysis to the pattern of scores these individual patterns can be generalized to produce a classification of areas based on social variations. The resultant social taxonomy, real world or empirical 'social regions' in the terminology of Figure 9.4, provides a shorthand description of the distinctive social areas produced by the specific combination of social structures in particular areas. Figure 9.7 shows the results of the application of this method to component scores obtained from a study of 541 enumeration districts and 27 variables in Cardiff (Davies 1983). The specific techniques used in the study were a Principal Axes procedure followed by Direct Oblimin rotation. Ward's Hierarchical grouping method was followed by a re-allocation produced by Discriminant Analysis. Nine basic taxonomic types were recognized in Cardiff and the types were described by the location of the areas and the size of the mean scores on the various clusters (Table 9.11). In theory, of course, there can be a multitude of different combinations of factor scores in the 6x541 matrix which formed the basis of the study. As in most investigations of this type, one or at most three axes had high enough scores in particular areas to produce sufficient distinctiveness for these units to emerge as a separate type. If these areas were locationally contiguous they formed a separate region. The important point about this set of results is, of course, the fact that a precise measurement of the social characteristics of the areas is obtained, one that is, of course, restricted to the data set originally factor analyzed. Although the original data set could have been grouped directly using some Area Taxonomy method, whether using Cluster Analysis or more recently developed procedures such as Q analysis (Atken 1974; Gatrell 1981), the advantage of the factorial approach is that it produces an initial simplification of the data, and the isolation of the general patterns of variation. The disadvantage is that unique highly localized patterns are eradicated as error variance.

If the study of intra-urban differentiation is to progress beyond the description of individual cities, generalizations of the relevance of city-specific typologies to other centres must be made. In the history

Table 9.11
Cluster groups in Cardiff

Short Title of Cluster	Importance of Individual Dimensions to Each Group			2. Postulated Types
	Very High Scores (> ± 1.5)	High Scores (± 1.5 to 1.0)	Fairly High Scores (± 0.99 to 0.5)	
1. Inner City: Decay and Immigrant		(4) Substandard Housing and Immigrant	+(1)Low Status (Includes Manuf. and Welsh Born)	Inner City: a) Decay b) Immigrant
2. Inner City: Non Family and Transient	(2)Young Family and Non Family		+(4)Substandard Housing and Immigrant +(5)Migrants +(1)Low Status	
3. Inner City: Public Housing			(4)Public Housing +(1)Low Status	Inner City: Old Public Housing
9. Inner City: Aging Status and Non Family	(3)Young Adult and Non Family	+Old Aged Females(2)	(1)High Status (Includes English born and Tertiary workers)	Inner: Aging Status
4. Middle City Areas			Old Aged and Females (3)	Middle City: Established Middle Status
6. Public Housing & Mixed Family, Middle City		(4)Public Housing (5)+Late Family (3)+Young Family	+Low Status (1)	Middle City: Established Public Housing
7. High Status, Established Family Areas		(1)High Status	+Old Aged and Females (2) +(5)Late Family	Middle City: Established High Status
8. New, Peripheral High Status Areas	(6)Urban Fringe +(1)High Status		+Owner Occupied(2)	Outer City: a) New High Status b) Rural Urban Fringe
5. New, Peripheral, Public Housing Estates	(5)Migrants	+(3)Young Family +(4)Public Housing	+Low Status (1)	Outer City: New Public Housing; Outer: New Middle Status

of investigations in this area a variety of very different regional typologies have been proposed for western cities, ranging from Burgess (1925), through Shevky and Bell (1955) and Mann (1965), to the schema suggested by investigators who have used factorial methods such as Rees (1972) and Robson (1975). Such typologies or models are complemented by the social regions proposed for non-western cities (McGee 1967). Davies (1983) has shown that despite the differences between the various western schemes they can be generalized and extended into a single model or theoretical pattern of social regions. This is composed of 18 types of social areas outside the central business district that seem relevant for at least the case of the medium sized British city. Figure 9.8 shows the distribution of these types which are arranged in four broad distance bands from the central area, called inner, middle, outer and fringe. No attempt is made in this idealized construction to add the location of subsidiary commercial, industrial or recreational areas since interest is only in social regionalization. Moreover, it has previously been emphasized that like all models it is not

> 'intended to incorporate the specific patterns of the real world city; there are always distortions caused by roads, rivers, parks or truncations caused by rivers, docks, industrial areas, etc.' (Davies 1983 p.168)

Before describing the distribution and characteristics of these areas it is worth noting that nine of the eighteen types identified in the model were isolated in the case of Cardiff, although Figure 9.8 shows that the specific social topography of Cardiff ensures that the systemetrical distributional pattern of the model is not duplicated in the real world example. Like all models Figure 9.8 is a composite generalization based on the individual patterns in Figure 9.6, only parts of which are applicable to most individual cities. In addition, it must be noted that the patterns are based on studies almost exclusively based on census information, not the quality-of-life, behavioural or attitudinal domains investigated by workers such as Knox and Maclaren (1978).

(b) Characteristics of the social region model

(i) <u>Inner city</u>. Figure 9.8 shows that the inner city is considered to be the most complex area because of the variety of ecological forces that operate. The area often has an old, decaying housing stock due to the age of the inner city and frequent absence of renewal (because more affluent groups have moved out). It is occupied by the lowest economic groups who are only able to afford such property. This leads to an areal type called <u>Substandard, Working Class</u>. The additional processes of ethnic segregation create an area called <u>Substandard, Immigrant</u> area (or areas, if several ethic groups are present) caused by the influx of low income ethnic groups to the only area they can afford. The high accessibility of the inner city has led some middle class groups to move back into these areas in the last decade, refurbishing the original property to so-call 'gentrify' the area. The resultant type is called <u>Renewal:Gentrification</u>. Many parts of the inner city area had become so run-down in the past that local authorities razed the area and created council or public authority housing areas. Elsewhere housing has been brought up to modern

A MODEL OF SOCIAL REGIONS IN BRITISH CITIES

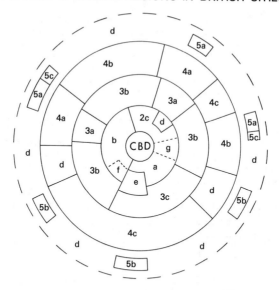

Figure 9.8 A model of social regions for British cities (Davies 1983)

1. CENTRAL BUSINESS DISTRICT
   (RETAIL, OFFICES, WHOLESALING)

2. INNER CITY
a. SUBSTANDARD, WORKING CLASS
b. SUBSTANDARD, IMMIGRANT
c. BEDSIT, TRANSIENT
d. AGING STATUS
e. OLD COUNCIL
f. RENEWED : COUNCIL
g. RENEWED : GENTRIFICATION

3. MIDDLE CITY
   (LATE AND ESTABLISHED FAMILY)
a. HIGH STATUS
b. MIDDLE INCOME
c. COUNCIL HOUSING

4. OUTER CITY
a. HIGH STATUS
b. MIDDLE INCOME
c. COUNCIL HOUSING
d. UNDEVELOPED

5. FRINGE
a. HIGH STATUS
b. MIXED STATUS SATELLITES
c. OLDER VILLAGES
d. RURAL

standards by the active sponsorship of local authorities in renewal rather than re-building, creating the Renewed:Council (public housing) type. In North America in particular the addition of an alternative type is often required, one linked to the way new houses and apartments have been created in the inner city by private developers. Such a type can be called Renewed:Private, although it is not typical enough to be part of the British model.

341

Two other types complete this inner city social mosaic. Some areas originally occupied by higher income groups have retained part of their status but are described as 'Aging Status' because of the higher average age of the residents, and the tendency for the areas to become run-down, as houses are not kept in repair or are turned into small residential hotels, etc. The influx of young people in search of cheap, one or two-room flats and apartments has created another distinctive region called Bedsit-Transient because of the constant flow of young adults in and out of the region. The larger houses, homes of the middle and upper class who are moved out of the area, provide relatively desirable property since they can be subdivided into many flats.

(ii) Middle city. In the middle city the aging process usually leads to a doughnut shaped zone of late or established family groups usually interrupted by sectors of commercial or industrial use cutting across the area alongside major roads. Low levels of migration characterize these areas, which have high proportions of mature to middle aged adults. Variations within the area are associated primarily with income differences. The extremes are occupied by council housing estates and high status private areas and can be called High Status, Middle City and Council Housing, Middle City with a type in between described as Middle Income, Middle City. The high status areas are frequently found on the edge of the inner city status areas, creating a sectoral pattern of high status residential development.

(iii) Outer city. In the outer city young family groups dominate the area, for these are the newest suburbs. Again, socio-economic status variations lead to at least a threefold division linked to council housing, low and middle and low income private estates and high status residential areas. However, many areas of undeveloped land are also intermingled with these types. The titles shown in Figure 9. reflect these four types.

(iv) Fringe. Outside the confines of the continuously built-up zone rural areas and pre-existing villages may survive and can be described as Older Villages, or Rural areas. In addition High Status Fringe communities are found. These are the high status residential enclaves created by the differential migration of higher income groups to larger houses in rural areas. These areas are located either in isolation or are attached to pre-existing villages. Satellite commercial or industrial centres are also found in the area. Usually they have mixed status so are described as such. Such communities are more typical around the largest cities. Again, in North America less strict land use regulations and higher levels of affluence have led to the growth of acreages and hobby farms producing an Acreage, Urban Fringe type. This is not a characteristic of most British cities.

Many elements of this model of the types of social regions found in medium sized British cities can be parallelled in other western countries, although without the emphasis on council housing, whilst the major additions needed in the North America case have been noted. It is obvious that Family Status and Socio-Economic Status variations provide the basic sources of differentiation, with the addition of Migrant and Non-Migrant dimensionality in the inner and middle city, as well as Ethnic or Immigrant variations. Through time, the greater complexity of Family Status variations has created a more differentiated structure, as has the effect of such massive interpretation in the housing market by successive British governments, which has led to distinctive areas of public housing for low status groups.

As with all generalizations, the model of social regions shown in Figure 9.6 is proposed as a summary, or idealized representation of the major social areas that could be identified from census based variables. Modifications associated with particular cities, whether due to the presence of specific social groups, topographic, economic or land use variations, or even historical associations that have perpetuated certain characteristics, can be expected. Like all generalizations, therefore, the model is designed as a _framework_ for the spatial interpretation of social variations in the medium sized British city. As such it may be useful in its own right, but may have further uses in providing a basis for differentiating areas for more behavioural investigations of social groups. Obviously, however, the limitations of the construction to the variations identifiable in census data must be noted – a conclusion that applies to the results of the dimensions or structures of cities described here. It goes without saying that the variations produced by what Booth (1896) referred to as the 'sights, sounds and life' of individual areas are _not_ taken into account. To do so requires a very different methodology based on humanistic or even phenomenonological considerations.

5. TIME

So far this whole discussion of intra-urban differentiation has been based on a very static interpretation of social character. Like most human ecological studies the information base that has been used relates to the residences of people: after all, census authorities count and record people by their home. Human ecologists have long recognized the various behavioural rhythms associated with cities from journey-to-work in city centre to particular concentrations for entertainment. Chapin (1968) has classified the range of activities and shown specific time-budget relationships (Chapin and Hammer 1972; Szalai 1972). Taylor and Parkes (1975), however, have demonstrated how factorial methods can help define the activity structures of the various parts of a city. They used a hypothetical city divided up into ten districts, a central zone and nine other units composed of three socio-economic sectors and three distance bands linked to family status. Eight time periods, 6-9 a.m., 9 a.m.-12 noon, 12-2 p.m., 2-4:30 p.m., 4:30-7 p.m., and 10 p.m. to midnight, were identified producing a set of eighty space-time units. A set of twenty-two variables were considered to define the basic character of the city in terms of land use, demographic, social status and transport character. The short titles are shown in Table 9.11. The data set is more comprehensive than the typical factorial ecology because of the need to take the interpretation beyond the purely static or residential framework used in most ecologies.

Table 9.12 shows the results of the application of a Principal Axes common factor solution with communality estimated by iteration, followed by Varimax rotation. A seven factor solution accounted for 91% of the total variance. The results show that the largest factor is a White Collar Workday Cycle which has its highest scores in the city

Table 9.12
Loadings for space:time factorial ecology

Variables	Factors						
	1	2	3	4	5	6	7
(1) Residential	-41	24	-42	42	-35	-32	39
(2) Retail and service	84	10	03	-15	-13	23	-16
(3) Other commercial	96	11	36	-06	05	09	-04
(4) Industrial	06	-33	-11	-06	24	24	-85
(5) Entertainment	13	-00	83	-08	07	00	03
(6) Open space	14	38	-01	10	-68	-11	-07
(7) Population	78	05	23	15	31	37	12
(8) Persons/dwelling	-43	-13	-37	57	-16	-34	35
(9) Male adults	12	-07	14	-02	89	-08	-22
(10) Age 0-4	-22	-22	-47	28	-68	-23	15
(11) Age 5-17	-33	-08	-44	37	-65	-27	17
(12) Age 18-29	32	-02	82	25	30	21	06
(13) Age over 65	-07	24	-19	-92	10	04	-04
(14) Married adults	08	00	-60	72	-25	-12	-06
(15) Employed females	85	05	18	01	19	25	-32
(16) Upper class	12	92	00	-09	-05	-03	15
(17) Lower class	01	-88	10	07	13	04	-09
(18) Professional services	92	08	24	05	14	20	12
(19) Manual services	45	-11	34	-15	-11	65	-18
(20) Private transport	37	22	12	-20	11	82	06
(21) Public transport	27	-15	-02	01	15	83	-34
(22) Access	39	68	37	-33	04	32	02
Eigenvalues	9.05	3.42	2.82	1.90	1.23	0.93	0.63
Cumulative percentage of total variance	41.1	56.7	69.5	78.1	83.7	88.0	90.8

Decimal points are removed.
Source: Taylor and Parkes (1975 p.677)

centre during the working day. The second axis is linked to Class
Segregation. The third largest axis proved difficult to describe but
since it was associated with Entertainment and Young Adults it can be
described in these terms. As such it reinforces the type of Pre-Family
axis shown in the traditional factorial studies. The fourth factor is
an Old Age Dependency axis emphasizing the socio-temporal segregation
of the old, and again demonstrating the importance of the split in
family character in modern society. The fifth factor can be called
Workers-Young Dependents because it separates daytime central or inner
city with its high proportion of working males from the married female
- high children suburbs, demonstrating the fundamental city:suburb

split in the working day. Factor six is a Journey-to-Work factor whilst factor seven is a <u>Blue Collar Workday</u> cycle primarily linked to industrial areas.

Taylor and Parkes (1975) provided a series of maps to illustrate the relative importance of each space-time to these various factors. The authors admit the study is limited because it is a study of an artificially-created data set. However this does not really matter in methodological terms since the study was designed as an <u>experiment</u> to bring back conceptions of temporal variations into ecological studies. One can only hope that it will soon prove possible to provide real empirical data for a city or set of cities to determine whether the dimensions shown in Table 9.12 really are the significant or important time-space axes. In this way the daily temporal dimension will be added to our quantitative studies of urban social character.

6. SUMMARY

Despite the large number of applications of factorial methods in studies of intra-urban character it must be remembered that cities are enormously complex. As such the factorial approach can only help describe and interpret part of the social mosaic. Yet such a limitation is common to <u>all</u> approaches, for all extract only <u>part</u> of the variation found in cities for study. The big advantage of the factorial approach is that it has provided evidence on social structural character, clarified the various stages in the process of urban social description, and provided explicit measurements of the variations based on particular data sets. Since these can be generalized to deal with the structures, patterns and regions typical of the internal social differentiation of sets of cities, the findings can be formalized as structures. These represent the constructs that still have to be explained by deep seated theories of social change. As yet investigators are still more concerned with identifying the constructs, despite the theoretical efforts in the field. In this context it must be admitted that factorial ecologists seem content to accept far lower level of loadings to identify axes of differentiation than is typical in the psychology literature. Whether experience with more carefully constructed variable sets will solve this problem, or whether it is a function of dealing with a non homogeneous population (the city areas are, after all, different) is still open to debate. Nevertheless despite this and other problems there does seem hope that an integrated body of literature is not far away to set alongside the behavioural, managerial and process oriented studies of residential variations.

10 CONCLUSION

It is just over twenty years since factor analysis methods have been extensively used by human geographers and human ecologists. In the first half of this period the factorial approach was viewed with great enthusiasm by its exponents and became one of the frontier areas of human ecological research. This led to a series of studies demonstrating the utility of the approach in various parts of the field culminating in two sets of essays edited by B.J.L. Berry (1971, 1972) which are still the basic reference sources to work in human geography. During the last decade it must be admitted that interest in the approach has waned considerably, curiously enough when one might have expected some consolidation of the research effort and integration of the previous set of results. Moreover, this is at a time when many of the more sophisticated factorial routines developed by statistically minded psychologists are widely available in computing centres and can be quickly and inexpensively run. No introduction to the field of factorial ecology is complete without understanding the reasons for this change in attitude to these methods. Three broad sets of explanation seem to be particularly relevant: the reaction against quantitative methods in the field; the general problems of the factorial approach; and the development of what can only be called new frontiers in techniques and methods. These issues provide a context for evaluating the utility of the factorial ecology approach now that the problems in the field have been introduced and some applications have been summarized.

1. REACTION AGAINST QUANTITATIVE METHODS

Part of the decline in interest in factorial methods undoubtedly stems from the reaction against the use, or more appropriately against the apparently dominating nature of quantitative methods in human geography in the early 1970s. This was linked to the anti-positivist philosophy displayed by many geographers who saw human geography as being a more pluralistic and wide ranging field than that envisaged by spatial scientists. Beginning, perhaps, with behavioural and perceptual studies the reaction against a strictly quantitative approach was carried furthest by those of more humanistic persuasions who preferred to look at the values and at the life-world of people, in which a number of alternative philosophies could profitably be applied to human geographical concerns. Added to these very different interests were the genuine concerns about the applicability of many statistical techniques (especially those with an inferential basis) to human geographical populations, a situation initially eloquently described by Gould (1970). Soon after these developments, which have been succinctly reviewed by Johnston (1979), critics pointed to the rather limited horizons displayed by most quantitatively minded researchers, in the sense that the causes of particular distributions, processes and organizations were only rarely dealt with, or were investigated at a superficial level. Geographers such as Harvey (1973) and Gregory (1978) were leaders in pointing to the need to search for the deep seated causes and meanings of empirical results, the approach eventually called the structuralist position. This interpretation of 'structure' conflicts, it must be emphasized, with the more usual meaning of a repetitive regularity displayed by empirical patterns (Blau 1976; Davies 1983). In addition some investigators saw in the extensive Marxist literature other keys to understanding, this time linked to desires to change the existing system. Yet such prescriptive concerns obviously go beyond the Marxists to those interested in planning in general and the managing of urban places.

There seems little point in denying the fact that adherents to all the positions described above have a set of perfectly reasonable arguments but it is important not to exaggerate the case for any one position. Many of the issues are based on a particular philosophical position, so the challenge is really against the sets of values held by factorial ecologists. In a liberal pluralistic society it is a personal decision as to which approach is considered to provide the most advantage. In a field such as human ecology with many, many possible alternative positions, it is difficult to accept that a single approach can claim a premium on truth, or more appropriately the various truths. This means that all approaches should be cultivated as rigorously as possible. In this context it seems difficult to deny that factorial methods have proved to be useful in synthesizing information and in isolating certain descriptive structures in our multivariate world. These structures or repetitive patterns are of intrinsic interest but could also be used as building-blocks for more comprehensive theories. As always, however, there may be other ways of arriving at these generalizations. One advantage of the factorial approach is that the mechanism used to identify these patterns is usually carefully specified, or should be. In this study it has been shown that a number of objectives may be fulfilled by factor analysis but the most important use of the factorial approach has been to synthesize relationships from a set of variables. It is rather as if an investigator is presented with an atlas of single variable distributions. Many of these indicators can be synthesized into some basic clusters of variation and associated spatial patterns. Certainly something is lost in the generalizing process but this is true of all generalizations. When viewed against the situation of the human ecology field before factorial methods were applied, it is clear that the approach has provided a liberalizing approach in at least part of the field of human ecology. To claim more would be presumptuous. For example the method has raised the sights of human ecologists from individual case studies to societal linkages and to the identification of constructs and the precise measurement of sources of variation. In the future, of course, other approaches may provide equally useful for these tasks. It must be admitted, however, that the factorial approach adds little to our understanding of the causes of dimensionality or spatial patterning. Yet by the same token the recent structuralist literature tells us little about the repetitive patterns or spatial regularities.

2. CRITIQUES OF FACTORIAL METHODS IN PARTICULAR

Many human ecologists are not particularly interested in quantitative methodology or the production of generalizations as opposed to ideographic study. In addition there are those critical of the use of factorial methods in particular. However, many of these criticisms apply as much to the factorial procedures in general (Ehrenberg 1962; Armstrong 1962; Williams 1972), not simply to their application in human ecology. Six major problem areas exist.

The first is that factorial methods are based on a specific set of assumptions which are not always relevant to the ecological or areal based data sets that are the basis of factorial ecology. These issues have been discussed in Chapter VI and need not be repeated in detail. But it is worth emphasizing that the major thrust of the factorial approach is to fit a linear model to a data set. There is no doubt that this type of model will not fit all circumstances, so it is important that the investigator determines the suitability of the model before such an analysis is completed or reported. Nevertheless, evidence has accumulated in many fields, as shown by repeated refer- ences to Cattell (1978) and Rummell (1970), that factor models are robust and produce acceptable results in a wide variety of circum- stances, whilst the future application of non linear factor methods (McDonald 1967) is likely to expand the horizons of the procedures. In the last resort, however, it must be stressed that factor analysis is not a universal panacea for all problems of synthesizing or understand- ing large data sets or the complex spatial mosaic of human distribu- tions. No technique, method or philosophy for that matter ever is. All approaches have some limitations.

The second major problem area relates to the invariance of the solutions, the fact that a number of alternative results can be obtained from some data sets. Yet if strong linear structures are present in a data set the factorial methods have been shown to uncover the structures. If very different results are obtained, either the procedures applied are inappropriate, or there really are a set of weak structures involved. There is no doubt that a large number of decisions have to be made in any factor analysis, but this is true for all empirical work, although in most cases these decisions are made subjectively and are rarely the object of much comment. As always in technical analyses it is important to understand the basis for the various decisions, rather than depending only upon the default options of the standard computer programme. Indeed the debate in the factorial literature about the suitability of various correlation coefficients, improved factoring or rotation designs or alternative results is a strength not a weakness, since it provides a cumulative evaluation of quantitative work. To use a quotation from a speech of Aneurin Bevan in another context: 'better the conflict of minds than the unanimity of the grave.' Moreover the ability of ecologists to learn from the parallel work of psychologists is another advantage.

A third set of criticisms relate to the generality of the results, where it is argued that despite the vast data output only very general or superficial conclusions are provided, as when the Shevky-Bell axes of differentiation are discovered in yet another empirical study. Such a criticism does seem unfair. Until the addition of factorial methods to the battery of research techniques urban ecologists were arguing about the relevance of the three axis model and the competition between concentric zones or sectors, etc., in spatial patterns. Any evaluation of the history of ideas will demonstrate that the factorial approach has made an important contribution. Moreover additional axes of differentiation are still being uncovered and the reasons why variations occur between different cities is still inadequately understood. It can be argued that the additional generalizations identified in this study in relation to urban system structure both in social character or interactions, or in terms of intra-urban differentiation, are less important than those defined a decade ago by Berry et al (1972). However it seems difficult to deny there has been an incremental addition to the substantive literature.

It is possible, of course, that the factorial approach has helped swing the pendulum of interest too far in the direction of generality as opposed to particularism. Yet it must be remembered that in the early 1970s Rees (1971) was able to claim that one of the major advantages of the factorial approach was to raise the sights of ecologists above the case study level. This demonstrates that any evaluation of the impact of the procedures must be based as much upon the state of the field at the time of the application of the methods as upon the level of current knowledge. What seems obvious today was not obvious ten or twenty years ago. In this context factorial methods can be used to emphasize the general:specific dichotomy found in all ecological distributions, rather than only the individual sets of structures. However the search for generality does not mean one has to fall into the trap of the universalist fallacy. For example, cities in particular societies may have their own general features, whilst ethnocentric considerations will ensure that variables may be measuring different things.

The fourth major criticism made about factorial methods in general is that they are superficial, in the sense that little thought goes into the creation of the results. Again there seems little point in denying that some of the earliest ecological applications may have been exercises in number crunching, not carefully designed experiments to test some hypothesis or to explore the dimensionality of data sets. But such problems occur in many first approaches, whether quantitative or qualitative, where particular perceptual biases may occur. In part, of course, the criticisms link back to the basically inductive approach of the factorial method as used by many people. This is not one that finds much favour in the scientific community. But Cattell (1978 p.9) has emphasized that the hypothetico-deductive method does not exist in isolation; the progress of this type of research really is a synthesis. The hypothesis is usually achieved by prior reading, analogy, or by intuition. But in some (not all!) areas of human ecology it can also be based on several exploratory factor analyses of variable sets. Subsequently, of course, the final set of hypotheses may be tested by appropriate factorial designs, or investigators may decide to go back to the original variable, having used the factorial approach simply to isolate the separate sources of variation as a set of data-screening device. So the methods can be used at the beginning and the end of studies, to sort out or confirm the basic sets of interrelationships. Moreover the results of factor analysis studies cannot be lightly dismissed, especially when one considers the continued utility of the approach in allied fields such as psychology. Nevertheless, one cannot deny that factorial ecologists have had less experience with the approaches ensuring that far less substantive results have been obtained. Indeed, compared with the psychology literature it is clear that factorial ecologists have a long way to go. Whether all the battery of inferential test procedures used by psychologists can ever be applied is a moot point, given the particular problems based on areal distributions. Nevertheless, even if used in a descriptive context the procedures have still identified the dimensions and patterns of regions, intra and inter-urban systems.

A fifth point is that the methods do not need to be so complex and there is no need for the new 'jargon' that is created. It can be argued that these unnecessarily complicated procedures produce results that could be obtained by much simpler methods. This is always a difficult issue to rebut, since it is always easy to be wise after the event. In some cases, such as the study of flow patterns, a nodal structure technique (Davies and Robinson 1968) may work equally as well on well structured data sets because the data falls into defined rank order relationships. When it does not, other structure seeking methods may be needed to tease out the relationships. Factor methods are only one of the means for abstracting the generality. Yet they allow investigators to go back to the original data set to describe the flow patterns left out by the analysis, because they cannot be accommodated in a linear model of the component or common factor approach. Similarly the results of factor methods can be used to identify key variables 'standing for' separate sources of variation. Certainly these could always be used. The point is that it is only after the factorial results that we can have some confidence these indicators are separate sources of variation, rather than depending on intuition to make this decision.

Finally, far more telling criticisms relate to the utility of factorial methods in identifying causes or causal relationships. Throughout this study it has been emphasized that factor axes can be viewed as new constructs or new sources of measurement identifying what the variables have in common. They do not necessarily establish anything about the causal links between the original variables. Applied factor analysts have long been aware that their results cannot stretch this far and Meyer (1971) has carefully documented the contrast between correlation and factorial methods. Yet it seems similarly inappropriate to criticize factorial ecologists for failing to describe the reasons why social dimensions or spatial patterns occur. The factorial approach is not designed to a universal methodology, or theoretical structure; rather it provides one mechanism for isolating or confirming the structures found in the multivariate ecological reality, structures that have to be causally explained.

3. NEW FRONTIERS

Many of the points of criticism raised so far about factorial methods seem to have been based on misunderstandings of the objectives of the procedures or upon the inevitably limited areas in human ecology to which the methods apply. However another set of criticisms derive from the addition of new techniques to the battery of multivariate procedures applied to human geography and ecology. Some, such as Multi-dimensional Scaling (Golledge and Rushton 1972) certainly have their place, especially in circumstances where only non metric data sets are appropriate. But such approaches seem to be more satisfactory in the purely taxonomic approach rather than in the search for dimensionality where interval scale data can be obtained. Other techniques have not yet been applied to the ecological field in any comprehensive way but studies have been carried out on the relationships between Factor Analysis and procedures such as Latent Structure and Latent Profile Analysis (Gibson 1959) or Order Analysis (Krus and Weiss 1976). More rigorous competition seems likely to be provided by Q Analysis (Atkin 1974).

The basis of Q analysis is to define the _variety_ of structures in data sets such as those composed of areas and variables rather than imposing a single classification upon the information as in the classical numerical taxonomy. Although alternative taxonomies may be produced if different techniques are employed (Johnston 1968) the end product is still usually one final result. Advocates of Q analysis suggest that the social variations in cities are often continuous and that the application of the classical taxonomic procedures imposes a _single_ regionalization which loses much of the structure existing in the data. In other words Alihan's (1938) important criticism of the concentric zone model, namely that zones are imposed on a continuous distribution, is revived. This is hardly the place to discuss the specifics of Q analysis in detail. Gould (1980) and Gatrell (1981) have provided introductions to the ideas. Yet is is necessary in this review to appreciate the basics of the approach. If a matrix of areas and variables is considered, some critical value is defined to create an incidence matrix of 1 (above the critical level or slice) and 0. Geometrically those variables above this critical value can be viewed as a series of nodes connected by lines which define a polyhedron or simplex for each area. Q analysis identifies the simplices that are connected at different levels, for example the areas or places that are connected by three variables above a level, or two variables, etc. The result of the procedure (which Gatell warns is a highly abstract and sophisticated approach) is to define the _variety_ of areal structures present in a data set. Like all classifications the procedure can be reversed to classify the variables on the basis of the areas.

In the context of intra-urban and inter-urban systems, Gatrell (1981) and Beaumont (1981) have argued that Q analysis provides a more satisfactory methodological approach because all the relationships in a data set are explored, without the necessity of imposing a single classification on the data or of making assumptions about the characteristics of the data. Although the arguments are persuasive it must be emphasized that the complexity of so many alternative structural relationships will undoubtedly cause serious problems in interpretation, although Gatrell (1981) has claimed significant benefits in isolating various structures. Unlike factor analysis there seems to be no equivalent procedure for isolating the importance of each dimension or pattern or the contribution of single variables to the overall structure that is displayed. In addition, the fact that the method is dependent upon incidence matrices, admittedly with the flexibility of different slices or levels of the variables, means that the whole range of data variation does not seem to be simultaneously dealt with. More importantly, in comparison with factorial ecology it is difficult to determine the importance of each dimensional structure or the general-unique dichotomy between cities or internal urban patterns that is at the heart of the common factor approach.

Although the Q approach can be applied to classification of variables as opposed to areas, at this time it seems likely to be more useful in the classificatory process of areas rather than dimensional structures of variables in which all the data (not just the common and specific variance) is being dealt with. However, much more comparative work on the respective utilitiy of the Q technique compared to all other methods in various circumstances is needed before a final conclusion can be drawn. If the history of other multivariate procedures is anything to go by it is likely that the factorial method will still claim its adherents, but perhaps in more restricted areas. It is difficult to imagine that the whole methodology that has proved so profitable in areas such as psychology will be abandoned wholesale. Only careful data analysis will reveal whether Q analysis will become the principal method for isolating and exploring structures in ecological circumstances.

Finally, it must be pointed out that it has been argued that advances in the field of human ecology, helped, in part, by the results of factorial methods may have created a new phase of enquiry which Davies (1983) has decribed as the Multivariate-Structural approach to ecology in which the alternative definitions of 'structure' and the limits of the approach have been described. It is obvious that the term factorial is not comprehensive enough to cover all the techniques that have been applied to problems such as the definition of urban social structure. The typical addition of cluster techniques (Sokal and Sneath 1973), or canonical correlation analysis (Berry 1978; Clark 1973), even apart from recent advances in Q analysis, ensures that the term 'factorial' is too restrictive to cover the alternative procedures. Yet all are similar in having a multivariate basis. Moreover, since the primary objective of the procedures is with structures (interpreted as the patterning or the repetitive features of areas, not some deep seated cause) it is more appropriate to use this term as a description of the essentials of the approach, since human ecology covers a wider field including social and behavioural consider-ations. If this argument proves acceptable then it is possible that factorial ecology may be a passing phase of enquiry in human ecology, perhaps with some built-in limitations, much like social area analysis is viewed today. But even if this proves to be an accurate statement of the future, there seems little doubt that factor analysis will still be one of the families of useful multivariate technique whilst factorial ecology has provided an important methodological key to unlocking some of the secrets of the complex areal human mosaic that is the subject matter of human geography and ecology. It has been and will continue to be useful in abstracting generalizations from appropriate data sets. Yet as with all methods it is only <u>one</u> of the paths to understanding, one that has provided important advances over older methodologies and techniques, and judging from its use in psychology, will still continue to play an important role in human geography and ecology in the identification of general or common sources of patterns of variation.

APPENDIX A DERIVATION OF PRINCIPAL AXES

This appendix describes how a factor matrix (specifically a Principal Axes method using the component model) can be derived from a correlation matrix such that the first factor vector extracted contains the largest amount of variance, and the second and subsequent vectors contain progressively smaller variance values. The method that is described here is called the 'iteration and exhaustion' procedure because of the way it extracts the roots (eigenvalues) of the correlation matrix. There are, of course, many different ways of extracting roots from a correlation matrix as described in standard statistical sources (Harman 1976). This particular method is relatively simple in that it does not necessarily require a prior understanding of matrix algebra. The method assigns arbitrary values to a test eigenvalue and iterates until the test eigenvectors converge. The residual correlation matrix is calculated and the whole process begins again.

This example is designed to be self-instructive to provide confidence to newcomers to the field and can be easily translated into a computer algorithm which carries out all the procedures that are described. It is worth noting that not all correlation matrices are suitable for factoring in this particular way.

Step 1

Derive a correlation matrix COR(J,K) and assume a test eigenvector or column of values Y(J) such that all the entries in the test vector are 1.0.

Step 2

Post-multiply the correlation matrix COR (J,K) by the trial vector Y(J) to obtain another vector X(J). The first value of this vector X(J) is the first trial eigenvalue EIG(L). Post-multiplication is a matrix algebra term in which each of the entries in the first <u>row</u> of the correlation matrix are multiplied by the corresponding entries in the first <u>column</u> of the Y matrix. The sum of these values is the appropriate entry for the product matrix X(J). For example: the calculation for Row 1 of the correlation matrix Column 1 of the vector is $(1.00 \times 1.00) + (-0.855 \times 1.00) + (0.367 \times 1.000) - 0.512.$

355

$$
\underset{COR(J,K)}{\begin{bmatrix} 1.000 & -0.855 & 0.367 \\ -1.855 & 1.000 & -0.516 \\ 0.367 & -0.516 & 1.000 \end{bmatrix}} \times \underset{\underset{Y(J)}{1}}{\begin{bmatrix} 1.00 \\ 1.00 \\ 1.00 \end{bmatrix}} = \underset{\underset{X(J)}{2}}{\begin{bmatrix} 0.512 \\ -0.371 \\ 0.851 \end{bmatrix}} \underset{EIG(L)}{\begin{bmatrix} 0.512 \end{bmatrix}} = \underset{\underset{VEC(J,L) = X(J) \div EIG(L)}{3}}{\begin{bmatrix} 1.000 \\ -0.724 \\ 1.662 \end{bmatrix}}
$$

Step 3

Standardize the entries in the new trial vector X(J) by dividing its entries by the trial eigenvalue EIG(L) to produce the new standardized trial vector VEC(J,L).

Step 4

Calculate the sum of the absolute difference (ignoring the sign) between the trial vectors Y(J) and VEC(J,L). Establish the test (or root convergence) criterion for controlling the number of iterations necessary to extract the root or eigenvalue. It is usual to assign 0.00001 for TEST.

e.g.   First Iteration

$$\left| Y(J) \ - \ VEC(J,L) \right|$$

1.000 - 1.000   =   0.0

1.000 - (-0.724) =   +1.724

1.000 - (1.662) =   -0.662

Sum of the
absolute differences
= Sum 1   2.386   This is the result
of Iteration 1-1
(i.e., Iteration 1
for Eigenvector 1)

Second Iteration

1.000 - (1.000)   =   0.000

-0.724 - (-1.093) =   +0.369

1.662 - (1.078)   =   0.584

Sum 1   =   0.953   Iteration 2-1

Step 5

If SUM 1 is greater than 0.00001 replace the original trial vector Y(J) with the new calculated trial vector VEC(J,L) and repeat Stage 2 using Y(J) for VEC(J,L):

6   7   8

COR(J,K)   Y(J)   X(J)   EIG(L)   VEC(J,L)

$$
\begin{bmatrix} 1.000 & -0.855 & 0.367 \\ 0.855 & 1.000 & -0.516 \\ 0.367 & -0.516 & 1.000 \end{bmatrix}
\times
\begin{bmatrix} 1.00 \\ -0.724 \\ 1.662 \end{bmatrix}
=
\begin{bmatrix} 2.228 \\ -2.436 \\ 2.402 \end{bmatrix}
\begin{bmatrix} 2.228 \end{bmatrix}
\begin{bmatrix} 1.00 \\ -1.093 \\ 1.078 \end{bmatrix}
$$

Steps 6, 7, 8

Steps 6, 7 and 8 are basically repetitions of Steps 2, 3 and 4. New values for X(J), EIG(J) and VEC(J,L) are obtained as in the example above.

Step 9

Calculate the absolute sum of the difference between the new Y(J) and VEC(J,L) vectors as in Step 4:

$$1.00 \quad - (+1.000) = \quad 0.0$$

$$-0.750 - (-1.103) = \quad +0.353$$

$$2.2 \quad - (+1.213) = \quad \underline{+0.987}$$

Sum 2   = $\underline{1.340}$ Result of Iteration 2-1 (Iteration 2 for Root 1)

Step 10

Test whether the difference between Sum 1 and Sum 2 is greater than 0.00001. If it is, then replace vector Y(J) from Step 5 with the new vector VEC(J,L) from Step 8. Repeat Steps 6 to 9. Continue the iterations until the test shows a value smaller than 0.00001. When this occurs the largest eigenvalue has been found. In this example stability, or the convergence of the test Y(J) and VEC(J,L), occurs on the 12th iteration when the following values are found:

X(J)   EIG(L)   VEC(J,L)

$$
\begin{bmatrix} 2.185 \\ -2.309 \\ 1.681 \end{bmatrix}
\begin{bmatrix} 2.185 \end{bmatrix}
\begin{bmatrix} 1.00 \\ -1.056 \\ 0.769 \end{bmatrix}
$$
    Sum 1 = 0.000007 Iteration 12

Hence the first Eigenvalue is 2.11.   The first unstandardized
Eigenvector VEC(J,L) is:

$$\begin{bmatrix} 1.000 \\ -1.056 \\ 0.769 \end{bmatrix}$$

It may be noted that convergence is frequently speeded up if the
original correlation matrix is squared a number of times (Harman 1968
p.147)

Step 11

Normalize the eigenvectors so that the sum of their squares equal
1.0. This is achieved by: (a) squaring the vector entries and summing
these values; (b) taking the square root of their sums; (c) dividing
this value into each of the standardized vector entries:

(a) Square the vector entries, i.e., $1.00 \times 1.00 =$ 1.000

$-1.056 \times -1.056 =$ 1.115

$0.769 \times 0.769 =$ 0.591

Sum totals $\sum =$ 2.706

(b) Sum the squares and take
     its square root, i.e. $\sqrt{2.706}$ $=$ 1.645

(c) Divide VEC(J,L) by the square
     root in (b) to obtain the
     normalized vectors, i.e. $1.00 \div 1.645 =$ $\begin{bmatrix} 0.608 \\ -0.642 \\ 0.467 \end{bmatrix}$

$-1.056 \div 1.645$

$0.769 \div 1.645$

The last column is the final eigenvector 1

Step 12

These calculations can be checked by making use of the fact that
$(R - \lambda_i I)v_L = 0.0$ (Harman 1968 p.144; Cooley and Lohnes 1971 p.99)
where: R is the correlation matrix; λ_L is the eigenvalue L; I is
the identity matrix (a matrix with 1.0 in the diagonals and zero
entries elsewhere); V_L is the eigenvector L.   The $(R - \lambda_L I)$ matrix
is calculated by taking the eigenvalues away from the diagonal entries
in the correlation matrix.

$$(R - \lambda_L I) \qquad \times \qquad V_L \quad = \quad 0.0$$

$$\begin{bmatrix} (1.00 - 2.18) & -0.855 & 0.367 \\ -0.815 & (1.00 - 2.18) & -0.516 \\ 0.367 & -0.516 & (1.00 - 2.18) \end{bmatrix} \times \begin{bmatrix} 0.608 \\ -0.642 \\ 0.467 \end{bmatrix} = \begin{bmatrix} 0.0 \\ 0.0 \\ 0.0 \end{bmatrix}$$

e.g.   Row 1 and Column 1

$((1.00 - 2.18) \times 0.608) + (-0.855 \times -0.642) + (0.367 \times 0.467)$

$= (0.717) + (+0.549) + (0.171) = 0.00$

N.B.   Rows will sum to 0.00 except for rounding errors.

Step 13

Calculate the Residual Correlations (CORR), namely the correlations left after the effect of the first eigenvalue has been removed. This is obtained by multiplying the eigenvector VEC(J,L) by its transpose, VEC(K,L), multiplying this product by the eigenvalue E(L) and taking this away from the original correlation matrix:

$\text{CORR} = \text{COR}(J,L) - (\text{VEC}(J,L) \times \text{VEC}(K,L)) \times E(L)$

$$\begin{bmatrix} 1.00 & -0.855 & 0.367 \\ -0.815 & 1.000 & -0.516 \\ 0.367 & -0.516 & 1.000 \end{bmatrix} - \begin{bmatrix} 0.608 \\ -0.642 \\ 0.467 \end{bmatrix} \times \begin{bmatrix} 0.608 & - 0.642 & + 0.467 \end{bmatrix} \times 2.185$$

For the first row (columns 1 to 3)

$1.00 - ((0.608 \times 0.608) \times 2.185) \quad = \quad 1.000 - (0.370 \times 2.185)$

$$= \quad 1.000 - (0.808)$$

$$= \quad +0.192$$

The final matrix, a Residual Correlation matrix, after the extraction of the first eigenvector is:

$$\begin{bmatrix} 0.192 & -0.002 & -0.254 \\ -0.002 & +0.099 & 1.140 \\ -0.254 & 0.140 & 0.522 \end{bmatrix}$$

Calculate the second eigenvalue and eigenvector by the procedures described above in Steps 2 to 11. Begin by assuming a value of 1.00 for all entries in the second eigenvector using the residual correlation matrix (CORR) instead of the original correlations (COR).

$$
\begin{array}{ccccc}
\text{CORR(J,K)} & & Y(J) & = & X(J) \quad E(L) \quad \text{VEC(J,L)} \\
\begin{bmatrix} 0.193 & -0.002 & -0.254 \\ -0.002 & 0.099 & 0.140 \\ -0.254 & 0.140 & 0.522 \end{bmatrix} & x & \begin{bmatrix} 1.00 \\ 1.00 \\ 1.00 \end{bmatrix} & = & \begin{bmatrix} -0.063 \\ 0.237 \\ 0.408 \end{bmatrix} \begin{bmatrix} -0.063 \end{bmatrix} \begin{bmatrix} 1.000 \\ -3.741 \\ -6.435 \end{bmatrix}
\end{array}
$$

Check the convergence of the eigenvectors by Step 10 above:

$$1.00 - 1.000 = 0.000$$

$$1.00 - (-3.741) = 4.741$$

$$1.00 - (-6.435) = \underline{7.435}$$

Sum of absolute difference =
Sum 1   12.176   Result of
                                          Iteration 1-2

Repeat Steps 5-10 above:

$$
\begin{array}{ccc}
X(J) & E(L) & \text{VEC(J,L)} \\
\begin{bmatrix} 1.837 \\ -1.276 \\ -4.138 \end{bmatrix} & \begin{bmatrix} 1.837 \end{bmatrix} = & \begin{bmatrix} 1.000 \\ -0.695 \\ -2.253 \end{bmatrix}
\end{array}
$$

Sum of absolute difference =
Sum 2   7.227   Result of
                                          Iteration 2-2

Iterate until convergence. This is achieved on the 10th iteration when Eigenvalue 2 equals 0.687.

The standardized values for Eigenvector 2 are:

$$
\begin{bmatrix} 0.448 \\ -0.209 \\ -0.869 \end{bmatrix}
$$

Check the calculations of the second Eigenvector as in Step 12, calculate the residual correlation matrix after the extraction of the second Eigenvalue (from the first residual matrix) as in Step 13. Calculate the third Eigenvector and Eigenvalue as in Step 14.

In this example, convergence for the third eigenvector is achieved after the third iteration when:

Eigenvalue 3 = 0.128 and Eigenvalue 3 = 0.656

0.738

0.161

The final results of the calculation of the eigenvalues and standardized eigenvectors are, therefore:

Eigenvectors			Eigen-values	Square Root of Eigen-values	Recip. of Sq.Ft. of Eigen-values	Percent Variance Explana-tion	Determin-ant of the Correla-tion Matrix
1	2	3					
0.608	0.448	0.656	1) 2.185	1.478	0.677	72.83%	0.1922
0.642	−0.209	0.738	2) 0.687	0.829	1.206	22.90%	
0.468	−0.869	0.161	3) 0.128	0.358	2.793	4.27%	

Step 16

Several derivations of these results can now be obtained.

(a) The amount of variance extracted by each eigenvalue can be calculated by dividing the eigenvalue by the number of variables in the original matrix, and multiplying by 100: i.e., (2.185 ÷ 3) x 100 = 72.83%. This is the variance for Eigenvalue 1.

(b) The determinant of the matrix can be obtained by multiplying the eigenvalues together: 2.185 x 0.687 x 0.128 = 0.1922.

Step 17

The Factor Structure matrix (the correlation of the variables with the factor or component axes) is derived from standardized eigenvector matrices as follows:

(a) Take the square root of the eigenvalues and put these in the diagonals of a matrix that otherwise contains zero entries.

(b) Post multiply the eigenvector matrix by this new matrix to obtain the Component Structure matrix.

	Eigenvectors			Square Root of Eigenvalues in diagonal				Component Structure	

$$
\begin{bmatrix} 0.608 & 0.448 & 0.656 \\ -0.642 & -0.209 & 0.738 \\ 0.468 & -0.869 & 0.161 \end{bmatrix} \times \begin{bmatrix} 1.478 & 0.00 & 0.00 \\ 0.00 & 0.829 & 0.00 \\ 0.00 & 0.00 & 0.358 \end{bmatrix} = \begin{bmatrix} 0.898 & 0.371 & 0.235 \\ -0.949 & -0.173 & 0.264 \\ 0.691 & -0.720 & 0.058 \end{bmatrix}
$$

For example, Row 1 of the Factor (Loading) Matrix is calculated as follows:

$(0.608 \times 1.478) + (0.448 \times 0.00) + (0.656 \times 0.00) = 0.898$

$(0.608 \times 0.00) + (0.448 \times 0.829) + (0.656 \times 0.00) = 0.371$

$(0.608 \times 0.00) + (0.448 \times 0.00) + (0.656 \times 0.358) = 0.235$

The Factor (Component) Coefficient matrix can be calculated in a similar manner to Step 17. This time the reciprocals of the square root of the eigenvalues ($\dfrac{1}{\sqrt{EIG_L}}$) replace the square root of the eigenvalues in the diagonals of the matrix. The calculations are therefore:

	Eigenvectors		x	Square Root of Reciprocals of EIG(L)			=	Factor Coefficients	

$$
\begin{bmatrix} 0.608 & 0.448 & 0.656 \\ -0.642 & -0.209 & 0.738 \\ -0.468 & -0.869 & 0.161 \end{bmatrix} \times \begin{bmatrix} 0.677 & 0.0 & 0.0 \\ 0.0 & 1.206 & 0.00 \\ 0.0 & 0.0 & 2.793 \end{bmatrix} = \begin{bmatrix} 0.411 & 0.541 & 1.832 \\ -0.434 & -0.252 & 2.062 \\ 0.316 & -1.049 & 0.449 \end{bmatrix}
$$

Step 18

The Factor Score matrix can be calculated by multiplying the matrix of standard scores $S(I,J)$ by the factor coefficient matrix according to the normal row by column multiplication rules. This example began with the correlation matrix so we have to produce the standard scores from the original data matrix to demonstrate the procedures involved. For example:

	Standard Scores			x	Factor Coefficients			=	Factor Scores
Variables:	1	2	3		1	2	3		
Place 1)	1.892	−1.102	0.514		0.411	0.541	1.832		
2)	−0.277	0.297	−1.149	x	−0.434	−0.252	2.062		
3)	−1.609	1.362	−0.506		0.316	−1.049	0.449		
4)	−1.231	0.381	0.391						
.									
.									
.									
.									
.									
.									
.									
28)	1.148	−0.463	1.376						

		Factor Scores		
		1	2	3
Place	1)	1.419	0.761	1.42
	2)	−0.606	0.981	−0.411
	3)	−1.413	−0.683	−0.368
	4)	−0.549	−1.172	−1.290
	.			
	.			
	.			
	.			
	.			
	.			
	.			
	28)	1.108	−0.706	1.767

For example:

for Place 1, Factor 1, the score is:

$$1.892 \ (0.411) + -1.102 \ (-0.434) + 0.514 \ (0.316)$$

$$= \ 0.778 + 0.478 + 0.163 \ = \ 1.419$$

Step 19

This is the end of this particular simple example. It is usual to go further than the calculation of the initial component solution. Typically one makes some decision about the number of meaningful components, then one rotates these axes to a simpler structure followed by the calculation of the rotated component scores. Alternatively one can adopt a common factor testing framework and use some of the many common factoring procedures described in Chapter V.

Now that the arithmetic associated with a simple example has been completed the next step is to learn the 'language' of matrix algebra in which these calculations are usually expressed. In this way an investigator will quite quickly obtain a working knowledge of the statistical foundation of factor analysis methods. This, of course, is beyond the objectives of this study.

APPENDIX B
TRANSFORMATION
PROCEDURES

The assumption of normality is basic to the use of parametric statistics such as the standard Pearson's Product Moment correlation coefficient. This means that if variables are normally distributed a greater range of statistical procedures can be applied to data sets. As a result one of the first procedures to be carried out in data screening prior to the adoption of multivariate methods such as factor analysis is the inspection of the frequency distribution of variables to see if they are normally distributed. Although it is assumed in this study that investigators have at least a basic course in statistics it is worth reviewing the procedures for measuring the character of distributions before showing how transformations can be carried out.

The simplest approach is measuring the character of a distribution of a variable is to divide the distribution into a set of equal intervals. The number of occurrences in each of these categories is counted and placed in a table. Table B.1 shows this approach for four different distributions called N, R, L and F. These values can also be portrayed visually by a histogram or graph. Figure B.1 shows the distribution for N. Figure B.3a, b c shows the distributions for the other variables.

Most variables have very different means, may have dissimilar variabilities and are often measured in various measurement units. So it has been found convenient to use a standardization procedure to remove these sources of variation. Standard scores represent one example of the approach. The standard score (z_i) is calculated by taking each observation (x_i) from the mean for that variable (x_i), and dividing by its standard deviation (σ_x). In symbolic terms:

$$z_i = \frac{x_i - \bar{x}}{\sigma_x} \quad \text{where:} \quad \sigma_x = \sum_{i=1}^{n} \sqrt{\frac{(x - \bar{x})^2}{n}}$$

σ_x = the square root of the mean of the sum of the squared deviations from the mean.

One advantage of the conversion of observations to standard scores is that the original values are expressed in deviation units away from a zero mean rather than in the original units of measurement. This makes it easier to compare individual values. Another advantage is that since statisticians have calculated the frequency with which individual

Table B.1
Four distributions: number of values per category

Categories	Number of Observations in Categories			
Midpoints	N(C)	R	L	F
35	1(1)	5	0	5
45	2(3)	10	1	6
55	4(7)	22	2	7
65	9(16)	22	4	9
75	15(31)	16	4	11
85	19(50)	9	5	12
95	19(69)	5	9	12
105	15(84)	4	16	11
115	9(93)	4	22	9
125	4(97)	2	22	7
135	2(99)	1	10	6
145	1(100%)	0	5	7
Mean	90	69.7	110.3	90

(C) shows the cumulative distribution of N.

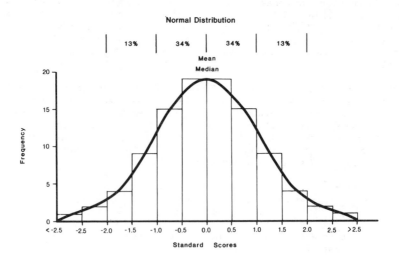

Figure B.1   Graph of distribution N

cases should occur in a normal distribution, it is easier to compare actual distributions with the theoretical ones. Standard statistical texts in geography (King 1969; Gregory 1978) show the so-called areas under the Normal Curve, which display the frequency of values in a normal distribution as probabilities. The whole occurrence, or the area under the normal curve, is given a value of 1.00 and the probability of an observation occurring in any of the various score categories is given as a frequency, or decimal of 1.00. For example, the probability of an observation falling between the mean, 0.0, and +0.5 Standard Deviations is 0.1915. In other words 19.15% of the observations should lie in this category in a perfect normal distribution. Taking into account both sides of the curve it means that 38.30% of the observations in a normal distribution lie within +0.5 standard deviation of the mean, whilst the probability of occurrence within +2.5 Standard Deviation of the mean is 80.9876. From these probabilities the frequency of occurrence above or below any level can also be calculated. For example, 3% of the cases (0.03 probability) are greater than +2.0 Standard Deviations from the mean and 3% greater than −2.0 Standard Deviation, whilst 68% of the cases are within +1.0 Standard Deviations from the mean. If one allows for the whole number rounding errors in Table B.1 distribution N is a normal curve, as can be seen grpahically by the familiar bell-shaped curve shown in Figure B.1. Here the mean (or average), median (middle value) and mode (highest class) all coincide at the apex of the curve.

In practice, distribution curves come in various degrees of regularity. If we ignore for a moment the J shaped curves, or irregular curves in which two or more humps occur (bi-modal curves), any bell-shaped curve can be altered. It can vary either by the mode being pushed to one side and the tail drawn out, or by being squashed or flattened, or by any combination of these processes. This means that additional measurements other than the mean and standard deviation are needed to take these possible characteristics into account. These are the so-called third and fourth moments of distribution, namely the skewness and kurtosis.

Skewness measures the degree to which a distribution is 'off-centre' and is measured by the division of the differences between the mean and median by the standard deviation:

$$\text{Skewness} = \frac{\text{Mean} - \text{Median}}{\text{Standard Deviation}}$$

In a normal distribution skewness will be zero. If the skew is negative (left hand) in a unimodal distribution, as in Figure B.2b or L in Table B.1, the median must be a bigger value than the mean, and this can only occur if the mean is on the left of the median, and the tail of the distribution is drawn out to the right. Not surprisingly, negatively skewed distributions are referred to as showing right skew. If the skew is positive, the distribution is the opposite to the one described. The tail is drawn out to the left and is called a right or right hand skew, or positive skew.

Kurtosis measures the degree of 'peak' or the flatness of the distribution.

In a normal distribution the Kurtosis will attain a value of 3.0. Values less than 3.0 indicate a distribution that is more peaked than would be the case in a normal curve. Values greater than 3.0 describe a flat or spread-out distribution.

A preliminary indication of the degree of normality of any distribution is provided by studying the relationships between the mean and median and calculating the values for Skewness and Kurtosis.

Table B.2

Measures of the four moments

Moments	Distributions			
Median	N	R	L	F
1. Mean	90.0	69.7	110.3	90.0
2. Standard Deviation	20.8	27.7	21.7	29.9
3. Skew	0.0	0.8	−0.8	0.0
4. Kurtosis	3.0	2.5	2.5	3.8

In Table B.2 the distribution of R displays a right hand or negative skew, and L shows a left hand or positive skew. Distribution F, on the other hand, is a flat distribution, one that has a very small peak.

A more precise way of working out the degree of normality of a curve is to make use of the cumulative frequency of the values of a distribution. A normal distribution will produce an S shaped curve when plotted on an arithmetic scale. But when the cumulative distribution is plotted on probability paper in which the vertical scale is logarithmic, the result is a straight line. Distribution N in Figure B.2 shows this feature, whereas regular distributions such as R and L displaying right skews and left skews respectively by being located on either side of this straight line. The reason for such a relationship is easily understood. Positive, or right skewed distributions have too many values on the negative side of the mean, and too few on the positive side or right side of the mean so that they will lie below the normal curve. The opposite applies to negative distributions.

These two methods only provide a shorthand, or visual portrayal of the departure of any distribution from normality. In the last resort any investigator should always test the degree of normality of the distribution in probability terms. An accepted approach is to use the chi squared statistic (χ^2) which is expressed as:

$$\chi^2 = \sum_{i-1}^{n} \frac{(O_i - E_i)^2}{E_i}$$

Where   n   =   number of categories
         n-1 =   number of degrees of freedom
         O   =   Observed frequency in category i
         E   =   Expected frequency in category i

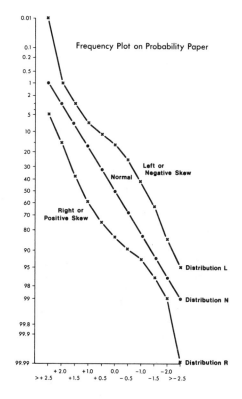

Figure B.2   Frequency plot of distributions

The expected distribution used in the statistic are the values obtained from the table of areas under the normal curve. The appropriate calculations for distribution R are shown in Table B.3 where it can be seen that the value of chi squared is 159.86. Tables for chi-squared can be found in most elementary statistics books. They show that with eleven degrees of freedom (the number of categories n, minus 1) the probability of the distribution of values of R having occurred by a chance deviation from the normal curve is less than 0.1%. So in 99.9% of the cases distribution R cannot be considered to represent a normal distribution. Similar calculations can be made for distributions N, R and F. In 99% of the cases N can be considered as a normal distribution.

	Frequency in Categories											
	35	45	55	65	75	85	95	105	115	125	135	145
1. Observed(O)	5	10	22	22	16	9	5	4	4	2	1	0
2. Expected(E)	1	2	4	9	15	19	19	15	9	4	2	1
3. Difference (O–E)	4	8	18	13	1	10	14	11	5	2	1	1
4. Squared Difference $(O-E)^2$	16	64	324	169	1	100	196	121	25	4	1	1
5. $\dfrac{(O-E)^2}{E}$ for each cell	$\dfrac{16}{1}$	$\dfrac{64}{2}$	$\dfrac{324}{4}$	$\dfrac{169}{9}$	$\dfrac{1}{15}$	$\dfrac{100}{19}$	$\dfrac{196}{19}$	$\dfrac{121}{15}$	$\dfrac{25}{9}$	$\dfrac{4}{4}$	$\dfrac{1}{2}$	$\dfrac{1}{1}$

$$\sum_{i-1}^{n} \frac{(O_i - E_i)^2}{E_i} = 159.86 \text{ for Distribution R}$$

After carrying out these tests an investigator may find that some of his variables are not normally distributed. This does not necessarily mean that the variables should be excluded. The assumption of normality simply makes it possible to use more statistical procedures in the study. When normality is not present the investigator is confronted with three choices:

(i) The variables could be used in the raw state but the investigator accepts that correlations, as measures of linear association will underestimate the relationships, and inferential statistics associated with many techniques should not be used. For example, this would exclude the use of tests of significance of factor loading in factor analysis. It should be recognized that the use of the raw data in a linear model means that much of the information contained in the distributions will be lost.

(ii) A second alternative is that the variables can be transformed to a different measurement scale. This may have the effect of moving the distribution closer to normality. For example, the standard score curve could be divided into an equal number of frequencies containing a fixed percentage of the cases. If 10 categories are used, each containing 10% of the cases and every section is given a rank value, 1, 2, 3, 4, 5, ... 9, 10, all the values of the actual distribution falling into these categories could be replaced by the rank value. If the first ten percent of the cases should occur between −1.282 Standard Deviations from the mean and the negative infinity, all cases should be given a value of 1. Every observation in the second ten percent, between −1.282 and −0.842 Standard Deviations will be given a value 2. Completing this process will produce a transformation of the interval scale data into a set of normally distributed rank order values. The obvious problem is that the transformed data set does not reproduce the original gaps between the original values, so part of the information of the set is lost.

(iii) The third alternative is another type of transformation. This time the same scale of measurement is kept. The procedure is most easily understood by referring back to the distribution R, the one with a right hand skew. Figure B.3 shows that if a way can be found of squeezing the tail of high values towards the middle the distribution could be altered into one that is closer to a normal curve. Similarly, the squeezing together of the low values of the left hand tail in positively skewed distributions, and the drawing-out of the bunching of high values on the right would produce the same effect for distribution L.

In Figure B.3 a set of different transformations are applied to distributions L, R and F. In the case of R the transformation to a distribution closer to the bell-shaped curve can be attained by transforming the original values of the right or positive skew to a logarithmic scale (Log. 10). This comes closest though the fit is not exact, given the 'tail' to the right, but it is closer than the square root, cube root, or reciprocal transformations. The load of calculation in the example has been lightened by assuming that all the values for distribution R are in fact concentrated at the midpoint of the categories. So there are 5 examples of 35, 10 examples of 45, etc. Hence, the values 35 and 45 will be replaced by their logarithm, namely 1.544 and 1.653, whereas 135 and 145 will be replaced by 2.130 and 2.161. The key in Table B.3 shows the new values. So instead of the constant arithmetic difference of ten units between the two sets of numbers the differences become 0.109 and 0.03. This means that the high values of any distribution are squeezed together and the low values are spread out when various root transformations are used. The reverse effect occurs when the numbers are powered, for example by changing to the square, or the cube. Figure B.3b shows the effect of the squeezing together of the low values and the pulling apart of the higher number by squaring in the case of distribution L.

Figure B.2 shows the effect of a series of different types of transformation on the positive and negatively skewed distributions. In the case of R, the right skewed distribution, it is apparent that the effect of successive transformations, by the square root, cube root, logarithm, and reciprocal, progressively shifts the mode of the distribution to the right and 'draws-in' the tail of the frequency curve. The investigator is able to choose which of these transformations is either closer to normality, perhaps by using the chi-squared test, or is at least near enough to normality to be satisfactory for the analysis. One point worth noticing in Figure B.2a is that the reciprocal of the values has proved to be too powerful in the case of distribution R. In this case it has changed the right hand skew to an almost left hand skew. Similarly, the square and the cube of the values progressively move the negative or left hand skew to the left by

371

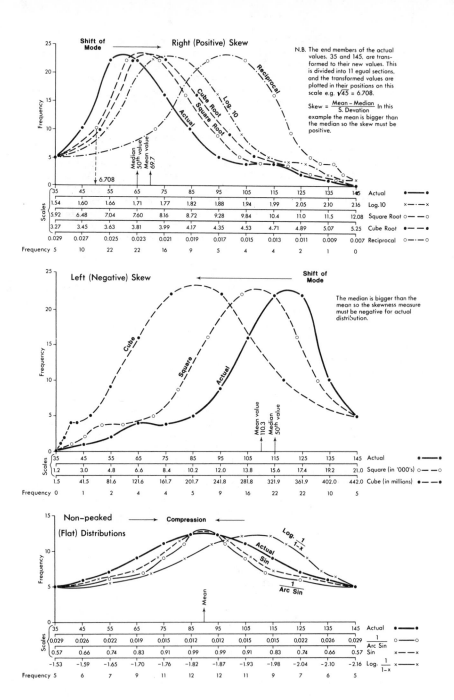

Figure B.3 Effect of various transformations

drawing in the tail of the distribution. Higher powering moves the distribution even further in this direction.

These two examples take care of skewed distributions. But the problem is rather different where the departure from normality is one of more or less Kurtosis. In the case of flat distributions the need is to squeeze the middle range of values upwards. Figure B.2c shows the example of a flat or non-peaked distribution called F in Table B.1. In this case the appropriate transformation involves the use of sin or cos curves. Any trigonometry text will show the positive unimodal character of the sin curve up to 90°, and its negative unimodal from 91° to 180°. So the result of transforming the midpoint values to their sin, and plotting the frequency distribution of F, is to squeeze in the middle range of values. This would produce a curve closer to the normal one. An even better compression occurs from the use of the reciprocal of the arc sin (1/arc sin). A third value, Log 1/1-x, was calculated to show how more complicated types of transformation can be used. This squeezes the low values of the original curve together but produces an overall skew to the left. Obviously flat distributions with negative skews could benefit from this type of transformation.

These examples demonstrate that the consequence of using various types of common transformations to a set of distributions. It is clear that extreme care has to be taken in transforming variables for trans-formations, if mis-applied, could take a distribution further from normality. Certainly it makes little sense to apply a single trans-formation type to a large set of different variables.

REFERENCES

Abiodun, J.O., (1967), 'Urban Hierarchy in a Developing Country', Economic Geography, 4, pp.347-67.

Abrahamson, (1974), 'The Social Dimensions of Urbanism', Social Forces, 52, pp.376-383.

Abu-Lughod, J., (1969), 'Testing the Theory of Social Area Analysis', American Sociological Review, 34, pp.198-212.

Abu-Lughod, J., (1969), 'Varieties of Urban Experience: Contrast, Coexistence and Coalescence in Cairo', in I.M. Lapidus (ed.), Middle Eastern Cities, University of California Press, Berkeley.

Adams, J.S., (1976), (ed.), Contemporary Metropolitan American: Twenty Geographical Vignettes, Ballinger, Cambridge, USA.

Adrian, C.R., (1972), 'A Comparative Typology of Urban Government Images, in B.J.L. Berry (ed.), City Classification Handbook, J. Wiley, New York.

Aigner, D.J., (1968), Principles of Statistical Decision Making, Prentice Hall, New Jersey.

Aigner, D.J., (1971), Basic Econometrics, Prentice Hall, New Jersey.

Aitchison, J.W., (1974), 'The Farming Systems of Wales', paper presented to I.G.U. Symposium on Agricultural Typology, Verona, Italy.

Aitchison, J.W., (1978), 'Classification and Mixed Mode Data', Cambria Welsh Geographical Journal, 5(2), pp.144-153.

Aitken, A.C., (1937), 'The Evaluation of the Latent Roots and Vectors of a Matrix', Proceedings of Royal Society of Edinburgh, 57, pp.269-304.

Alford, R.R., (1972), 'Critical Evaluation of the Principles of City Classification', in B.J.L. Berry (ed.), City Classification Handbook, J. Wiley, New York, pp.331-58.

Alfred, R.A., 'Critical Evaluation of the Principles of City Classification', in B.J.L. Berry and K.B. Smith (ed.), City Classification Handbook, J. Wiley, 1972, pp.331-359.

Alihan, M.A., (1938), Social Ecology, New York.

Alker, H.R., (1969), 'A Typology of Ecological Fallacies', in M. Dogan and S. Rokkan (ed.), Quantitative Ecological Analysis in the Social Sciences, MIT Press, Cambridge.

Alonso, W., (1964), Location and Land Use: Toward a General Theory of Land Rent, Harvard University Press, Cambridge.

Ames, H.B., (1898), The City Under the Hill, reprinted, University of Toronto Press, 1973.

Anderson, T.R. and Bean, L.L., (1961), 'The Shevky-Bell Social Areas: Confirmation of Results and Reinterpretation', Social Forces, 40, pp.119-124.

Anderson, T.R. and Egeland, J.A., (1961), 'Spatial Aspects of Social Area Analysis', American Sociological Review, 26, 1961, pp.392-398.

Armstrong, J.C., (1967), 'The Derivation of Theory by Factor Analysis, or Tom Swift and His Electronic Factor Analysis Machine, American Statistician, 21, p.17-21.

Arnold, D.S., (1972), 'Classification as Part of Urban Management', in B.J.L. Berry (ed.)., City Classification Handbook, J. Wiley, New York, pp.361-77.

Artibise, A.F.T., (1975), Winnipeg: A Social History, Queen's University Press.

Atkin, R.H., (1974), Mathematical Structure in Human Affairs, London.

Atkin, R.H., (1975), 'An Approach to Structure in Architectural and Urban Design', Envirionment and Planning, B, 2 pp.21-57.

Aurousseau, M., (1921), 'The Distribution of Population', Geographical Review, 11, p.563.

Baggaley, A.R. and Cattell, R.B., (1956), 'A Comparison of Exact and Approximate Linear Function Estimates of Oblique Factor Scores', British Journal of Statistical Psychology, 9, pp.83-86.

Bagley, C., (1965), 'Juvenile Delinquency in Exeter: An Ecological and Comparative Study', Urban Studies, 2, pp.35-50.

Bagley, C., Jacobson, S. and Rehin, A., (1976), 'Completed Suicide: A Taxonomic Analysis of Clinical and Social Data', Psychology and Medicine, 6, pp.429-438.

Balchin, W.G.V., (1972), 'Graphicacy', Geography, 57, pp.183-196.

Banks, A.S., and Gregg, P.M., 'Grouping Political Systems: Q-Factor Analysis of a Cross Polity Survey', American Behavioral Scientist, 9, 1965, pp.3-5.

Bargman, R. and Brock, R.D., (1966), 'Analysis of Covariance Structures', Psychometrika, 31, pp.507-534.

Barker, D., (1976), 'Hierarchic and Non Hierarchic Grouping Methods', Geographicka Annale B, pp.42-58.

Barker, D., (1978), 'A Conceptual Approach to the Description and Analysis of an Historical Urban System', Regional Studies, 12, pp.1-10.

Bartlett, M.S., (1951), 'A Further Note on Tests of Significance in Factor Analysis', British Journal of Psychology, 4, p.1.

Barton, K. and Cattell, R.B., (1975), 'An Investigation of the Common Factor Space of Some Well Known Questionnaire Scales', Journal of Multivariate, Experimental and Clinical Psychology, I, pp.268-277.

Bauer, R.A. (ed.), (1966), Social Indicators, M.I.T. Press, Cambridge, Massachusetts.

Beaumont, J.R. and Beaumont, C.D., (1981), A Comparative Study of the Multivariate Structure of Towns, Dept. of Geography, University of Keele, N. Staffs.

Bell, C.R., (1968), Middle Class Families, Routledge and Kegan Paul, London.

Bell, Daniel, (1974), The Coming of Post-Industrial Society, Heinemann, London.

Bell, Wendell and Force, Maryanne, (1967), 'Urban Neighbourhood Types and Participation in Formal Associations', American Sociological Review, 21, pp.23-33.

Bell, W., (1955), 'Economic, Family and Ethnic Status: An Empirical Test', American Sociological Review, 20, pp.45-52.

Bell, W., (1958), 'The Utility of the Shevky Typology for the Design of Urban Sub-area Field Studies', Journal of Social Psychology, 47, pp.71-83.

Bell, W. and Greer, S., (1962), 'Social Area Analysis and Its Critics', Pacific Sociological Review, 5, pp.3-9.

Bell, W. and Moskos, C.C., (1964), 'A Comment on Udry's Increasing Scale and Spatial Differentiation', Social Forces, 42, pp.414-417.

Berry, B.J.L. and Spodek, H., 'Comparative Ecologies of Large Indian Cities', in K.P. Schwirian (ed.), Comparative Urban Structure, D.C. Heath, Lexington, USA.

Berry, B.J.L., (1960), 'An Inductive Approach to the Regionalization of Economic Development', in Norton Ginsburg (ed.), Essays on Geography and Economic Development, Department of Geography, University of Chicago, Research Paper No. 62.

Berry, B.J.L., (1964), 'Approaches to Regional Analysis: A Synthesis', Annals Association American Geographers, 54, pp.2-11.

Berry, B.J.L. and Murdie, R.A., (1965),Socio-economic Correlates of Housing Condition, Metropolitan Planning Board, Toronto.

Berry, B.J.L. and Tennant, R.J., (1965), 'Socio-economic Classification of Municipalities in the Northeastern Illinois Metropolitan Area', Commercial Structure, Northeastern Illinois Metropolitan Area Planning Commission, Chicago.

Berry, B.J.L., (1966), 'Interdependency of Flows and Spatial Structure: A General Field Theory Formulation', in B.J.L. Berry, et al (ed.), Essays on Commodity Flows and the Spatial Structure of the Indian Economy, Department of Geography, University of Chicago, Research Paper No. 111.

Berry, B.J.L., et al, (1966), Indian Commodity Flows: Studies in the Spatial Structure of the Indian Economy, Department of Geography, University of Chicago, Research Paper No. 111.

Berry, B.J.L., (1968), 'Interdependency of Spatial Structure and Spatial Behavior: A General Field Theory Formulation', Papers of the Regional Science Association, 21, pp.207-227.

Berry, B.J.L. and Prakasa Rao, V.L.S., (1968), 'Urban-Rural Duality in the Regional Structure of Andhra Pradesh: A Challenge to Regional Planning and Development', Geograpische Zeitschrift, No. 21.

Berry, B.J.L., (1969), 'Relationships between Regional Economic Development and the Urban System: The Case of Chile', Tijdschrift voor Econ. Soc. Geografie, 60, pp.283-307.

Berry, B.J.L., and Rees, P.H., (1969), 'The Factorial Ecology of Calcutta', Amerian Journal of Sociology, 74, pp.445-491.

Berry, B.J.L. and Horton, F.E., (1970), Geographic Perspectives on Urban Systems, Prentice Hall, Englewood Cliffs, New Jersey.

Berry, B.J.L., (1971), 'Introduction. The Logic and Limitations of Comparative Factorial Ecology', Economic Geography, 4, June Supplement, pp.209-219.

Berry, B.J.L., (1972), The City Classification Handbook, J. Wiley, New York.

Berry, B.J.L., (1976), Urbanization and Counter Urbanization, Sage Publications, Beverly Hills, California.

Berry, B.J.L. and Kasarda, J., (1977), Contemporary Urban Ecology, Collier-MacMillan, New York.

Beshers, J., (1962), Urban Social Structure, Free Press, New York.

Bird, J.B., (1956), 'Scale in Regional Study', Geography, 41, pp.25-38.

Black, W., (1973), 'Towards a Factorial Ecology of Flows', Economic Geography, 49, pp.59-67.

Blalock, H.M., (1963), 'Correlated Independent Variables: The Problem of Multi-Collinearity', Social Forces, 42, pp.233-237.

Blalock, H.M. Jr., (1964), Causal Inferences in Nonexperimental Research, University of North Carolina Press, Chapel Hill.

Blalock, H.M., (1971), Causal Models in the Social Sciences, Aldine Atherton.

Blanchard, R., (1935), Grenoble: Etude de Geographie Urbaine, 2nd Edition, Grenoble, France.

Blau, P.M., (1974), 'Parameters of Social Structure, American Sociological Review, 39, pp.615-635.

Blau, P.M., (1976), Approaches to the Study of Social Structure, Open Books, London.

Bloom, L. and Shevky, E., (1949), 'The Differentiation of an Ethnic Group', American Sociological Review, 14.

Boal, F.W., (1976), 'Ethnic Residential Segregation', in D.T. Herbert and R.J. Johnston (ed.), Spatial Processes and Form, J. Wiley, New York.

Bogue, D.J., (1963), Skid Row in American Cities, The University of Chicago Press, Chicago.

Booth, C., (1893), 'Life and Labour of the People in London', Presidential Address, Journal of the Royal Statistical Society, 55(4), pp.557-591.

Booth, Charles, (1902-3), Life and Labour of the People in London, reprinted 1969, A. Kelley, New York.

Borgatta, E.R., (1965), 'Difficulty Factors and the Use of Phi Correlations', Journal of General Psychology, 73, p.321.

Bottomore, T., (1976), 'Structure and History', in P.M. Blau (ed.), op.cit., 1976, pp.159-171.

Bourne, L.S. and Barber, G.M., (1971), 'Ecological Patterns of Small Urban Centres', Economic Geography, 47(2), supplement, p.258-265.

Bourne, L.S., (1970), Dimensions of Metropolitan Land Use, University of Toronto Centre for Urban and Community Studies, No. 31.

Bourne, L.S. and Murdie, R.A., (1972), 'Interrelationships of Social and Physical Space in the City', Canadian Geographer, 16, pp.211-229.

Bowen, E.G., (1959), 'Le Pays de Galle', Transactions, Institute of British Geographers, 26, pp.1-25.

Bowen, E.G. and Carter, H., (1974), 'Preliminary Observations on the Distribution of the Welsh Language at the 1971 Census', Geographical Journal, 140(3), pp.432-440.

Briggs, R. and Leonard, W.A., (1977), 'Empirical Implications of Advances in Canonical Theory', Canadian Geographer, 21, pp.133-147.

Britton, J.N.H., (1973), 'The Classification of Cities: Evaluation of Q Mode Factor Analysis', Regional and Urban Economics, 2, pp.333-356.

Brown, L.A., and Horton, F.E., (1971), 'Social Area Change', Urban Studies, 7, pp.271-288.

Brown, L. and Holmes, J., (1971), 'The Delimitation of Functional Regions', Journal of Regional Science, 11, pp.57-72.

Browne, M.W., (1967), 'On Oblique Procrustes Rotation', Psychometrika, 32, pp.125-132.

Burgess, E.W., (1925), 'The Growth of the City', in R.E. Park et al. (1925), The City, University of Chicago Press, reprinted 1965, pp.37-44.

Burgess, E.W. and Bogue, D.J., (1964), Contributions to Urban Sociology, University of Chicago Press.

Burt, C.L., (1917), The Distribution and Relationships of Educational Abilities, P.S. King, London.

Burt, C.L., (1941), The Factors of the Mind: An Introduction to Factor Analysis in Psychology, MacMillan, London.

Burt, C.L., (1950), 'Tests of Significance in Factor Analysis', British Journal of Psychol. Statisticians, 5, pp.109-133.

Burt, C.L., (1966), 'The Appropriate Use of Factor Analysis and Analysis of Variance', in R.B. Cattell (ed.), Handbook of Multivariate Experimental Psychology, Rand McNally, Chicago, pp.267-287.

Buss, A., (1975), 'An Inferential Strategy for Determining Factor Invariance across Different Individuals and Different Variables', Multivariate Behavioural Research, 10, pp.365-372.

Buttimer, A., (1971), Society and Milieu in the French Geographical Tradition, Rand McNally.

Buttimer, A. and Seamon, D., (1980), The Human Experience of Space and Place, St. Martin's Press, New York.

Cant, R.G., (1971), 'Changes in the Location of Manufacturing in New Zealand 1957-68: An Application of Three Mode Factor Analysis', New Zealand Geographer, 27, pp.38-55.

Cant, R.G., (1975), 'Three Mode Factor Analysis as Applied to Industrial Location Data', in L. Collins and D.F. Walker (ed.), Locational Dynamics of Manufacturing Activity, Wiley, London.

Carroll, J.B., (1957), 'Biquartimin Criterion for Rotation to Simple Structure in Factor Analysis', Science, 126, p.1114.

Carroll, J.B., (1961), 'The Nature of the Data, or How to Choose a Correlation Coefficient', Psychometrika, 26, pp.347-372.

Carter, H. and Davies, W.K.D., (1970), Urban Essays: Studies in the Geography of Wales, Longmans, London.

Carter, H., (1974), 'Scale and the Dimensions of Socio-economic Variation', Geoforum, 19, pp.467-474.

Carter, H. and Wheatley, S., (1978), Merthyr Tydfil in 1851, S.S.R.C. Project Report, Geography Department, U.C.W., Aberystwyth.

Carter, H., (1981), The Study of Urban Geography, Ed. Arnold, Third Edition.

Castells, M., (1977), The Urban Question, Ed. Arnold, Second Edition, London.

Cattell, R.B., (1949), 'The Dimensions of Culture Patterns by Factorization of National Characters', Journal of Abnormal and Social Psychology, 44, pp.443-469.

Cattell, R.B. and Adelson, (1951), 'The Dimensions of Social Change in the U.S.A. as Determined by P Technique', Social Forces, 30, pp.190-201.

Cattell, R.B., (1952), Factor Analysis, Harper, New York.

Cattell, R.B., (1952), 'The Three Basic Factor Analytical Research Designs', Psychological Bulletin, 49, pp.499-520.

Cattell, R.B., (1957), 'A Universal Index for Psychological Factors', Psychologia, 1, pp.74-85.

Cattell, R.B., (1957), Factor Analysis: An Introduction to Essentials, 21, I. pp.190-215, II pp.405-435.

Cattell, R.B., (1960), 'Evaluating Interacton and Non Linear Relations by Factor Analysis', Psychological Reports, 7, pp.69-81.

Cattell, R.B. and Baggaley, A.R., (1960), 'The Salient Variable Similarity Index for Factor Matching', British Journal of Statistical Psychology, 13, pp.33-46.

Cattell, R.B., (1965), 'Higher Order Factor Structures', in C. Banks and P.L. Broadhurst (ed.), Studies in Psychology Presented to Cyril Burt, University of London Press.

Cattell, R.B., (1966), Handbook of Multivariate Experimental Psychology, Rand McNally.

Cattell, R.B., (1966), 'The Scree Test for the Number of Factors', Multivariate Behavioural Research, I, pp.140-161.

Cattell, R.B., Balcar, K.R., Horn, J.L. and Nesselroade, J.R., (1969), 'Factor Matching Procedures', Educational and Psychological Measurement, 29, pp.781-792.

Cattell, R.B., (1972), 'Real Base, True Zero Factor Analysis', Multivariate Behavioural Research Monographs, No. 72-1, Texas Christian University Press, Fort Worth.

Cattell, R.B. and Vogelmans, (1977), 'An Empirical Test ... of the Accuracies of the Scree and K-A Test for the Number of Factors', Multivariate Behavioural Research, 12.

Cattell, R.B., (1978), The Scientific Use of Factor Analysis, Plenum Press, New York.

Chapin, F.S., (1968), Urban Land Use Planning, Univ. of Illinois Press.

Child, D., (1970), The Essentials of Factor Analysis, Holt, Rinehart and Winston, London.

Christaller, W., (1932), Central Places in Southern Germany, translated by C. Barkin (1966), Prentice Hall, New Jersey.

Christofferson, (1975), 'Factor Analysis of Dichotomized Variables', Psychometrika, 40, pp.5-22.

Clark, D., (1971), Urban Linkage and Regional Structure, Unpublished Ph.D. Thesis, University of Wales.

Clark, D., (1973), 'Normality, Transformation and the Principal Component Solution', Area, 5, pp.110-113.

Clark, D., (1973), 'The Formal and Functional Structure of Wales', Annals, Association of American Geographers, 63(1), pp.71-84.

Clark, D., Davies, W.K.D. and Johnston, R.J., (1974), 'The Application of Factor Analysis in Human Geography', The Statistician, 23, pp.327-357.

Cliff, A.D. and Ord, J.K., (1973), Spatial Autocorrelations, Pion, London.

Cliff, N., (1970), 'The Relation Between Sample and Population Characteristics Vectors', Psychometrika, 35, pp.163-178.

Clignet, R. and Sween, J., (1969), 'Accra and Abidjan: A Comparative Examination of the Theory of Increase in Scale', Urban Affairs Quarterly, 5, pp.297-324.

Cole, J.P. and King, C.A.M., (1968), Quantitative Geography, J. Wiley, London.

Comrey, A.L. and Levonian, E., (1958), 'A Comparison of Three Point Coefficients in Factor Analysis', Educational and Psychological Measurement, 18, pp.739-755.

Comrey, A.L., (1973), A First Course in Factor Analysis, Academic Press.

Conway, F., (1967), Sampling for Social Scientists, G. Allen and Unwin, London.

Cooley, W.W., and Lohnes, P.R., (1962), Multivariate Procedures for the Behavioural Sciences, J. Wiley, New York.

Coughlin, R.E., (1973), 'Attainment along Goal Dimensions in 101 Metropolitan Areas', Journal American Institute of Planners, 39 (November), pp.413-425.

Cox, K.R., (1969), 'Voting in the London Suburbs', in M. Dogan and S. Rokkan (ed.), Quantitative Ecological Analysis in the Social Sciences, M.I.T. Press, Cambridge, pp.343-369.

Crawford, L.B. and Koopman, P., (1973), 'A Note on Horn's Test for the Number of Factors', Multivariate Behavioural Research, 8, pp.117-126.

Crawford, L.B., (1975), 'Determining the Number of Interpretable Factors', Psychological Bulletin, 82, pp.226-237.

Cureton, E.E. and Mulaik, S.A., (1975), 'The Weighted Varimax Rotation and the Promax Rotation', Psychometrika, 40, pp.183-196.

Davie, M.R., (1938), 'The Pattern of Urban Growth', in G.P. Murdock (ed.), Studies in the Science of Society, New Haven, U.S.A., reprinted in G.A. Theodorson (ed.), Studies in Human Ecology, Evanston, Illinois, 1961, pp.71-92.

Davies, F.M.T., (1971), Town Expansion Schemes in Cheshire and Staffordshire, Unpublished M.A. Thesis, University of Keele, England.

Davies, W.K.D. and Healey, D.M., (1977), 'Consistency and Change in the Ecological Patterns of Calgary 1961-71, in B.M. Barr (ed.), Research Studies by Western Canadian Geographers: The Edmonton Papers, Tantalus Press, Vancouver, pp.77-97.

Davies, W.K.D., (1968), 'Replication in Human Geography', Tijdschrift voor Economishe en Sociale Geografie, 59, pp.145-155.

Davies, W.K.D. and Robinson, G.W.S., (1968), 'The Nodal Structure of the Solent Region', Journal of Town Planning Institute, 54, pp.18-23.

Davies, W.K.D., (1970), 'Introduction to the Study of Urban Geography', in H. Carter, W.K.D. Davies (ed.), Urban Essays: Studies in the Geography of Wales, Longmans, U.K.

Davies, W.K.D., (1971), 'Varimax and the Destruction of Generality', Area, 3, pp.112-118.

Davies, W.K.D., (1971b), 'Varimax and Generality: A Reply to Mather', Area, 3, pp.254-259.

Davies, W.K.D., (1972), 'Conurbation and City Region in an Administrative Borderland: A Study of Greater Swansea', Regional Studies, 6, pp.217-236.

Davies, W.K.D., (1972), 'Data Analysis in Urban Geography', Geography, 57, pp.196-206.

Davies, W.K.D., (1972), The Conceptual Revolution in Geography, University of London Press.

Davies, W.K.D., (1972), 'Varimax and Generality: A Second Reply', Area, 4, pp.207-208.

Davies, W.K.D. and Barrow, G., (1973), 'A Comparative Factorial Ecology of Three Canadian Prairie Cities', Canadian Geographer, 17(4), pp.327-357.

Davies, W.K.D., (1973), 'Problems in Applied Factor Analysis', Proceedings of Institute of British Geographers Study Group in Quantitative Methods Working Paper No. 3, Birmingham, U.K.

Davies, W.K.D. and Lewis, G.J., (1973), 'The Urban Dimensions of Leicester, England', in B. Clark and B. Gleave (ed.), Social Patterns in Cities, Institute of British Geographers Special Publication No. 5, pp.71-86.

Davies, W.K.D. and Lewis, G.J., (1974), 'The Social Patterning of a British City, Leicester, England', Tijdschrift v. Econ. en Soc. Geografie, 65, pp.95-107.

Davies, W.K.D., (1975), 'A Multivariate Description of Calgary's Community Areas', in B. Barr (ed.), Calgary, Western Geographical Series, No. 11, University of Victoria, B.C., pp.231-269.

Davies, W.K.D., (1975), 'Variance Allocation and the Dimensions of British Towns', Tijdschrift v. Econ. en Sociale Geografie, 66, pp.358-372.

Davies, W.K.D., (1977), Factorial Ecology of Large Prairie Cities, Paper read at Annual Meeting of Canadian Geographers, Regina.

Davies, W.K.D. and Welling, S.L., (1977), 'The Socio-Economic Differentiation of Alberta Towns in 1971', in B.M. Barr (ed.), Research Studies by Western Canadian Geographers: The Edmonton Papers, Tantalus Press, Vancouver, pp.77-97.

Davies, W.K.D., (1977), 'Towns and Villages of Wales', Chapter 7 in D. Thomas (ed.), Wales: A New Study, David and Charles, Newton Abbot, pp.191-225.

Davies, W.K.D., (1978), 'Alternative Factorial Solutions and Urban Social Structure', Canadian Geographer, 22, pp.273-297.

Davies, W.K.D., (1978), 'Charles Booth and the Measurement of Urban Social Structure', Area, pp.290-296.

Davies, W.K.D. and Musson, T.C., (1978), 'Spatial Patterns of Commuting in South Wales: 1951-71', Regional Studies, 12, pp.353-366.

Davies, W.K.D., (1978), 'The Social Taxonomy of Edmonton's Community Areas in 1971', in P. Smith (ed.), Edmonton, Western Geographical Series, University of Victoria, B.C., pp.161-197.

Davies, W.K.D., (1979), 'Urban Connectivity in Montana', Annals Regional Science, 13, pp.29-46.

Davies, W.K.D. and Tapper, S., (1979), Urban Dimensionality of the Major American S.M.S.A. System, 1970, Unpublished paper presented to Annual Meeting, Western Division of Canadian Association of Geographers, Calgary, 1979.

Davies, W.K.D., (1980), 'Higher Order Factor Analysis and Functional Regionalization', Environment and Planning, A(12), pp.685-701.

Davies, W.K.D. and Thompson, R.R., (1980), 'The Structure of Inter-Urban Connectivity: A Dyadic Factor Analysis of Prairie Commodity Flows', Regional Studies, 14(4), pp.297-312.

Davies, W.K.D., (1983), Urban Social Structure: A Multivariate Structural Study of Cardiff and its Region, University of Wales Press, Social Science Monograph 8.

Davis, K., (1972), The City: Essays from Scientific American, W. Freeman, San Francisco.

Dean, K.A. and James, H.D., (1981), 'Social Factors and Admission to Psychiatric Hospital', Transactions, Institute of British Geographers, New Series 6(1), pp.39-52.

Dear, M. and Scott, A.J., (1982), Urbanization and Urban Planning in Capitalist Society, Methuen, London.

Dent, O., and Sakoda, J.M., (1973), 'Potential Sources of Spuriousness in Factor Ecology Studies', Institute of British Geographers, Quantitative Methods Study Group Working Paper No. 1, Birmingham Meeting, January 1973, pp.37-44.

Dickinson, R.E., (1970), Regional Ecology, J. Wiley, New York.

Dixon, W.J., (1969), Biomedical Series of Computer Programmes, University of California, Los Angeles.

Dogan, M. and Rokkan, S., (1969), <u>Quantitative Ecological Analysis in the Social Sciences</u>, M.I.T. Press.

Downs, R.M., and Stea, D., (1977), <u>Maps in Minds</u>, Harper and Row, New York.

Dubos, Rene, (1968), <u>So Human an Animal</u>, Chas Scribner, New York.

Dudzinski, M.L. et al, (1975), 'Repeatability of Principal Components in Samples: Normal and Non Normal Data Sets Compared', <u>Multivariate Behavioural Research</u>, 10, pp.109-117.

Duncan, O., (1955), 'Review of Social Area Analysis', <u>American Journal of Sociology</u>, 61, pp.84-85.

Duncan, O., (1966), 'Path Analysis: Sociological Examples', <u>American Journal of Sociology</u>, 72, pp.1-16.

Dunn, E., (1970), <u>A Flow Network Image of Urban Structure</u>, 7(3).

Ehrenberg, A.S.C., (1962), 'Some Questions about Factor Analysis', <u>Statistician</u>, 12, pp.191-208.

Eisenstadt, S.N., (1973), <u>Tradition, Change and Modernity</u>, J. Wiley, New York.

Engels, F., (1845), <u>Die Lage der arbei, tenden Klasse in England</u>, translated into English for London edition (1892) as <u>The Conditions of the Working Class in England in 1844</u>. Reprinted 1962.

Evans, D.J., (1973), 'A Comparative Study of Urban Social Structures in South Wales', in B.D. Clark and M.B. Gleave (ed.), <u>Social Patterns in Cities</u>, Institute of British Geographers, pp.87-102.

Evans, D.J., (1973), 'Urban Social Structures in South Wales', in B. Clark and B. Gleave (ed.), <u>Social Patterns in Cities</u>, Institute of British Geapherse, Special Publication No. 5, pp.87-102.

Evans, G.T., (1971), 'Transformation of Factor Matrices to Achieve Congruence', <u>British Journal of Mathematical Statistics in Psychology</u>, 24, pp.22-48.

Eysenck, H.J., (1952), 'Uses and Abuses of Factor Analysis', <u>Applied Statistics</u>, 1, pp.45-49.

Feyman, R., (1965), <u>The Character of Physical Law</u>, Cox and Wyman, London.

Firey, W., (1948), <u>Land Use in Central Boston</u>, Harvard Univ. Press.

Fisher, C.A., (1970), 'Whither Regional Geography?', <u>Geography</u>, 55, pp.373-389.

Fisher, C.S., (1974), 'Toward a Subcultural Theory of Urbanism', <u>American Journal of Sociology</u>, 80(6), pp.1319-41.

Fisher, J.C., (1966), <u>Yugloslavia</u>, Chandler, San Francisco.

Foley, D.L., (1962), 'An Approach to Metropolitan Spatial Structure', in M.M. Webber et al, <u>Explorations in Urban Structure</u>, University of Pennsylvania Press.

Foreman, P.B., (1948), 'The Theory of Case Studies', <u>Social Forces</u>, 26, pp.408-419.

Forrest, R., Henderson, J. and Williams, P., (1982), <u>Urban Political Economy and Social Theory</u>, Gower, Aldershot, U.K.

Fortes, M. (ed.), (1949), <u>Social Structure: Studies Presented to A.R. Radcliffe-Brown</u>, Clarendon Press, Oxford.

French, J.W., (1972), 'Application of T Technique Factor Analysis to the Stock Market', <u>Multivariate Behaviour Research</u>, 3, p.279.

French, R.A. and Hamilton, F.E.I., (1979), <u>The Socialist City</u>, J. Wiley, New York.

Frichter, B., (1964), <u>Introduction to Factor Analysis</u>, Van Norstrand First Edition 1951, Revised 1964.

Friedmann, J., (1972), 'A General Theory of Polarized Development', in N.M. Hansen (ed.), <u>Growth Centres in Regional Economic Development</u>, Free Press, New York, pp.82-107.

Friedmann, J., (1973), <u>Urbanization, Planning and National Development</u>, Sage, Beverly Hills, California.

Gans, H., (1962), <u>The Urban Villagers</u>, Free Press, New York.

Garrison, W.L. and Marble, D.F., (1964), 'Factor Analytic Study of the Connectivity of a Transportation Network', <u>Papers of the Regional Science Association</u>, pp.231-238.

Gatrell, A.C., (1981), 'On the Structure of Urban Social Areas: Explorations Using Q Analysis', <u>Transactions of Institute of British Geographers</u>, Vol. 6(2), pp.228-245.

Gauthier, H.L., (1968), 'Transportation and the Growth of the Sao Paulo Economy', <u>Journal of Regional Science</u>.

Gibb, C.A., (1956), 'Changes in the Culture Pattern of Australia 1906-46 as Determined by the P Technique', <u>Journal of Social Psychology</u>, 43.

Gibson, W.A., (1959), 'Three Multivariate Models: Factor Analysis, Latent Structure Analysis and Latent Profile Analysis', <u>Psychometrika</u>, 24, pp.229-252.

Giggs, J.A., (1970), 'Socially Disorganized Areas in Barry', in H. Carter and W.K.D. Davies, <u>Urban Essays: Studies in Geography of Wales</u>, Longmans, London.

Giggs, J.A., (1973), 'The Distribution of Schizophrenics in Notting-ham', Transactions of the Institute of British Geographers, 59, p.55-76.

Giggs, J.A. and Mather, P., (1975), 'Factorial Ecology and Factorial Invariance', Economic Geography, 51(5), pp.366-382.

Gittus, E., (1964), 'An Experiment in the Identification of Urban Sub Areas', Transactions of Bartlett Society, 2, pp.109-135.

Gittus, E., (1964), 'The Structure of Urban Areas', Town Planning Review, 35, pp.5-20.

Glass, G.V. and Maguire, T.O., (1966), 'Abuses of Factor Scores', American Educational Research Journal, 3, pp.297-304.

Glass, G.V. and Hakstian, A.R., (1969), 'Measures of Association in Comparative Experiments', American Educational Research Journal, 6, pp.403-414.

Goddard, J.B., (1970), 'Functional Regions Within the City Centre', Transactions of Institute of British Geographers, 49, pp.161-182.

Goheen, P., (1970), Victorian Toronto, Department of Geography, University of Chicago Research Paper.

Golledge, R.G. and Rushton, G., (1972), Multidimensional Scaling, Association of American Geographers, Washington, D.C.

Goodman, L.A., (1953), 'Ecological Regressions and the Behaviour of Individuals', Americal Sociological Review, 18, pp.663-664.

Goodman, L.A., (1966), 'How to Ransack Social Mobility Tables and Other Kinds of Cross Classification Tables', American Journal of Sociology, 75, pp.1-40.

Gorsuch, R.L., (1970), 'A Comparison of Biquartimin, Maxplane, Promax and Varimax', Educational and Psychological Measurement, 30, p.861.

Gorsuch, R.J., (1973), 'Using Bartlett's Test to Determine the Number of Factors to Extract', Educational and Psychological Measurement, 33, pp.361-364.

Gorsuch, R.L., (1974), Factor Analysis, Saunders, Philadelphia.

Gould, P.R., (1967), 'Structuring Information on Spacio-Temporal Preferences', Journal of Regional Science, 7, pp.259-280.

Gould, P.R., (1968), 'On the Geographical Interpretation of Eigen-values', Transactions Institute of British Geographers, pp.53-86.

Gould, P.R., (1968), 'The Mental Maps of British School Leavers', Regional Studies, 2, pp.161-182.

Gould, P.R., (1970), 'Is Statistix Inferens the Geographical Name for a Wild Goose Chase?', Economic Geography, 46, pp.439-448.

Gould, P.R. and White, R., (1974), Mental Maps, Penguin, Harmondsworth, U.K.

Gould, P.R., (1980), 'Q Analysis', International Journal of Man-Machine Studies.

Gower, J.C., (1967), 'A Comparison of Some Methods of Cluster Analysis', Biometrics, pp.623-628.

Gravier, J.F., (1947), Paris et le Desert Francais, Flammarion, Paris.

Green, F.H.W., (1950), 'Urban Hinterlands in England and Wales', Geographical Journal, 116, pp.64-81.

Green, F.H.W., (1971), 'Back to the Heartland', Area, 3(2), pp.129-130.

Green, B.F., (1976), 'On the Factor Score Controversy', Psychometrika, 41(2), pp.263-266.

Gregory, D., (1978), Ideology, Science and Human Geography, Hutchinson, London.

Gregory, S., (1978), Statistical Methods for Geographers, Longmans, 4th Edition, London.

Grigg, D., (1965), 'The Logic of Regional Systems', Annals Association of American Geographers, 55, pp.465-491.

Guertin, W.H. and Bradey, J., (1970), Introduction to Modern Factor Analysis, Edwards, Ann Arbor, Michigan.

Guilford, J.P. and Hoepner, R., (1969), 'Comparisons of Varimax Rotations with Rotations to Theoretical Targets', Educational and Psychological Measurement, 29, pp.3-23.

Guilford, J.P., (1977), 'The Invariance Problem in Factor Analysis', Educational and Psychological Measurement, 37, pp.11-19.

Guttman, L., (1954), 'Some Necessary Conditions for Common Factor Analysis', Psychometrika, 19, p.149.

Guttman, L., (1955), 'A New Approach to Factor Analysis', in P.F. Lazarsfield (ed.), Mathematical Thinking in the Social Sciences, Columbia University Press.

Guttman, L., (1955), 'The Determinacy of Factor Score Matrices with Implications for Five Other Basic Problems of Common Factory Theory', British Journal of Statistical Psychology, 4, p.65.

Hadden, J.K. and Boryalta, E.P., (1965), American Cities, Rand McNally, New York.

Hagerstrand, T., (1968), 'Methods and Techniques in Current Urban and Regional Research in Sweden', Tijdschrift for Planering av Landsbygd och Tatorter, 22, pp.3-11.

Haggett, P., (1965), <u>Locational Analysis in Geography</u>, Ed. Arnold, London.

Hagood, M.J., (1943), 'Statistical Methods for Delineation of Regions Applied to Data on Agriculture and Population', <u>Social Forces</u>, 21, pp.287-297.

Haining, R., 'Spatial Correlation Problems', in D.T. Herbert and R.J. Johnston (ed.), <u>Geography and the Urban Environment</u>, Vol. III, J. Wiley, pp.1-44.

Hakstian, A.R., (1971), 'A Comparative Evaluation of Several Prominent Methods of Oblique Factor Transformation', <u>Psychometrika</u>, 36, p.175.

Hakstian, A.R. and Abell, R.A., (1974), 'A Further Comparison of Oblique Factor Transformation Methods', <u>Psychometrika</u>, 39, pp.429-444.

Hall, P., (1968), <u>The World Cities</u>, Weidenfeld and Nicolson, London.

Hall, P., (1971), 'Spatial Structure of Metropolitan England', in M. Chisholm et al (ed.), <u>Spatial Policy Problems of the British Economy</u>, Cambridge University Press.

Hammer, A.G., (1971), <u>Elementary Matrix Algebra</u>, Pengammon Press.

Hannan, M.T., (1970), <u>Problems of Aggregation and Disaggregation in Sociolocial Research</u>, University of North Carolina.

Hansen, N.M., (1968), <u>French Regional Planning</u>, Indiana.

Harary, J., (1965), <u>Structural Models: An Introduction to the Theory of Directed Graphs</u>, J. Wiley, N. York.

Harman, H.H., (1976), <u>Modern Factor Analysis</u>, University of Chicago Press, 1967, Revised 1976.

Harris, C. and Ullman, E., (1945), 'The Nature of Cities', <u>Annals, American Academy of Political and Social Science</u>, 142, pp.7-17.

Harris, C.D., (1970), 'Cities of the Soviet Union', <u>Association of American Geographers Monograph</u>, Rand McNally, New York.

Harris, C.W. (ed.), (1956), <u>Problems in Measuring Change</u>, University of Wisconsin Press.

Harris, C.W., (1967), 'On Factors and Factor Scores', <u>Psychometrika</u>, 32, pp.363.

Harvey, D., (1973), <u>Social Justice and the City</u>, Ed. Arnold, London.

Hay, (1975), 'On the Choice of Methods in the Factor Analysis of Connectivity Matrices', <u>Transactions Institute of British Geographers</u>, 66, pp.163-167.

Haynes, K.E., (1971), 'Spatial Change in Urban Structure', Economic Geography, 47(2) Supplement, pp.324-335.

Haynes, K.E. and Conway, D., (1977), 'Advances in Comparative Ecological Analysis', Environment and Planning, 9(10).

Henderickson, A.E. and White, P.O., (1964), 'Promax: A Quick Method for Rotation to Oblique Simple Structure', British Journal of Statistical Psychology, 17, p.65.

Henshall, J.D., (1966), 'The Demographic Factor in the Structure of Agriculture in Barbados', Transactions Institute of British Geographers, 38, pp.183-195.

Herbert, D. and de Silva, (1974), 'Factorial Ecology of Colombo', Cambria, Vol. 4, p.147.

Herbert, D.T., (1967), 'Social Area Analysis: A British Study', Urban Studies, 4, pp.41-60.

Herbert, D.T., (1970), 'Principal Components Analysis and Urban Social Structure: A Study of Cardiff and Swansea', in H. Carter and W.K.D. Davies (ed.), Urban Essays: Studies in Geography of Wales, Longmans, U.K., pp. 79-100.

Herbert, D.T. and Evans, D.J., (1973), Urban Environment and Juvenile Delinquency, Report for U.K. Home Office Research Unit, Geography Department, University College, Swansea.

Herbert, D.T., (1977), 'An Areal and Ecological Analysis of Delinquency Residence: Cardiff 1966 and 1971', Tijdschrift v. E.S. Geografie, 68, pp.83-99.

Hirst, M.A., (1971), 'Hierarchical Aggregation Procedures for Inter-action Data', Environment and Planning, A9, pp.99-103.

Hirst, M.A., (1975), 'Dimensions of Urban Systems in Tropical Africa Geographical Analysis', Transactions Institute of British Geographers, pp.441-449.

Hitt, W.D., (1968), 'Two Models of Man', American Psychologist, 24, pp.651-658.

Hoch, I., (1973), 'Income and City Size', Urban Studies, 9, pp.299-328.

Hodge, G., (1968), 'Urban Structure and Regional Development', Papers and Proceedings of the Regional Science Association, 21, pp.101-123.

Hofstaetter, P.R., (1952), 'Your City Revised: A Factorial Study of Cultural Patterns', American Catholic Sociological Review, 13, pp.159-168.

Hohn, F.E., (1958), Elementary Matrix Algebra, MacMillan, New York.

Horn, J.L. and Miller, W.C., 'Evidence on Problems in Estimating Common Factor Scores', Educational and Psychological Measurement, 26, p.617.

Horn, J.L., (1965), 'An Empirical Comparison of Methods for Estimating Factor Scores', Educational and Psychological Measurement, 25, pp.313-322.

Horn, J.L., (1969), 'Factor Analysis with Variables of Different Metric', Educational and Psychological Measurement, 29, p.753.

Horst, P., (1963), Matrix Algebra for Social Scientists, Holt, Rinehart and Winston, New York.

Horst, P., (1965), Factor Analysis of Data Matrices, Holt, Rinehart and Winston.

Hotelling, H., (1933), 'Analysis of a Complex of Statistical Variables into Principal Components', Journal of Educational Psychology, 24, pp.417-441, 498-520.

Houghton, D.S., (1975), 'City Size and Social Differentiation', Tijdschrift v. Econ. en Sociale Geografie, 66(4), pp.217-230.

Hoyt, H., (1939), The Structure and Growth of Residential Neighbourhoods in American Cities, F.H.A., Washington, D.C.

Huff, D., (1973), 'The Delimination of a National System of Planning Regions', Regional Studies, 7, pp.323-329.

Hughes, J.A. and Carey, G.W., (1972), 'Factorial Ecology: Oblique and Orthogonal Solutions', Environment and Planning, 4, pp.147-162.

Hunter, A.A., (1972), 'Factorial Ecology: A Critique and Some Suggestions', Demography, 9, pp.107-118.

Hunter, A.A. and Latif, A.H., (1973), 'Stability and Change in the Ecological Structure of Winnipeg', Revue of Canadian Sociology and Anthropology, 6, pp.167-178.

Hunter, A.A., (1974), Symbolic Communities, University of Chicago Press.

Illeris, S., and Pedersen, P.O., (1968), 'Central Places and Functional Regions in Denmark', Land Studies in Geography, Series B., Sweden.

Imbrie, J. and Purdy, E.G., (1962), 'Classification of Modern Bahavrian Carbonate Sediments', in W.E. Harm (ed.), American Associate Petroleum Geology Memoirs, 1, pp.253-272.

Imbrie, J. and Andel, V., (1964), 'Vector Analysis of Heavy Mineral Data', Geological Society of America, Bulletin 75, pp.1131-1156.

Isserman, A.M., (1977), 'The Location Quotient Approach to Estimating Regional Economic Impacts', American Institute of Planners, pp.33-41.

Jackson, J.N., (1972), The Urban Future, Allen and Unwin, London.

Jackson, L.E. and Johnston, R.J., (1972), 'Structuring the Image', Environment and Planning, 4, pp.415-427.

Jansen, E., (1974), 'The Spatial Structure of Newark', Environment and Planning, 6, pp.273-290.

Janson, C.A., (1968), 'The Spatial Structure of Newark, New Jersey', Acta Sociologia, 11, pp.1441-1469.

Janson, C.G., (1971), 'A Preliminary Report on Swedish Urban Structure', Economic Geography, 47, pp.240-257.

Jennrich, R.I. and Sampson, P.F., (1966), 'Rotation for Simple Loadings', Psychometrika, 31, pp.313-323.

Jennrich, R.I., (1974), 'Standard Errors for Oblique Factor Loadings', Psychometrika, 39, pp.593-604.

Johnston, R.J., (1968), 'Choice in Classification: The Subjectivity of Objective Methods', Annals, Association of American Geographers, 58, pp.575-589.

Johnston, R.J., (1970), 'On Spatial Patterns in Urban Residential Structure', Canadian Geographer, 14.

Johnston, R.J., (1971), 'Some Limitations of Factorial Ecologies and Social Area Analysis', Economic Geography, 47(2) Supplement, pp.314-323.

Johnston, R.J., (1971), Urban Residential Patterns: An Introductory Review, G. Bell, London.

Johnston, R.J., (1972), 'Towards a General Model of Intra-Urban Residential Patterns', in C. Board, P. Haggett et al, Progress in Geography, 4, pp.83-115.

Johnston, R.J., (1973), 'Possible Extensions to the Factorial Ecology Method', Environment and Planning, 5, pp.719-734.

Johnston, R.J., (1977), 'Principal Components Analysis and Factor Analysis in Geographical Research', South African Geographical Journal, 59, pp.30-44.

Johnston, R.J., (1977), 'Regarding Urban Origins, Urbanization and Urban Patterns', Geography, 62, pp.1-8.

Johnston, R.J., (1978), Multivariate Statistical Analysis in Geography, Longmans, London.

Johnston, R.J., (1978), 'Residential Area Characteristics: Research Methods for Identifying Urban Sub-Areas', in D.T. Herbert and R.J. Johnston (ed.), Social Areas in Cities: Processes, Patterns and Problems, J. Wiley, London, New York, pp.193-236.

Johnston, R.J., (1979), Geography and Geographers, Ed. Arnold, London.

Johnston, R.J., (1979), 'On the Characterization of Urban Social Areas', Tijdschrift voor Economische en Sociale Geografie, 70(4), pp.232-238.

Jollifee, I.J., (1972), 'Discarding Variables in Principal Component Analysis, Applied Statistics, 21, pp.160-173.

Jones, F.L., (1968), 'Social Area Analysis: Some Theoretical and Methodological Comments Illustrated with Australian Data', The British Journal of Sociology, 14, pp.424-444.

Jones, F.L., (1969), Dimensions of Urban Social Structure, Australian National University Press, Canberra.

Jones, P.N., (1978), 'The Distribution of Coloured Immigrants in the U.K.', Transactions of Institute of British Geographers, 3(4), pp.515-592.

Joreskog, K.G., (1967), A Computer Programme for Unrestricted Maximum Likelihood Factor Analysis, Research Bulletin, Princeton Educational Service.

Joreskog, K.G., (1967), 'Some Contributions to Maximum Likelihood Factor Analysis', Psychometrika, 32, pp.443-482.

Joreskog, K.G. and Goldberger, A.S., (1972), 'Factor Analysis by Generalized Least Squares', Psychometrika, 37, pp.243-260.

Joshi, T.R., (1972), 'Toward Computing Factor Scores', in W.P. Adams and F. Helleiner (ed.), International Geography, (I.G.U. 1972), University of Toronto Press, pp.906-909.

Kaiser, H.F., 'The Varimax Criterion for Analytical Rotation in Factor Analysis', Psychometrika, 23, pp.187-200.

Kaiser, H.F. and Caffrey, (1965), 'Alpha Factor Analysis, Psychometrika, 30, pp.1-14.

Kaiser, H.F., (1963), 'Image Analysis', in C.W. Harris (ed.), op. cit.

Kaiser, H.F., Hunka, S., and Bianchini, J., (1971), 'Relating Factors Between Studies Based on Different Individuals', Multivariate Behavioural Research, 6, p.409.

Kaiser, H.F., (1974), 'An Index of Factorial Simplicity', Psychometrika, 39, pp.31-36.

Kansky, K.G., (1976), Urbanization under Socialism: Czechoslovakia, Praegor, New York.

Kendall, M.G., (1939), 'The Geographical Distribution of Crop Productivity in England', Journal Royal Statistical Society, 102, pp.21-48.

Kendall, M.G. and Stuart, A., (1966), The Advanced Theory of Statistics, Griffin, London.

King, L.J., (1966), 'Cross Sectional Analysis of Canadian Urban Dimensions', Canadian Geographer, 10, pp.205-224.

King, L.J., (1969), Statistical Analysis in Geography, Prentice Hall, Englewood Cliffs.

King, L.J., (1972), 'City Classification by Oblique Factor Analysis of Time Scores Data', in B.J.L. Berry (ed.), City Classification Handbook, op. cit.

Klovan, J.E., (1966), 'The Use of Factor Analysis in Determining Depositional Environments', Journal of Sedimentary Petrology, 36(1), pp.115-125.

Klovan, J.E. and Joreskog, K.G., (1976), Geological Factor Analysis, Elsevier, New York.

Kmentra, J. and Oberhoffer, W., (1974), 'A General Procedure for Obtaining Maximum Likelihood Estimates in Generalized Regression Estimates', Econometrika.

Knox, P.L., (1975), Social Well Being, Oxford University Press.

Knox, P.L. and MacLaran, P., (1978), 'Values and Perceptions in Descriptive Approaches to Urban Social Geography', in D.T. Herbert and R.J. Johnston (ed.), Geography and the Urban Environment: Progress in Research and Applications, J. Wiley, London, pp.197-248.

Korth, B. and Tucker, L., (1975), 'The Distribution of Chance Congruence Coefficients from Simulated Data', Psychometrika, 40, pp.361-372.

Krus, J.K. and Weiss, D.J., (1976), 'Empirical Comparison of Factor and Order Analysis on Pre-Structured and Random Data', Multivariate Behavioural Research, 11, pp.95-118.

Kruskal, J.B. and Shepard, R.N., 'A Non Metric Variety of Linear Factor Analysis', Psychometrika, 39, pp.123-158.

Kuh, E. and Meyer, J.E., (1955), 'Correlation and Regression Estimates When the Data are Ratios', Econometrika, 23, pp.400-416.

Kuh, E. and Belsley, (1980), Regression Diagnostics, J. Wiley, New York.

Langton, J., (1977), 'Late Medieval Gloucester', Transactions Institute of British Geographers, 2(3), pp.259-278.

Lapidus, M. (ed.), (1969), Middle Eastern Cities, University of California Press.

Latif, A.H., (1974), 'Structure and Change Analysis of Alexandria, Egypt', in K.P. Schwirian (ed.), Comparative Urban Structure.

Lausen, J.R., (1969), 'On Growth Poles', Urban Studies, 6(2), pp.137-161.

Laut, P., (1974), A Geographical Analysis and Classification of Canadian Prairie Agriculture, University of Manitoba Geographical Series Number 2, (Chapter 7).

Lawley, D.N., (1943), 'The Application of the Maximum Likelihood Method to Factor Analysis', British Journal of Psychology, 33, pp.172-175.

Lawley, D.N. and Maxwell, A.E., (1971), Factor Analysis as a Statistical Method, Butterworths, London, First Edition 1963.

Lawton, R., (1968), 'The Journey to Work in Britain', Regional Studies, 2, pp.27-40.

Levi-Strauss, C., (1952), 'Social Structure', in A.L. Kroeber (ed.), Anthropology Today, University of Chicago Press, pp.321-390.

Liu, B-C, (1977), Quality of Life Indicators in U.S. Metropolitan Areas: A Statistical Analysis, Praeger, New York.

Lovejoy, A.O., (1929), The Revolt Against Dualism, Open Court Publishing Company, La Salle, Indiana, Reprinted 1955.

Lovejoy, A.O., (1953), 'The Meanings of Emergence and its Modes', in P.P. Weiner (ed.), Readings in the Philosophy of Science, Scribern's, New York, pp.119-147.

Lynch, K., (1960), The Image of the City, M.I.T. Press.

Mabogunje, A., (1968), Urbanization in Nigeria, University of London Press.

Mackinder, H., (1902), Britain and British Seas.

MacRae, D., (1960), 'Direct Factor Analysis of Sociometric Data', Sociometry, 23, pp.360-370.

Maddala, G.S., (1977), Econometrics, McGraw Hall.

Magee, A., (1971), 'Problems of Economic Development and Migration in Southern Europe', in R.J. Johnson and J.M. Soons (eds.), Proceedings of the Sixth New Zealand Geography Conference, Christchurch, N.Z., pp.179-185.

Malinvauch, E., (1980), Statistical Methods in Econometrics, North Holland, New York, Third Edition.

Marx, K., (1936), Selected Works, Reprinted by International Publishers, New York.

Marx, K., (1967), Capital, Three Volumes, International Publishers, New York, 1967 Edition.

Masser, I. and Brown, P.J.B., (1975), 'Hierarchical Aggregation Procedures for Interaction Data', Environment and Planning, A7, pp.509–523.

Mather, P.M., (1971), 'Varimax and Generality: A Comment', Area, 3, pp.252–254.

Mather, P.M., (1972), 'A Further Comment', Area, 4, pp.27–30.

Mather, P.M., (1976), Computational Methods of Multivariate Analysis in Geography, J. Wiley, London.

Mayhew, H., (1862), London Labour and London Poor, 1–4, 1862, Griffin and Bohn, London, Reprinted Dover Publications, London, 1968.

McDonald, R.P. and Burr, E.J., (1967), 'A Comparison of Four Methods of Constructing Factor Scores', Psychometrika, 32, p.381.

McDonald, R.P., (1967), 'A General Approach to Non Linear Factor Analysis', Psychometrika, 27, pp.397–415.

McDonald, R.P., (1967), 'Non Linear Factor Analysis', Psychometrics, Monograph 15.

McElrath, D.C., (1968), 'Social Differentiation and Societal Scale', in S. Greet et al, The New Urbanization, St. Martin's Press, New York.

McElrath, D.C., (1968), 'Societal Scale and Social Differentiation, Accra, Ghana', in S. Greer et al (ed.), The New Urbanization, St. Martin's Press, New York, pp.33–52.

McGee, T., (1967), The South East Asian City, Bell, London.

Meyer, D.R., (1971), 'Factor Analysis Versus Correlation Analysis', Economic Geography, 47(2) Supplement, pp.336–343.

Moseley, E.C. and Klett, C.J., (1964), 'An Empirical Comparison of Factor Scoring Methods', Psychological Reports, 14, p.179.

Moser, C.A. and Scott, W., (1961), British Towns: A Statistical Study of the Social and Economic Differences, Oliver and Boyd.

Mulaik, S.A., (1972), The Foundations of Factor Analysis, McGraw-Hill, New York.

Murdie, R.A., (1969), The Factorial Ecology of Toronto, Department of Geography, University of Chicago Research Paper No. 116.

Murdie, R.A., (1976), 'Spatial Form in the Residential Mosaic', in D.T. Herbert and R.J. Johnston, Social Areas in Cities, Vol. I, J. Wiley, London, pp.237–272.

Muth, R.F., (1969), Cities and Housing, University of Chicago Press.

Nagel, E., (1961), The Structure of Science, Harcourt, Brace and Wild, New York.

Nelson, H.J., (1955), 'A Service Classification of American Cities', Economic Geography, 31, pp.189-201.

Neuhaus, J.O. and Wrigley, C., (1954), 'The Quartimax Method', British Journal of Statistical Psychology, 7, pp.81-91.

Newton, P.W. and Johnston, R.J., (1976), 'Residential Area Characteristics and Residential Area Homogeneity', Environment and Planning, A5, pp.543-552.

Nicholson, T.G. and Yeates, M.H., (1969), 'The Factorial Structure of Winnipeg 1961', Canadian Review of Sociology and Anthropology, 6, pp.162-178.

Nie, N., Hull, C.H., Jenkins, J.G., Steinbrenner, K. and Brent, D., (1975), Statistical Package for the Social Sciences, McGraw Hill, New York.

Norman, P.C., (1968), 'A New Typology of London's Districts', in M. Dogan and S. Rokkan (ed.), Quantitative Ecological Analysis in the Social Sciences, M.I.T. Press, Cambridge, 1969.

Nosal, M., (1977), 'A Note on the Minres Method', Psychometrika, 42, pp.149-151.

Nystuen, J.D. and Dacey, M., (1961), 'A Graph Theory Interpretation of Nodal Regions', Papers of the Regional Science Association, 7, pp.29-42.

Oldershaw, S., (1973), 'An Empirical Study of Scale or Principal Component and Factorial Studies of Spatial Amount', Proceedings of Quantitative Methods Study Group, Institute of British Geographers, Birmingham, U.K.

Openshaw, S., (1977), 'A Geographical Solution to Scale and Aggregation Problems in Region Building', Transactions Institute of British Geographers New Series, 2, pp.459-472.

Openshaw, S., Cullingford, D. and Gillard, A., (1980), 'A Critique of the National Classifications of OPCS PRAG', Town Planning Review, 51, pp.421-439.

Oppenheim, A.N., (1966), Questionnaire Design and Attitude Measurement.

Osgood, C.E. et al, (1957), A Measurement of Meaning, University of Illinois Press.

Palm, R. and Caruso, D., (1972), 'Factor Labelling in Factorial Ecology', Annals, Association of American Geographers, 62, pp.122-133.

Palm, R., (1973), 'Factorial Ecology and the Community of Outlook', Annals, Association of American Geographers, 63, pp.341-346.

Palmer, C.J., Robinson, M.E. and Thomas, R.W., (1977), 'The Countryside Image', Environment and Planning, A9, pp.739-749.

Park, R.E., (1925), 'The City as a Social Laboratory', in T.V. Smith and L. White (ed.), Chicago: An Experiment in Social Science Research, University of Chicago Press, pp.46-63.

Parkes, D.N., (1973), 'Timing the City', Royal Australian Institute Journal, 12, pp.130-135.

Parsons, Talcot, (1960), Structure and Process in Modern Societies, Free Press, New York.

Parsons, Talcot, (1964), 'A Functional Theory of Change', in A. an E. Etzioni (ed.), Social Change, Basic Books, New York, pp.83-97.

Pawlick, K. and Cattell, R.B., (1964), 'Third Order Factors in Objective Personality Tests', British Journal of Psychology, 55, pp.1-18.

Peach, Ceri (ed.), (1975), Urban Social Segregation, Longman, U.K.

Pearson, K., (1893), 'Contributions to the Mathematical Theory of Evolution', Journal of the Royal Statistical Society, 55(4), pp.675-679.

Pearson, K., (1901), 'On Lines and Planes of Closest Fit to Systems of Points in Space', Philosophical Transactions of Royal Society of London, pp.559-572.

Pennal, R., (1972), 'Routinely Compatible Confidence Intervals for Factor Loadings Using the Jack Knife', British Journal of Mathematical Statistical Psychology, 25, pp.107-114.

Perle, E.D., (1977), 'Scale Changes and Impacts on Factorial Ecology Structures', Environment and Planning, A, pp.549-558.

Piaget, J., (1970), Structuralism, Basic Books, New York.

Pigozzi, J., (1975), 'Spatial-Temporal Structure of Inter-Urban Economic Impulses', Tijdschrift voor Economishe en Sociale Geografie, 66(5), pp.272-276.

Pinneau, S.R. and Newhouse, A., (1964), 'Measures of Invariance and Comparability in Factor Analysis for Fixed Variables', Psychometrika, 29, p.271.

Poole, M.A. and O'Farrell, P.N., (1971), 'The Assumptions of the Linear Regression Model', Transactions Institute of British Geographers, 52, pp.145-158.

Poor, D.D.S., and Wherry, R.J., (1976), 'The Invariance of Multi-dimensional Configurations', British Journal of Mathematical and Statistical Psychology, 29, pp.114-125.

Potter, R.B., (1981), 'The Multivariate Functional Structure of the Urban Retailing System: A British Case Study', Transactions of Institute of British Geographers, New Series 6(2), pp.188-214.

Pred, A., (1974), 'Major Job Providing Organizations and Systems of Cities', Commission on College Geography, Resource Paper 27, A.A.G., Washington, D.C.

Pred, A., (1977), City-Systems in Advanced Economies, Hutchinson, London.

Prescott, J.R.V., (1972), Political Geography, St. Martin's Press, New York.

Price, D.O., (1942), 'Factor Analysis in the Study of Metropolitan Centres', Social Forces, 20, pp.449-455.

Pringle, D., (1976), 'Normality, Transformations and Grid Square Data', Area, 8, pp.42-45.

Przeworski, A. and G. and Soares, G.A.D., (1971), 'Theories in Search of a Curve: A Contextual Interpretation of Left Vote', American Political Science Review, 65, pp.51-68.

Przeworski, A., (1974), 'Contextual Models of Poliltical Behaviour', Political Methodology, 1, pp.27-60.

Radcliffe-Brown, A.R., (1940), 'On Social Structure', Journal Royal Anthropological Institute, 61, pp.1-19.

Rao, C.R., (1955), 'Estimation and Tests of Significance in Factor Analysis', Psychometrika, 20, pp.93-111.

Rao, C.R., (1965), Linear Statistical Inference and Its Applications, J. Wiley, New York.

Ray, D.M. and Murdie, R.A., (1972), 'Canadian and American Urban Dimensions', in B.J.L. Berry (ed.), City Classification Handbook, J. Wiley, New York.

Rees, P.H., (1970), 'The Factorial Ecology of Metropolitan Chicago', in B.J.L. Berry and F.L. Horton (ed.), Geographic Perspectives on Urban Systems, Prentice-Hall, New Jersey, pp.319-365.

Rees, P.H., (1972), 'Problems of Classifying Sub-Areas Within Cities', in Berry, op. cit. (1972), pp.265-330.

Richardson, H.W., (1973), The Economics of Urban Size, Saxon House.

Rippe, D.D., (1953), 'Application of a Large Sampling Criterion to Some Sampling Problems in Factor Analysis', Psychometrika, 18, pp.191-205.

Robinson, W.S., (1950), 'Ecological Correlations and the Behaviour of Individuals', American Sociological Review, 15, pp.351-357.

Robson, B.T., (1969), Urban Analysis, Cambridge University Press.

Robson, B.T., (1973), 'A View on the Human Scene', in M. Chisholm and B. Rodgers (ed.), Studies in Human Geography, Heinemann, London, pp.203-241.

Robson, B.T., (1975), Urban Social Areas, Oxford University Press, London.

Rodwin, L., (1970), Nations and Cities, Houghton Mifflin.

Roff, A., (1977), 'The Importance of Being Normal', Area, 9, pp.195-201.

Romsa, G., Hoffman, W. and Brozowski, (1972), 'A Test of the Influence of Scale in Factorial Ecology on Windsor, Ontario', Ontario Geographer, 7, pp.87-92.

Royce, J.R., (1963), 'Factors as Theoretical Constructs', American Psychologist, 18, pp.522-528.

Royce, J.R., (1976), 'Factors of Mouse Emotionality at the Second, Third and Fourth Order', Multivariate Behavioural Research, 11, pp.63-78.

Rumley, D., (1979), 'The Study of Structural Effects in Human Geography', in Tijdschrift vor Economische en Sociale Geografie, 70(6), pp.350-360.

Rummel, R.J., (1967), 'Understanding Factor Analysis', Journal of Conflict Resolutions, 11, pp.444-480.

Rummel, R.J., (1970), Applied Factor Analysis, Northwestern University Press, Evanston, U.S.A.

Rummel, R.J., (1971), Applied Factor Analysis, North Western University Press.

Rummel, R.J., (1972), The Dimensions of Nations, Sage, Beverly Hills, California.

Russett, B.M., (1967), International Regions and the International System, Rand McNally.

Schmid, J. and Leimann, J., (1957), 'The Development of Hierarchical Factor Solutions', Psychometrika, 22, pp.53-61.

Schmidt, C.F. et al, (1958), 'The Ecology of the American City', American Sociological Review, 23, pp.392-401.

Schonemann, P.H., (1966), 'The Generalized Solution of the Orthogonal Procrustes Problem', Psychometrika, 31, p.1.

Schonemann, P.H. and Wong, M.M., (1972), 'Some New Results on Factor Indeterminacy', Psychometrika, 37, p.61.

Schuesster, K., (1973), 'Ratio Variables and Path Models', in A.S. Goldberger and O.D. Duncan, Structural Equation Models in Social Sciences, Seminar Press, New York.

Schwirian, K.P. and Smith, R.K., (1974), 'Primary, Modernization and Urban Structure', in K.P. Schwirian, Comparative Urban Structure, D.C. Heath, pp.324-337.

Shaw, C.R., Zorbaugh, F.M., McKay, H.D. and Lottrell, L., (1929), Delinquency Areas, Chicago.

Shaw, M., (1977), 'The Ecology of Social Change Wolverhampton 1851', Institute of British Geographers, New Series 2(31), pp.332-348.

Shevky, E. and Williams, M., (1949), The Social Areas of Los Angeles, Berkeley.

Shevky, E. and Bell, W., (1955), Social Area Analysis, Stanford University Press.

Siegal, S., (1956), Nonparametric Statistics for the Behavioural Sciences, McGraw Hill, New York.

Simmey, T.S., (1969), 'Charles Booth', in T. Raison (ed.), The Founding of Social Science, Penguin, pp.92-99.

Sjorberg, G., (1962), The Pre-Industrial City, Free Press, Glencoe.

Slater, P.B., (1976), 'A Hierarchical Regionalization of Japanese Prefections', Regional Studies, 10, pp.123-132.

Smith, C.J., (1980), 'Neighbourhood Effects on Mental Health', in D.T. Herbert and R.J. Johnston, Geography and the Urban Environment, III, J. Wiley, London.

Smith, David, (1976), 'Myth and Meaning in the Literature of the South Wales Coalfield: The 1930s', Anglo-Welsh Review, 25, pp.21-42.

Smith, D.M., (1972), The Geography of Social Well Being, McGraw-Hill, New York.

Smith, R.H.T., (1965), 'Method and Purpose in Functional Town Classification', Annals, American Geographers, Vol. 55(3), pp.539-548.

Sneath, P.H.A. and Sokal, R.R., (1973), Numerical Taxonomy, W.H. Freeman, San Francisco.

Snider, J.G. and Osgood, C.E., (1969), Semantic Differential Technique, Aldine, Chicago.

Soja, E.W., (1968), 'Communications and Territorial Integration in East Africa', East Lakes Geographer, 4, pp.39-59.

Sokal, R.R. and Sneath, P.H., (1963), Principles of Numerical Taxonomy, Freeman, San Francisco.

Spearman, C., (1904), 'General Intelligence, Objectively Determined and Measured', American Journal of Psychology, 15, pp.201-293.

Spearman, C., (1927), The Abilities of Man, MacMillan, New York.

Spence, N.A., (1968), 'A Multi-Factor Regionalization of British Counties on the Basis of 1961 Employment Data', Regional Studies, 2, pp.87-104.

Spence, N.A. and Taylor, P.J., (1970), 'Quantitative Methods in Regional Taxonomy', Progress in Geography, 2, pp.3-50.

Sprout, H. and Sprout, M., (1965), The Ecological Perspective on Human Affairs, Princeton University Press.

Sugden, D. and Hamilton, P., (1971), 'Scale, Systems and Regional Geography', Area, 3(4), pp.139-144.

Suttles, G.D., (1968), The Social Order of the Slum, University of Chicago Press.

Suttles, G.D., (1972), The Social Construction of Communities, University of Chicago Press.

Sweetser, F.L., (1965), 'Factorial Ecology: Helsinki 1960', Demography, 2, pp.372-386.

Sweetser, F.L., (1965), 'Factor Structure as Ecological Structure in Helsinki and Boston', Acta Sociologica, 8, pp.205-225.

Taafe, E.J. and Gauthier, H.L., (1973), The Geography of Transportation, Prentice Hall, New Jersey.

Taylor, P.J., (1971), 'Distance Transformation and Distance Decay Function', Geographical Analysis, 3, pp.221-238.

Taylor, P.J., (1975), 'A Kantian View of the City: A Factorial Ecology Experiment in Space and Time', Environment and Planning A, pp.671-688.

Taylor, P.J., (1977), Quantitative Methods in Geography, Houghton, Mifflin.

Theodorson, G.A. (ed.), (1961), Studies in Human Ecology, Row Paterson.

Thomas, E.H. and Anderson, D.L., (1965), 'Additional Comments on Weighting Values in Correlation Analysis of Areal Data', Annals, Association of American Geographers, 55, pp.492-505.

Thompson, I.B., (1970), Modern France: A Social and Economic Geography.

Thompson, W.R., (1965), A Preface to Urban Economics Resources for the Future, John Hopkins Press.

Thorndike, E.L., (1938), Your City, Harcourt Brace and Co., New York.

Thorndike, R.M. and Weiss, D.J., (1973), 'A Study of the Stability of Canonical Correlations and Components', Educational and Psychological Measurement, 33, pp.123-134.

Thurstone, L.L., (1931), 'Multiple Factor Analysis', Psychological Review, 38, pp.406-427.

Thurstone, L.L., (1938), Primary Mental Abilities, University of Chicago Press.

Thurstone, L.L., (1947), Multiple Factor Analysis, University of Chicago Press.

Timms, D.W.G., (1971), The Urban Mosaic, Cambridge University Press.

Tinkler, K.J., (1975), 'On the Choice of Methods in the Factor Analysis of Connectivity Matrices: A Reply', Transactions Institute of British Geographers, 66, pp.168-170.

Tinkler, K.J., (1972), 'The Physical Interpretation of Eigenfunctions of Dichotomous Matrices', Transactions Institute of British Geographers, 55, pp.17-46.

Tucker, L.R., (1966), 'Some Mathematical Notes on Three Mode Factor Analysis', Psychometrika, 31, pp.279-311.

Tucker, L.R., (1971), 'Relations of Factor Score Estimates to Their Use', Psychometrika, 36, p.427.

Tucker, L.R., (1972), 'Relations Between Multidimensional Scaling and Three Mode Analysis Factor Analysis', Psychometrika, 37, pp.3-27.

Tucker and Lewis, (1973), 'A Reliability Coefficient for Maximum Likelihood Factor Analysis', Psychometrika, 38, pp.1-10.

Tyron, R.C., (1955), Identification of Social Areas by Cluster Analysis, University of California Press, Berkeley.

Udry, J.R., (1964), 'Increasing Scale and Spatial Differentiation: New Tests of Theories from Shevky and Bell', Social Forces, 42, pp.403-413.

Ullman, E.L. and Dacey, M.F., (1960), 'Minimum Requirements Approach to the Urban Economic Base', Papers Regional Science Association, 6, pp.175-194.

Van Arsdol, M., Camillieri, S.F. and Schmid, C.F., (1958), 'The Generality of Urban Social Area Indices', American Sociological Review, 23, pp.277-284.

Vance, J.E., (1964), Geography and Urban Evolution in the San Francisco Bay Area, Institute of Government Studies, Berkeley, California.

Vance, J.E., (1977), This Scene of Man, Harper's College Press, U.S.A.

Velicer, W.F., (1974), 'An Empirical Comparison of the Stability of Factor Analysis, Principal Component Analysis and Image Analysis', Educational and Psychological Measurement, 34, pp.563-572.

Velicer, W.F., (1975), 'The Relationships Between Factor Scores, Image Scores and Principal Component Scores', Educational and Psychological Measurement, 35, pp.

Velicer, W.F., (1977), 'An Empirical Comparison of the Similarity of Principal Components, Image and Factor Patterns', Journal of Multivariate Behavioural Research, 12, pp.3-22.

Ward, J.H., (1963), 'Hierarchical Grouping to Optimize an Objective Function', Journal of the American Statistical Association, 58, pp.236-243.

Warnes, A.M., (1973), 'Residential Patterns in Emerging Industrial Towns', in B. Clark and B. Gleave, Social Patterns in Cities, Institute of British Geographers Special Publication No. 5, (March), pp.169-190.

Webber, M.M., (1964), 'The Urban Place and Non-Place Urban Realms', in M. Webber et al (ed.), Explorations in Urban Structure, University of Pennsylvania Press, pp.79-153.

Webber, R.J., (1977), The National Classification of Residential Neighbourhoods, Planning Research Applications Group, Paper No. 23, Centre for Environmental Studies, London.

Webber, R.J., (1980), 'A Response to the Critique of the OPCS PRAG, National Classifications', Town Planning Review, 51, pp.440-450.

Weber, M., (1921), The City, Reprinted, Free Press, Glencoe, 1958.

Weclawowicz, A., 'The Structure of Socio-Economic Space in Warsaw 1931 and 1970: A Study in Factorial Ecology', in French and Hamilton op.cit., pp.387-423.

Wheatley, R., (1974), Pivot of the Four Quarters, Aldine Press, U.S.A.

Williams, K., (1971), 'Do You Sincerely Want to be a Factor Analyst?', Area, 3, pp.228-229.

Wilson, A. et al, (1977), Models of Cities and Regions, J. Wiley, New York.

Wilson, A.G. and Clarke, M., (1981), The Dynamics of Urban Spatial Structures, Geography, University of Leeds, No. 313.

Wilson, G. and Wilson, M., (1945), The Analysis of Social Change, Cambridge, U.K.

Wiltshire, R., Murdie, R. and Greer-Wootten, B., (1973), Comparative Factorial Ecology, Unpublished Paper, Geography Department, York University, Toronto.

Wirth, L., (1938), 'Urbanism as Way of Life', American Journal of Sociology, 44, pp.1-24.

Wirth, L., (1945), 'Human Ecology', American Journal of Sociology, 50, pp.483-488.

Wishart, D. and Pocock, D.C.D., (1969), 'Methods of Deriving Multifactor Uniform Regions', Transactions Institute of British Geographers, pp.73-98.

Wishart, D., (1975), CLUSTAN IC, Package Program Manual, University of London Computing Centre, U.K.

Wright, S., (1960), 'The Treatment of Reciprocal Interaction in Path Analysis', Biometrics, 16, pp.423-425.

Wrigley, C.S. and Neuhaus, J.O., (1955), 'The Matching of Two Sets of Factors', American Psychologist, 10, p.418.

Yeates, M.H. and Garner, B.J., (1976), The North American City, Harper and Row, New York.

Zelinsky, W., (1973), The Cultural Geography of the U.S.A., Prentice Hall, New Jersey.

Zorbaugh, H.W., (1926), 'The Natural Areas of the City', Publications of American Sociological Society, 20, pp.128-197.

INDEX